INDUSTRIAL INSTRUMENTATION

INDUSTRIAL INSTRUMENTATION

DONALD P. ECKMAN
Professor of Mechanical Engineering
Case Institute of Technology
Consultant, Conoflow Corporation

CBS

CBS Publishers & Distributors Pvt. Ltd.

New Delhi • Bengaluru • Chennai • Kochi • Kolkata • Mumbai
Hyderabad • Uttarakhand • Nagpur • Patna • Pune • Jharkhand

ISBN: 81-239-0810-5

First Indian Reprint: 2002
Reprint: 2003, 2004

This edition has been reprinted in India by arrangement with
The Author

Published by **Satish Kumar Jain** and produced by **Varun Jain** for
CBS Publishers & Distributors Pvt. Ltd.,
4819/XI Prahlad Street, 24 Ansari Road, Daryaganj, New Delhi - 110002
delhi@cbspd.com, cbspubs@airtelmail.in • www.cbspd.com
Ph.: 23289259, 23266861, 23266867 • Fax: 011-23243014

Corporate Office: 204 FIE, Industrial Area, Patparganj, Delhi - 110 092
Ph: 49344934 • Fax: 011-49344935
E-mail: publishing@cbspd.com • publicity@cbspd.com

Branches:
- ***Bengaluru:*** 2975, 17th Cross, K.R. Road, Bansankari 2nd Stage, Bengaluru - 70 • Ph: +91-80-26771678/79 • Fax: +91-80-26771680 E-mail: cbsbng@gmail.com, bangalore@cbspd.com
- ***Chennai:*** No. 7, Subbaraya Street, Shenoy Nagar, Chennai - 600030 Ph: +91-44-26681266, 26680620 • Fax: +91-44-42032115 E-mail: chennai@cbspd.com
- ***Kochi:*** Ashana House, 39/1904, A.M. Thomas Road, Valanjambalam, Ernakulum, Kochi • Ph: +91-484-4059061-65 Fax: +91-484-4059065 • E-mail: cochin@cbspd.com
- ***Kolkata:*** 6-B, Ground Floor, Rameshwar Shaw Road, Kolkata - 700014 Ph: +91-33-22891126/7/8 • E-mail: kolkata@cbspd.com
- ***Mumbai:*** 83-C, Dr. E. Moses Road, Worli, Mumbai - 400018 Ph: +91-9833017933, 022-24902340/41 • E-mail: mumbai@cbspd.com

Representatives:
- Hyderabad: 0-9885175004
- Nagpur: 0-9021734563
- Patna: 0-9334159340
- Pune: 0-9623451994
- Jharkhand: 0-9811541605
- Uttarakhand: 0-9716462459

Printed at:
J.S. Offset Printers, Delhi (India)

PREFACE

This book is an introduction to the science of measurement. Rapid development of new industrial methods and techniques has brought the subject of measurement into a science on its own physical foundation and has extended its scope far beyond the elementary concepts of temperature and pressure determinations. New measurement methods based on both new and old physical phenomena are becoming common in industrial operations.

The purpose of this book is to review the principles of those methods of measurement employed in industrial processing and manufacturing. Primary emphasis is given to the method rather than to the mechanism. Attention is paid, however, to important practical details. In most cases it has been deemed necessary to review the fundamentals of physics (mechanics, thermodynamics, and electricity) pertaining to the problems of measurement. This serves two purposes. First, it focuses attention on the fundamentals of engineering applications, and, second, it induces a more analytical and less empirical approach to the problems of measurement.

The book is intended primarily for the undergraduate in engineering. With the exception of a few sections containing differential equations, it is assumed that the student is familiar with the elements of calculus, mechanics, thermodynamics, and fluid mechanics beyond elementary courses in physics. The problems are mainly elementary. They may serve to suggest to the instructor other problems of similar character.

The author wishes to acknowledge the valuable assistance of Mr. A. E. Krogh of The Brown Instrument Company, particularly in regard to data on temperature measurement. Dr. R. K. Munch of the Monsanto Chemical Company made many valuable suggestions on the importance of various spectrographic and high vacuum methods. The suggestions of Professor C. R. Otto and Dr. D. Dropkin of Cornell University were much appreciated. Mr. Kermit Fischer of the Fischer and Porter

Company aided considerably in the preparation of the discussion of rotameters. The co-operation of the manufacturers of industrial measuring equipment in providing information and operating data is especially appreciated. Much credit is due Mrs. D. P. Eckman for her typing of the manuscript and consideration of grammatical details.

DONALD P. ECKMAN

Ithaca, New York
June 22, 1950

CONTENTS

1· Qualities of Measurement

The progress of measurement is the progress of science. Likewise advancement of industrial progress must be preceded by advancement in the art and science of measurement, for the whole foundation of industrial processing and manufacturing lies in the measurement of materials entering into manufacture, measurement of product, and, not the least, measurement of cost. One important phase of measurement is the utility of measuring instruments designed for guiding the progress of various steps in manufacturing and processing. Of these, this book will be primarily concerned with instruments for measurements in temperature, pressure, rate of flow, and composition. Such factors are important in the scheme of things, for it was very early pointed out that—

"If a man would be a painter, navigator, or architect, it is not enough that he be advised to be one, that he is convinced by the arguments of his adviser that it would be for his advantage to be one, and that he resolves to be one; but he must also be taught the principles of the art; be shown all the methods of working and acquire the habit of using properly all the instruments."—*Benjamin Franklin*

The progress of instrumentation in industry as it is known today took place largely in the 1930's but began with the introduction of a reliable instrument for recording temperatures. With the growth of continuous manufacturing, the need for continuous measurement of temperature, pressure, and flow became urgent. There are continual demands not only for improvements in quality of measurement but also for development of new methods based on newly found physical and chemical laws and effects.

The Meaning of Measurement

The fundamental and underlying purpose of measurements in industrial manufacturing and processing is to aid in the economics of industrial operations by improving either the quality of product or the efficiency of production. In view of this purpose, one should always bear in mind these questions in regard to measurement:

1. Why is the measurement being made?
2. What should the measurement mean?
3. What measurement is made?
4. What is the result of making this measurement?

In an industrial operation these questions may be difficult to answer. Nevertheless a measurement that has a definite purpose and a definite meaning and accomplishes a specific result is one that should be made.

Direct measurement is a tremendously important approach in modern instrumentation. Direct measurement can best be illustrated by an example. In manufacturing a mechanical part in a lathe section of a machine shop, the purpose of processing is to produce a part with given physical dimensions. During and after the operation, measurement of its dimensions (called inspections) are usually made. This is a direct measurement in which the meaning of the measurement (determining a physical dimension) and the purpose of the processing operation (producing a physical dimension) are identical. Indirect measurement is illustrated by the measurement of temperature in a milk pasteurizer. The purpose of the pasteurizing operation is to eliminate bacteria. This is an indirect measurement, wherein the meaning of the measurement (temperature of the milk) and the purpose of processing (eliminating bacteria) are not the same but are related to each other. A direct measurement in this example would be a bacteria count.

The point of these examples is to bring out the meaning and intent of a measurement. *If the primary purpose of making a measurement is to determine a quality of a product, then, one should measure that quality directly.*

Direct measurement is not always possible, however, and then indirect measurement must be resorted to. In indirect measurement an empirical relation is generally established between the measurement actually made and the results that are desired. There are many factors, however, that result in a change in the relations in indirect measurement, and a breakdown in these relations will cause the measurements to become meaningless.

The Elements of Instruments

A measuring instrument is simply a device for determining the value of a quantity or condition. The purpose served by the instrument is, first, to determine or ascertain the value (magnitude) of some particular phenomena. The instrument does not necessarily have to indicate, signal, record, or otherwise make known what value it has ascertained. On the other hand, it may be required to indicate, record, register, signal, or perform some operation on the value it has determined. The value determined by the instrument is generally, but not necessarily, quantitative. An instrument for the measurement of the presence of acid in a chemical solution may simply answer yes or no.

The following functions may be fulfilled by an instrument.

Transmitting, in which the instrument is intended to convey information concerning the measured quantity over some distance to a remote point. The value (magnitude) of the measured quantity may never be made known, because it may be used for some other purpose. A homely example is the telephone.

Signaling, where the instrument only indicates the general value or a range of values of its measured quantity. Some grocer's scales, for example, show only that the weight is too little or too great.

Registering, in which the instrument merely indicates, by numbers or some other symbol of discrete increments, the value of some quantity. A cash register and certain water meters only register incremental quantities.

Indicating, in which an instrument provides some kind of calibrated scale and pointer. The value of the quantity may be read on the scale to any fraction within the limitations of the instrument and the human eye. Most clocks are indicating instruments, for example.

Recording, in which an instrument makes a written record, usually on paper, of the value of the measured quantity against some other variable or against time. Although this function of an instrument is not fundamentally different from an indicating function, it is classed separately.

Any combination of these functions may be found in a measuring instrument, and no one of them is more important than the others. A great number of each type are used industrially.

One of the most important functions of an instrument is found in the automatic controller, every one of which includes a mechanism performing the function of measurement. The quality of automatic control is often determined by the quality of the measuring means.

Another function accomplished in an instrument is to perform various manipulations on a measured variable. This is generally incidental to the main purpose of the instrument but nevertheless is an important function. Several examples will illustrate these features. A type of instrument much used with combustion furnaces takes the *ratio* of fuel flow and air flow. A differential pressure instrument may measure the *difference* in pressure between two points in a process. A flowmeter indicating rate of flow usually *integrates* the flow rate to obtain quantity. Other examples are easily found showing addition, multiplication, and differentiation. In fact, instruments are used in which the solution of rather complex algebraic or differential equations is involved.

All instruments contain various parts that perform prescribed functions in converting a variable quantity or condition into a corresponding indication. The process of conversion in an instrument is necessary in order to change the measured variable, a temperature, a pressure, a flow, or a chemical composition, into a more useful quantity, such as displacement, pressure, force, or potential. In most cases an instrument ultimately converts the measured variable into a displacement.

The parts of an instrument are indicated by Fig. 1–1, together with a simple example of an indicating thermometer. The *primary element* is the part of the instrument that first utilizes energy from the measured medium to produce a condition representing the value of the measured variable. In this case the thermometer bulb is the primary element, because it first converts energy in the form of heat into a fluid displacement, which is proportional to the temperature at the bulb.

The *secondary element* merely converts the condition produced by the primary element into a condition useful to the function of the instrument. In the example the secondary element is the pressure spring, which converts the fluid displacement into a displacement of a link.

The *manipulation element* performs given operations on the condition produced by the secondary element. In Fig. 1-1 the

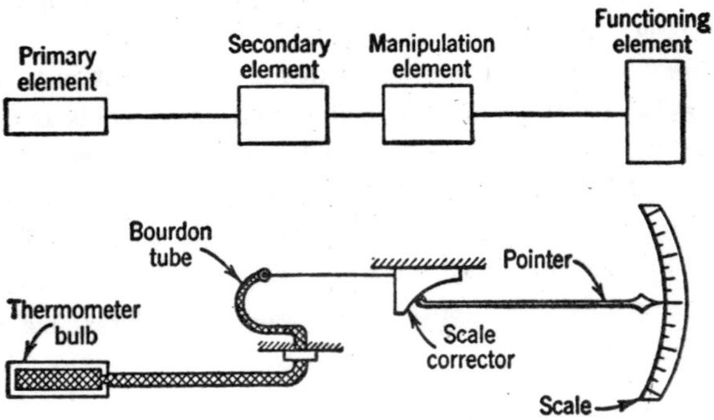

FIG. 1-1. Elements of an Instrument

motion of the pressure spring is modified by the cam in order to correct for nonlinearity in the preceding conversion processes. The manipulation element sometimes precedes the secondary element.

The *functioning element* simply denotes the part of an instrument used for transmitting, signaling, registering, indicating, or recording.

The parts of an instrument named previously do not necessarily appear in all instruments. Some instruments like the mercury-in-glass thermometer are much simpler and require no dissection such as this. On the other hand, an instrument like the automatic balanced potentiometer thermocouple pyrometer is much more complex than the example in Fig. 1-1.

A distinction is sometimes made between *automatic* instruments and *manual* instruments. The automatic instrument does not require the services of an operator in fulfilling its function whereas the latter does. A mercury-in-glass thermometer is an automatic instrument because it indicates temperature without requiring any manual assistance. One type of resistance thermometer with a Wheatstone bridge requires manual operation of the bridge in order to indicate the temperature being measured and is therefore a manual instrument.

Another classification of instruments is made according to their source of power. The *self-operated* instrument like the mercury-in-glass thermometer derives its power wholly from the thermal expansion of mercury. The *power-operated* instrument requires a source of auxiliary power, such as compressed air, electricity, hydraulic supply, or a mechanical source of power.

A further classification of instruments is according to their arrangement. In the *self-contained* instrument such as a room-temperature thermometer, all parts of the instrument from primary element to indicating element are contained in one physical assembly. On the other hand, there are instruments so used that the primary element is located hundreds of feet from the secondary element. The indicating element may also be remotely located.

Static Characteristics

The static characteristics of an instrument are, in general, those that must be considered when the instrument is used to measure a condition not varying with time. The characteristics discussed here are:

Accuracy	Static error
Reproducibility	Drift
Sensitivity	Dead zone

The qualities on the left are desirable, and their opposite qualities are undesirable. It must be stated, however, that there are many existing definitions for these characteristics, and in some cases the definitions are unrelated. We have attempted to select the most generally used definitions. It is suggested that caution be employed in handling specifications for static characteristics because of this confusion.

In an indicating or recording instrument the value of the measured quantity is indicated on a scale or a chart by a pointer or some similar means. Suppose that the highest point of cali-

ation is b units and the lowest point a units and that the calibration is continuous between these two points. Then the instrument *range* is from a to b. The instrument *span* is given by

$$\text{Span} = b - a$$

For a pyrometer calibrated 0 to 1000° the range is 0 to 1000° and the span is 1000°. For a thermometer calibrated 200 to 350° the range is 200 to 350°, and the span is 150°.

The *accuracy* of an instrument may be expressed in several ways. A common method is to specify "accurate to within x per cent." This is an abbreviation, which means "accurate to within $\pm x$ per cent of instrument span at all points of the scale, unless otherwise stated." This is the definition followed in this book whenever possible. Often, however, accuracy is based on instrument range and instrument reading instead of instrument span. Obviously these specifications may differ greatly. Suppose we have a pyrometer calibrated 1000 to 1800°, and the accuracy is stated as within 0.5 per cent. The accuracy of the pyrometer is ± 0.5 per cent times $(1800° - 1000°)$ or $\pm 4°$.

The *static error* of an instrument is the difference between the true value of a quantity not changing with time and the value indicated by the instrument. The static error is expressed as $+x$ units, or $-x$ units. For static error in units,

$$\text{True value} + \text{static error} = \text{instrument reading}$$

For plus static errors the instrument reads high, and for minus static errors the instrument reads low.

The *static correction* of an instrument reading is specified in units:

$$\text{True value} = \text{instrument reading} + \text{static correction}$$

Therefore, static error and static correction are related as follows:

$$\text{Static correction} = -\text{static error}$$

Error calibration means that an industrial instrument has been especially calibrated against a suitable standard and its static error determined at a number of points of its scale. These data form an error curve, which can be used for correcting instrument readings. A typical error curve is shown in Fig. 1–2.

In the specifications for industrial instruments it is often necessary to designate the over-all accuracy. Many industrial instruments are not self-contained but are composed of separate units, such as the primary element, secondary element, and manipulation element. These elements, as in a pressure-gage transmitter–receiver system, may be obtained separately with accuracy limits specified for each element of the system. Let the

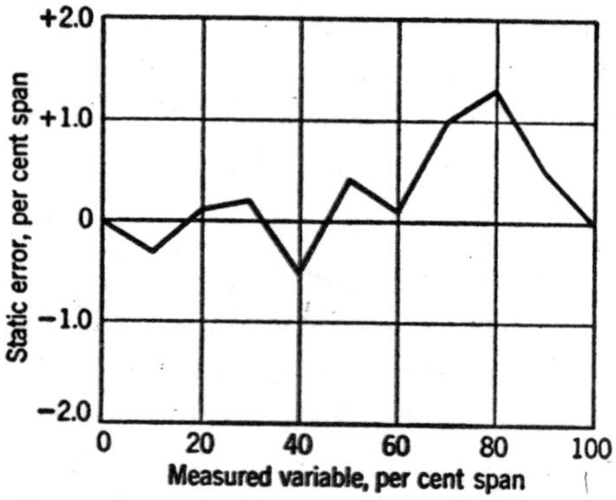

FIG. 1–2. Error Curve for an Instrument

accuracy of each unit be within $\pm a$, within $\pm b$, within $\pm c$, and so on. The *least accuracy* is

$$\text{Within } \pm(a + b + c + \text{etc.})$$

The *root-square accuracy:*

$$\text{Within } \pm\sqrt{a^2 + b^2 + c^2 + \text{etc.}}$$

Obviously the root-square accuracy is more justifiable in an instrument system, because it is not probable that all units of the system will have the greatest static error at the same point at the same time.

The *reproducibility* of an instrument is the degree of closeness with which a given value may be repeatedly measured. It may be specified in terms of units for a given period of time. Perfect reproducibility means that the instrument has no *drift:* that is, the instrument calibration does not gradually shift over a long period of time, such as a week, a month, or even a year.

Several kinds of drift may occur. The whole instrument calibration may gradually shift by the same amount. This is sometimes called a zero drift and is usually due to some kind of simple effect, such as permanent set or slippage. In an indicating or recording instrument it is easily corrected by shifting the pen or pointer position. A span drift involves a gradual change in which the calibration from zero upward changes a proportional amount. This may be caused by a gradual change in a spring gradient. A third kind of drift occurs when only one portion of a calibration changes. This may take place at the high end of an instrument scale when some portion of the instrument is at high stress.

Drift may occur in thermocouples and resistance-thermometer elements because of contamination of the metal and changes in metallurgical or atomic structure. Drift occurs in orifice flowmeters because of wear and erosion of the orifice plate.

Drift is an insidious quality in industrial instruments, because it is rarely apparent and must be carefully guarded against by continuous inspection and maintenance.

Sensitivity unfortunately has two common meanings, which bear little relation to each other. It is a commonly used term, however. Usually sensitivity denotes the smallest change in value of a measured variable to which an instrument responds: that is, the instrument has a dead zone. On the other hand, sensitivity may mean that the instrument produces a given deflection for a given change in value of a measured quantity: that is, the instrument has a given span.

Dead zone is the largest range of values of a measured variable to which the instrument does not respond. This is sometimes called dead spot and hysteresis. Dead zone usually occurs with friction in an indicating or recording instrument, particularly the latter. It may also be found in certain kinds of mechanisms that can only indicate small and discrete changes in value of a measured variable.

Hysteresis caused by friction or backlash results in the characteristic loop when the instrument is calibrated first in one direction and then in the other. Hysteresis may also be caused by permanent set in highly stressed materials. This effect, as such, is not usually of primary importance in an instrument but is always considered important as dead zone. The reason is that industrial instruments used in continuous manufacturing or processing measure values that do not change a great deal over long periods of time. Consequently, the dead zone is noticeable, but the hysteresis may not appear directly.

Dynamic Characteristics

Instruments rarely respond instantaneously to changes in the measured variable. Instead, they exhibit a characteristic slowness or sluggishness due to such things as mass, thermal capacitance, fluid capacitance, or electric capacitance. Furthermore, pure delay in time is often encountered where the instrument "waits" for some reactions to take place. Industrial instruments, such as are discussed here, are nearly always used for measuring quantities that fluctuate with time. Dynamic and transient behavior of the instrument is therefore as important as and often more important than static behavior.

The dynamic behavior of an instrument is determined by subjecting its primary element to some known and predetermined variation in measured quantity. The three most common variations are: (1) *step change,* in which the primary element is subjected to an instantaneous and finite change in measured variable; (2) *linear change,* in which the primary element is "following" a measured variable, changing linearly with time; (3) *sinusoidal change,* in which the primary element follows a measured variable, the magnitude of which changes in accordance with a sinusoidal function of constant amplitude. The physical conditions of imposing these changes on an instrument may cause great difficulty. Nevertheless methods can usually be devised for each test, so that close approximation to each type of change can be had.

The dynamic characteristics are:

Speed of response.........	Lag
Fidelity.................	Dynamic error

where the qualities on the left are desirable and those on the right are undesirable.

Speed of response is the rapidity with which an instrument responds to changes in the measured quantity. *Measuring lag* is a retardation or delay in the response of an instrument to changes in the measured quantity. *Fidelity* is the degree to which an instrument indicates the changes in measured variable without dynamic error. *Dynamic error* is the difference between the true value of a quantity changing with time and the value indicated by the instrument if no static error is assumed.

The calibration of many instruments is nonlinear: that is, the plot of measured variable versus scale calibration in inches is not a straight line on ordinary graph paper. The orifice flow-

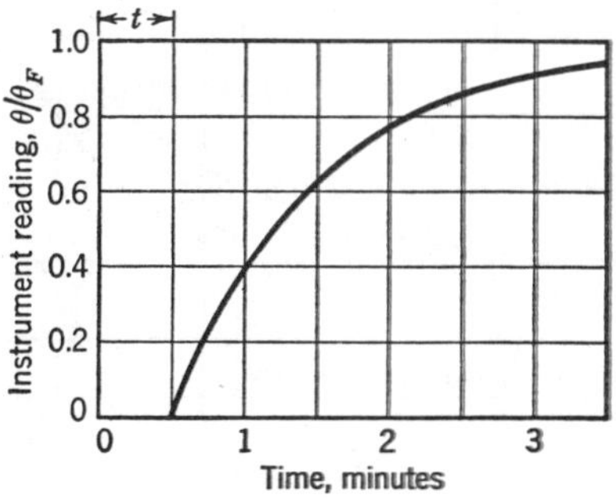

FIG. 1–3. Dead-Time Lag

meter, for example, is generally nonlinear because of the square-root characteristic. In such cases it is necessary to restrict the dynamic analysis to a limited portion of the range, selecting as many different portions as are required to express its dynamic characteristics.

The response of power-operated instruments run by an automatic-control device or a servomechanism is a specialized subject. Generally, for purposes of industrial instrumentation a simple approximation of their lag is sufficient.

Measuring lag may be of the retardation type, that is, where the response of the instrument begins immediately on a change in measured variable, or of the time-delay type called *dead time*, illustrated by t in Fig. 1–3. Dead time simply shifts the response of the instrument along the time scale and causes a corresponding dynamic error. Generally speaking, measuring lags of this type are very small and may be expressed in fractions of a second. Some types of instruments have periodic measuring systems, in which changes in the value of the measured variable are detected only at given intervals, such as every 1 sec, or every 10 sec. An instrument with appreciable dead time is generally not satisfactory for measuring a variable that fluctuates rapidly.

Finally, dead time is caused by a finite dead zone in the instrument. This effect is illustrated in Fig. 1–4, where the trend of the actual value of the measured variable is indicated. Suppose that the instrument possesses a small but finite dead zone indicated by the dotted lines. The dead zone may be caused by

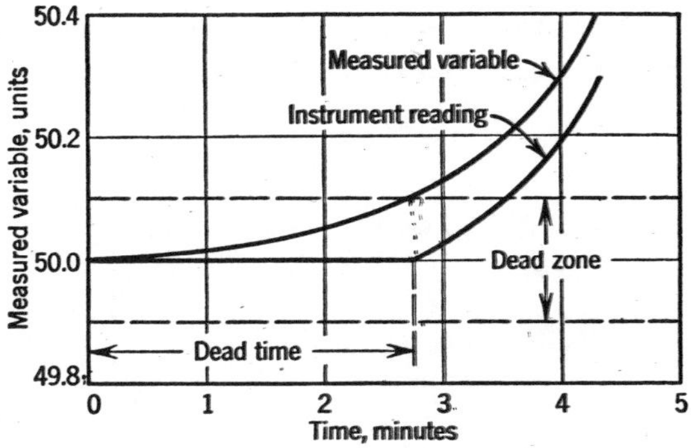

FIG. 1–4. Dead Zone and Dead Time

starting friction or a similar effect. The instrument does not respond for a certain delay time and acts only when the measured variable has changed sufficiently to overcome the starting friction. Obviously this type of delay depends on how fast the measured variable is changing and on the extent of the instrument dead zone.

Dynamic Response of First-Order-Type Instruments

An elementary temperature-measuring device such as a thermometer has a bulb or other primary element, which is mainly thermal capacitance, surrounded by a heat-transfer film, which is mainly resistance. As Fig. 1–5 shows, this is thermally equivalent to a resistance and capacitance in series. Also a pressure gage may have a lightweight bellows or diaphragm, which encloses a given volume, and a pressure inlet made by a small tube that provides resistance to fluid flow. Such instruments may be called first-order type.

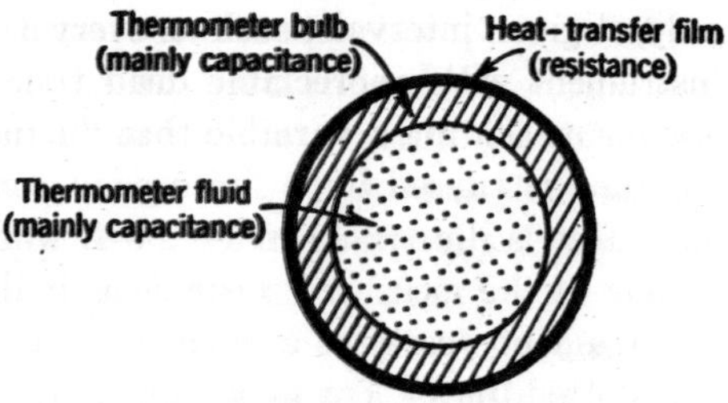

FIG. 1-5. Cross Section of Thermometer Bulb

Ideally the dynamic response of first-order-type instruments to a *step change* can be represented by [1]

$$T\frac{d\theta}{dt} + \theta = \theta_F \tag{1-1}$$

where θ = value indicated by instrument
θ_F = final steady value
t = time
T = time constant = RC sec or min

This linear first-order differential equation has the particular solution for given initial conditions,

$$\frac{\theta}{\theta_F} = 1 - e^{-t/T} \tag{1-2}$$

which represents a single exponential response of the kind shown by Fig. 1-6. This figure is plotted in dimensionless ratios so that

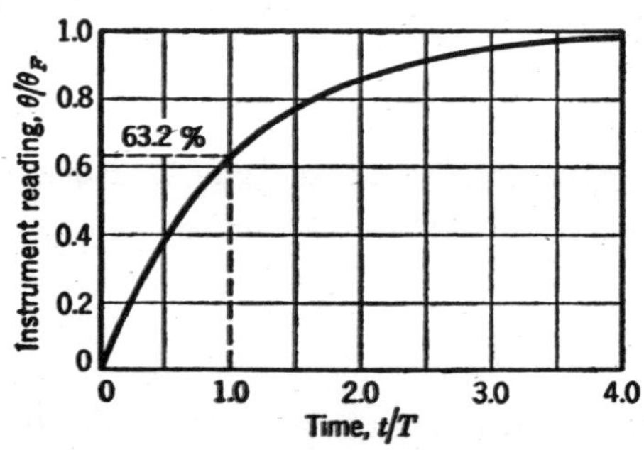

FIG. 1-6. Response of First-Order-Type Instrument to a Step Change

[1] In the transient analysis presented here we are using the approach of C. S. Draper and G. P. Bentley, "Design Factors Controlling the Dynamic Performance of Instruments," *ASME Trans.*, **62**, 5, 421 (1940).

a change of time constant T alters the time scale. Notice that, as the time constant T, which is the measuring lag, becomes larger, the response, while maintaining the same shape, becomes proportionately slower. The time constant T is the time required to indicate 63.2 per cent of the complete change. In many cases the measuring lag is specified by the time required to attain 90 per cent, 95 per cent, or 99 per cent of the full change. These specifications are easily converted to obtain the time constant T by use of tables for e^{-x} as shown in the following:

Time	To attain, per cent
T	63.2
$2.303T$	90
$2.996T$	95
$4.605T$	99

The measuring lag for first-order-type instruments is given by the time constant T. It is significant that the measuring lag of many thermometers, thermocouples, and pressure gages may be specified in this manner with accuracy sufficient for industrial-instrumentation purposes. The time constant T is numerically equal to the product of resistance and capacitance. In a thermometer bulb the resistance to heat flow is given in degrees temperature drop for each British thermal unit per minute heat flow, the capacitance is given in British thermal units per degree temperature rise, and their product is minutes.

The dynamic error following a step change is not so important as the dynamic error during a linear change or sinusoidal change, since a measured variable does not ordinarily change in this fashion.

The dynamic response of these instruments to a *linear change* can be represented by

$$T\frac{d\theta}{dt} + \theta = Kt \tag{1-3}$$

where K is the rate of change of the true value of measured variable and the last term, Kt, shows that the actual measured variable is changing linearly with time. A particular solution for given initial conditions is

$$\frac{\theta}{KT} = \frac{t}{T} - 1 + e^{-t/T} \qquad (1\text{–}4)$$

and the solution is plotted in Fig. 1–7, using dimensionless quantities. It is apparent that, after the transient has disappeared, the instrument lags behind by essentially a constant time and a constant value. The transient disappears after a time equal to about three time constants, since by then 95 per cent of the lag is attained.

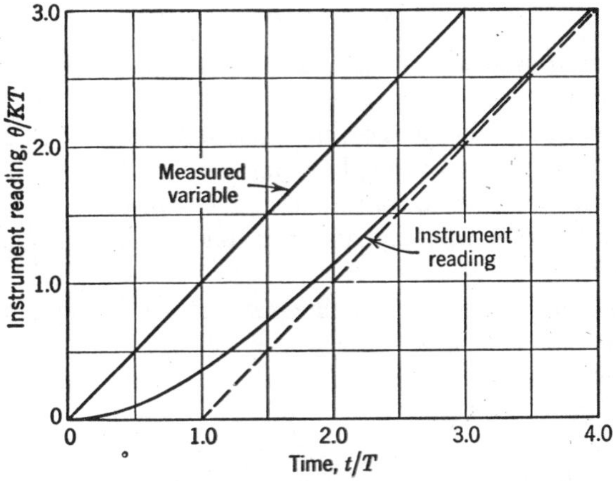

FIG. 1–7. Response of First-Order-Type Instrument to a Linear Change

The lag of the instrument is given directly by the time constant T, and we now see why this particular value was chosen. For example, a thermometer having a time constant of 1.0 min will ultimately lag behind a linear change by 1.0 min. The dynamic error is given in this case by the value KT. For the thermometer with 1.0-min lag and a linear change of 5.0° temperature per min, the dynamic error is 5.0°. By this we see also that the dynamic error increases directly with the rate of change of the measured variable. If a small dynamic error is desired (high fidelity), either the instrument must have small lag, or the measured quantity must change slowly.

The *dynamic response to a sinusoidal change* is found from

$$T\frac{d\theta}{dt} + \theta = A \sin \omega t \qquad (1\text{–}5)$$

where θ = value indicated by instrument
t = time
T = time constant, RC sec or min
A = amplitude of cycle of measured variable
ω = circular frequency of cycle

A solution with given initial conditions and with the transient terms omitted is

$$\left[\frac{\theta}{A}\right]_{ss} = \frac{1}{\sqrt{1 + (\omega T)^2}} \sin(\omega t - \phi) \tag{1-6}$$

where $\phi = \arctan \omega T$ = lag angle and the subscript ss denotes steady state.

This equation shows that the instrument lags the measured variable by a geometric angle ϕ and also that the amplitude is reduced or attenuated. Because the complete equation yields little information, the lag and dynamic error may be treated separately.

The instrument produces a sine wave, which is displaced by the lag angle ϕ varying from 0 towards 90 geometrical degrees. However, the phase lag in degrees is not so convenient as the lag of the instrument expressed in time units as follows:

$$\text{Lag} = \frac{1}{\omega} \arctan \omega T \tag{1-7}$$

and this equation is plotted in Fig. 1–8. Suppose that the thermometer with a time constant of 1.0 min measures a temperature cycling with a 10-min period. The lag is

$$\text{Lag} = \frac{10}{2\pi} \arctan\left(\frac{2\pi}{10}\, 1.0\right) = 0.89 \text{ min}$$

Figure 1–9 shows the ratio of amplitudes, which is

$$\left[\frac{\theta_m}{A}\right]_{ss} = \frac{1}{\sqrt{1 + (\omega T)^2}} \tag{1-8}$$

The instrument fidelity decreases rapidly when either its measuring lag increases or the fluctuations in actual measured variable become rapid. With the thermometer of 1.0-min lag and a cycle of temperature of 10-min period, the amplitude indicated by the instrument is only 85 per cent of the actual value. If the measuring lag of the thermometer increases to 5.0 min, which is entirely possible, the amplitude drops to 31 per cent. If the measuring lag is 1.0 min but the cycle increases to a 2-min

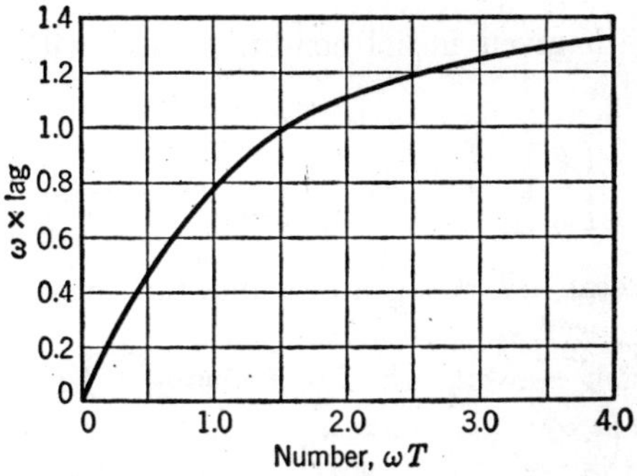

Fig. 1-8. Lag of First-Order-Type Instrument to a Sinusoidal Change

period, the amplitude drops to 31 per cent. Under such conditions the dynamic error of the instrument is so large that the instrument readings are meaningless. If the instrument must indicate a true value within 10 per cent, then the number ωT must be about 0.5 or less.

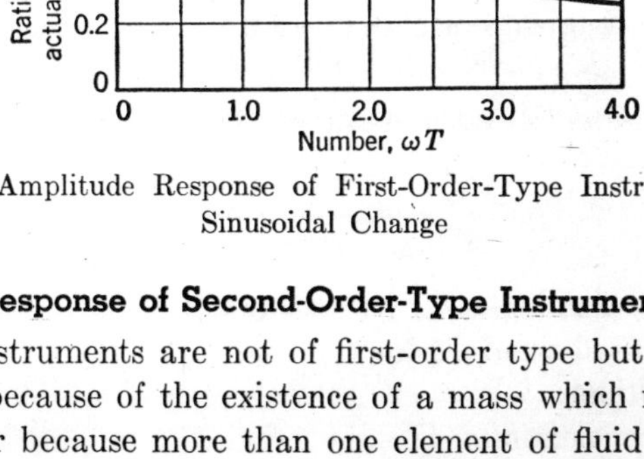

Fig. 1-9. Amplitude Response of First-Order-Type Instrument to a Sinusoidal Change

Dynamic Response of Second-Order-Type Instruments

Many instruments are not of first-order type but have additional lag because of the existence of a mass which must be accelerated or because more than one element of fluid capacity is involved. Usually a thermometer bulb is surrounded by a metal or ceramic tube for purposes of protection, as indicated in the diagram on the left in Fig. 1–10. The primary element can be assumed to be mainly thermal capacity, the air gap mainly resistance, the well mainly thermal capacitance, and the heat-transfer film resistance. Likewise, a pressure gage may have a similar construction involving separated resistances and capacitances, as indicated on the right in Fig. 1–10.

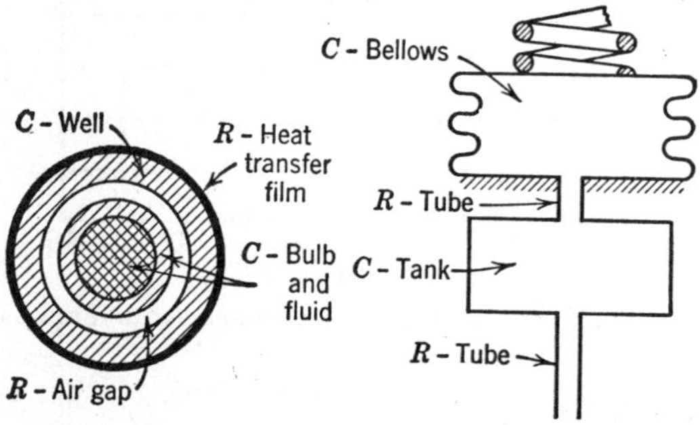

FIG. 1–10. Instruments of Second-Order Type

Most mechanical instruments such as pressure gages and manometers have some small but appreciable mass, which must move during instrument operation. The pressure gage of Fig. 1–11 has a mass in the system, a spring, and a viscous damping means. This last is often nothing but the air damping and viscous friction of the moving members. The liquid manometer of Fig. 1–11 also has these same components, with damping caused by viscous fluid friction. These instruments may be called second-order type.

All the instruments of Figs. 1–10 and 1–11 possess a dynamic response to a *step change* that may ideally be described by means of a second-order differential equation such as the following,

$$a_2 \frac{d^2\theta}{dt^2} + a_1 \frac{d\theta}{dt} + a_0\theta = a_0\theta_F \tag{1–9}$$

where θ = value indicated by instrument
θ_F = final steady value
t = time
a_2, a_1, a_0 = constants

It is necessary to derive the constants for equation 1–9 by analysis of each instrument. For instruments similar to those in Fig. 1–10, the constants are combinations of the resistances and capacitances. For the pressure gage of Fig. 1–11, the constants include the mass, the damping, and the spring gradient.

For instruments having rotating instead of linear motion, mass becomes moment of inertia. For the fluid manometer of Fig. 1–11, the constants include the length of the column and

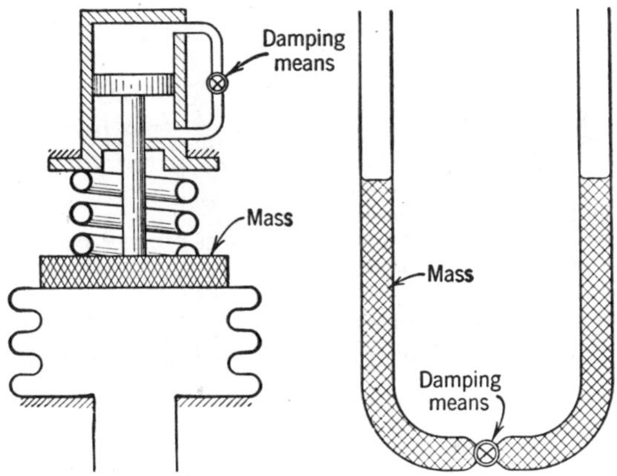

FIG. 1–11. Instruments of Mass-Spring-Damping Type

a damping factor. For a galvanometer-type electric instrument the constants become those for moment of inertia, electrical damping, and spring gradient.

The solutions to equation 1–9 are well known and involve three cases: an oscillatory condition in which the roots of the auxiliary equation are conjugate complex with negative real parts; critical damping in which the roots are negative, real, and equal; and an overdamped condition in which the roots are negative, real, and unequal.

The particular solutions for given initial conditions are:

1. Oscillatory Response $\zeta < 1$:

$$\frac{\theta}{\theta_F} = 1 - \frac{1}{\sqrt{1-\zeta^2}} e^{-t/T} \cos\left[\frac{\sqrt{1-\zeta^2}}{\zeta}\frac{t}{T} - \phi\right] \quad (1\text{–}10)$$

where $T = \frac{2a_2}{a_1} \doteq$ characteristic time

$\zeta = \sqrt{a_1^2/4a_2a_0}$ = damping ratio

$\phi = \arcsin(\zeta)$

2. Critically Damped Response, $\zeta = 1$:

$$\frac{\theta}{\theta_F} = 1 - e^{-t/T}\left[1 + \frac{t}{T}\right] \tag{1-11}$$

3. Overdamped Response, $\nu > 1$:

$$\frac{\theta}{\theta_F} = 1 - \frac{\nu}{\nu - 1} e^{\frac{-2}{\nu+1}\frac{t}{T}} + \frac{1}{\nu - 1} e^{\frac{-2\nu}{\nu+1}\frac{t}{T}} \tag{1-12}$$

where $\nu = \frac{1 + \sqrt{1 - (4a_2a_0/a_1^2)}}{1 - \sqrt{1 - (4a_2a_0/a_1^2)}}$ = damping number

These responses are given in Fig. 1–12 for various values of the parameters. The graph is dimensionless again, so that a change of characteristic time T alters the time scale. The measuring lag is first of all directly proportional to the characteristic

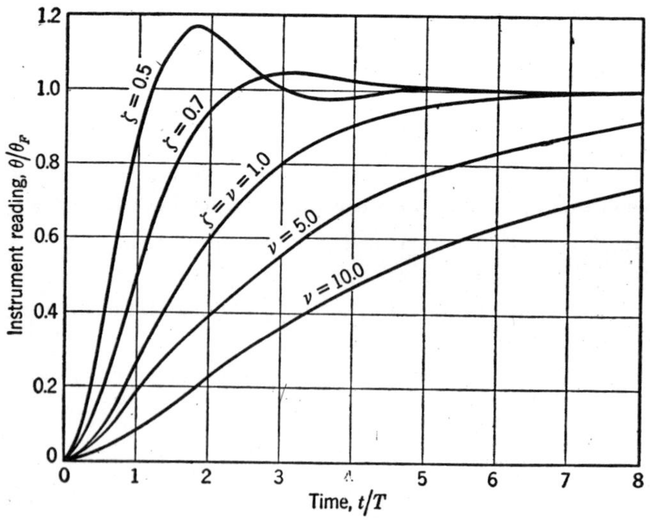

FIG. 1–12. Response of Second-Order-Type Instrument to a Step Change

time T; a larger characteristic time T produces a greater measuring lag and a slower response. In addition, an increase in damping ratio ζ or an increase in damping number ν also increases the measuring lag and makes the response slower. We see therefore that two numbers are required to express the measuring lag of such instruments.

Most pressure gages and most manometers have a slightly oscillatory response with a damping ratio ζ between 0.5 and 1.0, unless they are provided with special damping means. A greater damping obtained by closing a restriction between the legs of a fluid manometer can greatly increase the measuring lag. For these instruments the characteristic time T depends on the ratio of mass or moment of inertia to damping. A small mass is therefore highly desirable.

Most galvanometer-type electric instruments are very nearly critically damped, and their measuring lag can therefore be specified by the characteristic time T. Instruments, particularly temperature-measuring elements, like those in Fig. 1–10 are generally overdamped. Consequently, their measuring lag can be specified only by giving the characteristic time T, as well as the damping number ν. In general, an increase of either resistance or capacitance increases the characteristic time but may either increase or decrease the damping number.

The *dynamic response* of these instruments *to a linear change* is found from

$$a_2 \frac{d^2\theta}{dt^2} + a_1 \frac{d\theta}{dt} + a_0\theta = Kt \tag{1–13}$$

The solutions to this equation for given initial conditions are:

1. Oscillatory Response, $\zeta < 1$:

$$\frac{\theta}{KT} = \frac{t}{T} - 2\zeta^2 + \zeta\sqrt{4 + \frac{(2\zeta^2 - 1)^2}{(1 - \zeta^2)}}\, e^{-t/T} \cos\left[\frac{\sqrt{1 - \zeta^2}}{\zeta}\frac{t}{T} - \phi\right] \tag{1–14}$$

where $\phi = \arcsin\left[\frac{2\zeta\sqrt{1 - \zeta^2}}{2\zeta^2 - 1}\right]$

2. Critically Damped Response, $\zeta = 1$:

$$\frac{\theta}{KT} = \frac{t}{T} - 2 + e^{-t/T}\left[2 + \frac{t}{T}\right] \tag{1-15}$$

3. Overdamped Response, $\nu > 1$:

$$\frac{\theta}{KT} = \frac{t}{T} - \left(\frac{\nu+1}{2\nu}\right)^2 + \frac{\nu(\nu+1)}{2(\nu-1)}e^{\frac{-2}{\nu+1}\frac{t}{T}} - \frac{\nu+1}{2\nu(\nu-1)}e^{\frac{-2\nu}{\nu+1}\frac{t}{T}} \tag{1-16}$$

These equations are plotted in Fig. 1–13, again with dimensionless ratios. The instrument lags behind the true value by an

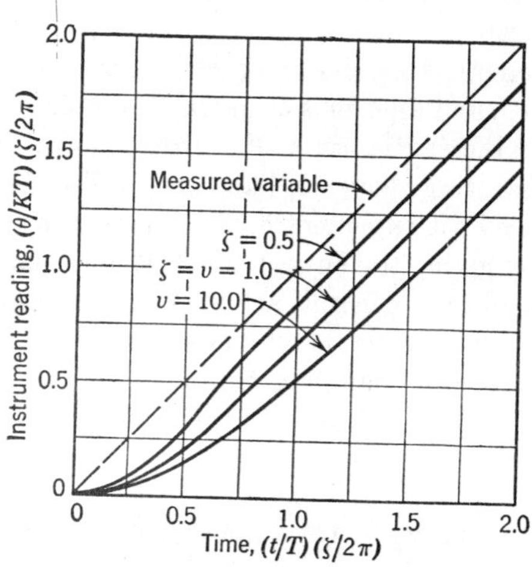

Fig. 1–13. Response of a Second-Order-Type Instrument to a Linear Change

amount depending on both its characteristic time and its damping ratio or damping number. In fact, a little study of the foregoing equations shows that the lag after the transient period can be expressed as follows:

1. Oscillatory Response, $\zeta < 1$:

$$\text{Lag} = 2\zeta^2 T \tag{1-17}$$

2. Critically Damped Response, $\zeta = 1$:

$$\text{Lag} = 2T \tag{1-18}$$

3. Overdamped Response, $\nu > 1$:

$$\text{Lag} = \frac{(\nu + 1)^2}{2\nu} T \tag{1–19}$$

As an example, a pressure gage with a damping ratio of 0.5 and a characteristic time of 1.5 sec has a lag of 0.75 sec. A thermal element with a characteristic time of 2.0 min and a damping number of 10.0, for example, shows a lag of 6.05 min.

The dynamic error during a linear change is given by:

1. Oscillatory Response, $\zeta < 1$:

$$\text{Dynamic error} = 2\zeta^2 KT \tag{1–20}$$

2. Critically Damped Response, $\zeta = 1$:

$$\text{Dynamic error} = 2KT \tag{1–21}$$

3. Overdamped Response, $\nu > 1$:

$$\text{Dynamic error} = \frac{(\nu + 1)^2}{2\nu} KT \tag{1–22}$$

In the pressure gage with a damping ratio of 0.5 and a characteristic time of 1.5 sec the dynamic error with a rate of change of 5 units per second is 3.75 units. For the thermal element with a characteristic time of 2.0 min and a damping number of 10.0 and with a rate of change of 2 units per minute the dynamic error is 12.1 units.

The dynamic response to a *sinusoidal change* is represented by

$$a_2 \frac{d^2\theta}{dt^2} + a_1 \frac{d\theta}{dt} + a_0\theta = A \sin \omega t \tag{1–23}$$

where A = amplitude of cycle of measured variable
ω = circular frequency of cycle

A solution with given initial conditions and with the transient terms omitted is

$$\left[\frac{\theta}{A}\right]_{ss} = \frac{1}{\sqrt{(1 - \beta^2)^2 + (2\zeta\beta)^2}} \sin (\omega t - \phi) \tag{1–24}$$

where $\beta = \zeta\omega T$ = frequency number

$$\phi = \arctan\left[\frac{2\zeta^2\omega T}{1 - (\zeta\omega T)^2}\right] = \text{lag angle}$$

The various constants for the three types of response are related by

$$\zeta T = \frac{(\nu + 1)T}{2\sqrt{\nu}} = \frac{1}{\omega_n} \tag{1-25}$$

where ω_n is the circular undamped frequency. We are interested in the amplitude of the cycle indicated by the instrument, and this is given by

$$\left[\frac{\theta_m}{A}\right]_{ss} = \frac{1}{\sqrt{(1 - \beta^2)^2 + (2\zeta\beta)^2}} \tag{1-26}$$

This function is plotted in Fig. 1–14, and it is evident that the instrument may indicate either a greater or smaller than true

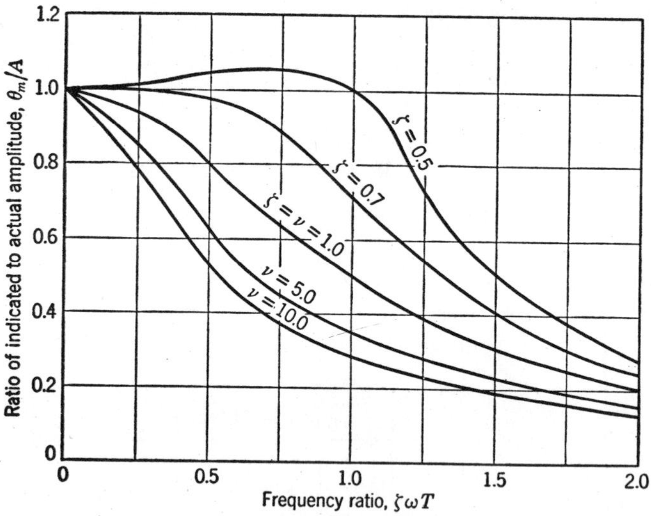

FIG. 1–14. Amplitude Response of Second-Order-Type Instrument to a Sinusoidal Change

amplitude. This is the well-known phenomenon of resonance in a vibrating system. It may occur in a pressure gage when the measured pressure fluctuates at a frequency for which $\zeta\omega T$ is near 1.0. For a pressure gage with a damping ratio of 0.5 and a characteristic time of 1.5 sec, if the frequency of cycle in pressure is 0.1 per second, then ω is 0.2π, and Fig. 1–14 shows the amplitude to be 1.04 or 4 per cent greater than the true value. With a temperature-measuring instrument having a characteristic time of 1.0 min, a damping number of 10, and measuring a temperature with a period of 10 min, the frequency ω is 0.63, and the amplitude is therefore 0.26 or only 26 per cent of the true value. Temperature-measuring elements always have an overdamped response; consequently, it may be seen from the graph that they must be operated at very low frequencies in order to maintain a small dynamic error.

The lag of the instrument can be expressed in time units as follows:

$$\text{Lag} = \frac{1}{\omega} \arctan \left[\frac{2\zeta^2 \omega T}{1 - (\zeta \omega T)^2} \right] \tag{1-27}$$

and this relation is plotted in Fig. 1–15.

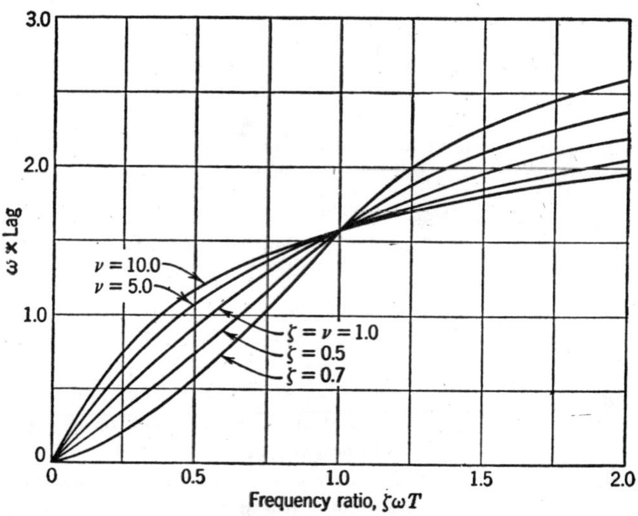

FIG. 1–15. Lag of Second-Order-Type Instrument in a Sinusoidal Change

For a pressure gage with a damping ratio of 0.5 and a characteristic time of 1.5 sec, the lag is 1.12 sec. The lag of a thermal element with a characteristic time of 1.0 min and a damping number of 10 is 2.57 min. The lag increases as the frequency of cycle of measured variable increases, as the damping ratio or damping number increases, and as the characteristic time increases.

Unfortunately the dynamic response of all instruments is not covered by the preceding analysis, although for most practical purposes it is sufficient. It is apparent that we might have instrument mechanisms that can be described only by third-, fourth-, or higher-order differential equations.

Problems

1-1. Suppose that your home has a thermostat which indicates the "temperature" of the room. Discuss the meaning of this measurement. Why is the temperature measured? Of what use is it? Suggest what measurements must be made if the "comfort" of the room is to be determined.

1-2. For an ordinary mercury-in-glass thermometer name the parts that might constitute the primary element and the indicating element.

1-3. A thermometer is calibrated 325 to 400°F. The accuracy is specified to be within ±0.25 per cent. What is the maximum static error in units?

1-4. A flowmeter is calibrated 0 to 10,000 cu ft per hr. The accuracy is specified as within ±0.75 per cent above 20 per cent of scale reading. What static error in units is possible if the instrument indicates 8000 cu ft per hr?

1-5. A transmitter, relay, and receiver are used in measuring a pressure at a remote point. The specified accuracies are:

Transmitter	Within ±0.25 per cent
Relay	Within ±1.0 per cent
Receiver	Within ±0.75 per cent

What is the root-square accuracy?

1-6. The dead zone of a certain pyrometer is 0.125 per cent of span. The calibration is 800 to 1800°F. What temperature change might occur before it is detected?

1-7. The time constant of a simple thermometer may be found by imposing a step change and noting the time required to achieve 63.2 per cent of the final value. Suppose this time is 0.23 min. If this thermometer is used in a bath, the temperature of which is cycling 3 times per minute, what per cent of actual amplitude would it indicate, and how many minutes is it delayed?

1-8. A pressure is cycling from 80 psi to 90 psi with a period of 5 sec. The attenuation of the cycle must be limited to 10 per cent.

(*a*) What maximum time constant is permissible in a pressure gage with a first-order-type response?

(*b*) What characteristic time is permissible in a critically damped pressure gage of the mass-spring-damping type?

1-9. The temperature of a furnace is raised at the rate of 300°F per hr. The instrument must not be in error by more than 10°F. What maximum time constant is permissible if the instrument has a first-order-type response?

1-10. The temperature of a tempering bath falls at a rate of 1°F per sec. The thermometer used for measuring the temperature of the bath has a second-order response with characteristic time of 0.1 min and a damping number of 10.

(*a*) How many minutes does the thermometer lag behind the bath temperature? (*b*) What is the maximum dynamic error?

1-11. A home thermostat for on-off control of room temperature produces a cycle with a 30-min period and 3°F amplitude. If a thermometer with a lag of 4.0 min is placed beside it, what cycle should it indicate? Assume the thermometer has first-order lag.

2· Expansion Thermometers

The thermometer is, without question, the most widely used instrument in industrial processing of all kinds. It is simple to use and economical and provides reliable performance under a variety of applicational conditions. Many forms of the expansion thermometer are in use; all of them are based on the thermal expansion of a solid, liquid, or gas, and each may be found in hundreds of varieties and forms. Of these many types, the mercury-in-glass thermometer, the bimetallic thermometer, and the pressure-spring thermometer are discussed in sufficient detail to give a thorough background of the fundamentals in applying thermometric measurements.

Thermometers are employed for the measurement of temperature. On the whole, temperature is the most important variable in industrial processing, primarily because it is a fundamental condition characteristic of the thermal state of a body. Speaking of industrial processes, it is this thermal state that determines whether desired chemical reactions will take place, as in certain chemical processes, or whether the physical conditions of a manufactured product are correct. Temperature and heat flow are related qualitatively by the second law of thermodynamics, which expresses the fact that heat flows of its own accord from a body at a higher temperature to a body at a lower temperature. It is therefore important in temperature measurement to remember that two bodies in intimate contact are at the same temperature only if there is no heat flow between them. Temperature measurement is based on the so-called zeroth law of thermodynamics, which states that two bodies in thermal equilibrium with a third body are in thermal equilibrium with each other.

Temperature Scales

A scale representing quantitatively the temperature of a body is arbitrary. Temperature scales are determined by choosing a particular physical property of a body, such as the thermal expansion of a gas at constant volume or the change of electrical resistance of a wire with temperature; designating two thermal equilibrium points by numbers; and subdividing the interval into a number of equal parts. However, since such scales depend on different physical properties, the *temperature* intervals need not be identical, even though the numbers on the scales agree.

Suppose, by way of simple example, that the thermal expansion of a solid-metal bar is directly proportional to temperature and that the thermal resistivity of a wire is proportional to the cube of temperature. If we place the rod and the wire in a heated enclosure, we can put a mark on the rod and measure the resistance of the wire. These points can be marked 0°. If we further heat the enclosure and mark new points and call them 100°, these intervals may be divided into 100 equal parts, and we have a temperature "scale." However, it is easy to see that a degree on the "expansion" temperature scale and a degree on the "resistance" temperature scale do not represent the same amount of change in thermal equilibrium—our scale is not independent of its working substance.

An ideal temperature scale is independent of the substance used in defining it. Such a scale was suggested by Lord Kelvin and is called the thermodynamic scale. This scale is based on the availability of energy and is arranged so that the ratio of the values of any two temperatures is equal to the ratio of the heat taken in to the heat rejected by a reversible heat engine working with a heat source at the higher temperature and a heat sink at the lower temperature. If a number of reversible heat engines were arranged to operate on the same Carnot cycle, all but the first receiving the heat rejected by its predecessor, then the temperature change through each engine would represent a temperature interval. This defines the thermodynamic scale independently of the working substance. Suppose we call the temperature at which heat is taken in by the first engine some fixed number and then number all the engines at progressively lower temperatures. The engine in the series that discharges no heat would end the series. This point is called absolute zero. On the Kelvin thermodynamic temperature scale the equilibrium temperature of ice and water at standard pressure is called the ice point and is 273.16°K. The equilibrium temperature of water and steam at standard pressure is called the steam point and is 373.16°K.

There are a number of other temperature scales, all based on the ideal thermodynamic scale but using different units. The most important industrially are: Fahrenheit, centigrade, Kelvin, Rankine, and Réaumur.

The *Fahrenheit* scale, abbreviated °F, was introduced about 1665 and is used in most English-speaking countries. This scale assigns 0°F to the lowest temperature of a certain salt-and-ice mixture, 32°F to the ice point, and 212°F to the steam point.

The *centigrade* scale, abbreviated °C, was introduced about 1740 and is commonly used in European countries, where it is called the Celsius scale. It is also commonly used in technical scientific literature. This scale assigns 0°C to the ice point and 100°C to the steam point.

The *Kelvin* scale, abbreviated °K, is also called the centigrade absolute scale. It is much used in technical literature. This scale assigns 273.16°K to the ice point and 373.16°K to the steam point.

The *Rankine* scale, abbreviated °R′ and also called the Fahrenheit absolute scale, is commonly used in engineering literature. It assigns 491.69°R′ to the ice point and 671.69°R′ to the steam point.

The *Réaumur* scale, abbreviated °R, was introduced about 1731 and is used in a few European countries. It assigns 0°R to the ice point and 80°R to the steam point. This scale is often used in the alcohol industries.

The relationship of the various temperature scales is shown in Fig. 2–1. The equation relating Fahrenheit temperature to centigrade temperatures is:

$$°C = \frac{5}{9}(°F - 32) \tag{2-1}$$

and the interrelation of all scales may be readily found.

An international temperature scale[1] has been set up, in order to define the temperature intervals on the basis of equilibrium of various substances, as shown in Table A-1 (appendix). From −310°F (−190°C) to 1220°F (660°C) the international temperature scale is defined by the resistance of a platinum re-

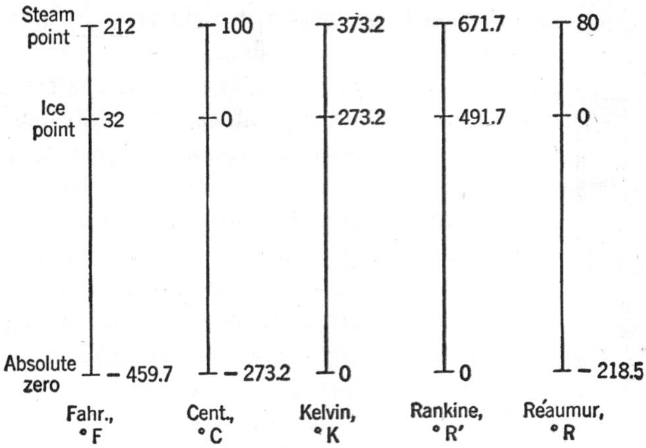

FIG. 2-1. Temperature Scales

sistance thermometer. From 1220°F (660°C) to 1945°F (1063°C) the scale is defined by the electromotive force (emf) of a platinum versus platinum–rhodium thermocouple. Above 1945°F (1063°C) the scale is defined by a formula for the radiation of a black body. For all practical purposes one degree Kelvin is the same as one degree centigrade on the international scale.

Constant-Volume Gas Thermometer

The reference standard in thermometry is the constant-volume gas thermometer employing helium below 32°F, hydrogen between 32 and 212°F, and nitrogen above 212°F. This thermometer is used in the laboratory for comparison calibration of other thermometers. The gas is maintained at constant volume,

[1] H. F. Stimson, *J. Research Natl. Bur. Standards* **42,** 209 (1949). G. K. Burgess, *Bur. Standards J. Research* **1,** 635 (1928).

and, by the ideal-gas law, the pressure is directly proportional to temperature. However, helium, hydrogen, and nitrogen are not ideal gases, and, consequently, the deviation from ideal behavior makes it necessary to correct the temperature indications.[2] The pressure reading of the thermometer can therefore be related to the ideal thermodynamic temperature scale.

Mercury-in-Glass Thermometer

The mercury-in-glass thermometer, widely used in both laboratory and industry, is one of the simplest temperature-measuring devices. It utilizes the volumetric expansion of mercury with temperature as a means of indicating the temperature. As illustrated in Fig. 2–2, the typical industrial thermometer has a bulb formed by a glass envelope, which contains mercury, enclosed in a metal well. As heat is transferred through the well and metal stem and into the mercury, the mercury expands, pushing the column of mercury higher in the capillary above.

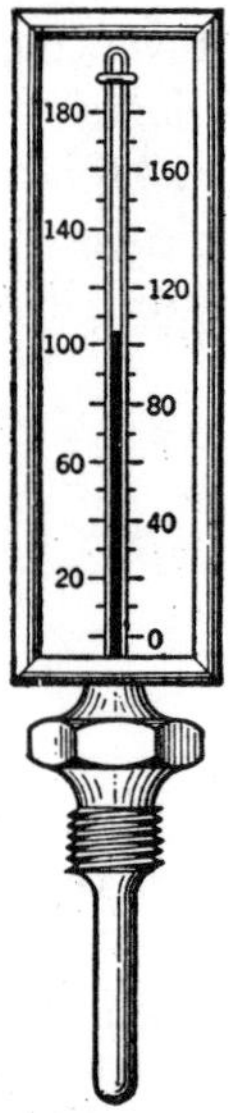

Fig. 2–2. Industrial Mercury-in-Glass Thermometer

The scale calibration of a mercury thermometer is not exactly linear, although the variation from linear is small. First, the glass envelope of the thermometer expands and contracts with temperature changes. This changes the volume inside the thermometer and must be taken into consideration in original calibration. Second, the coefficient of cubical expansion for mercury varies somewhat with temperature. Third, the space above the mercury in the thermometer bore is sometimes filled with dry nitrogen under pressure. The purpose of this gas is to elevate the boiling point of the mercury when the thermometer is used at high temperatures. The boiling point of the triple-distilled mercury, commonly used in thermometers, is about 675°F (357°C). Therefore, when the mercury column rises, the gas pressure increases, causing elastic expansion of the thermometer walls and compression of mercury volume.

[2] These corrections are given in a number of handbooks. For example, see *Handbook of Chemistry and Physics,* 1947, p. 2455.

There is sometimes a gradual change in volume of the thermometer, due to aging over long periods of time. Always, however, the glass in good-quality industrial thermometers is aged before use, in order to reduce this effect.

The industrial mercury-in-glass thermometer is constructed as shown in Fig. 2–2. It is contained in a metal case with the bulb inserted into a metal thermal well. A metal scale is mounted behind the upper end of the thermometer, and a glass covers the scale. This provides complete protection for the thermometer itself, making a rugged and compact instrument. The stem is generally protected and insulated by packing.

The thermal well is provided for the purposes of preventing breakage and providing a sealing means at the point of installation. It is generally of brass or steel, although cast iron, Monel, stainless steel, and aluminum are sometimes used. For use in ovens and air ducts, perforated metal guards are employed for protection against breakage.

The glass of the thermometer stem is generally shaped so as to magnify the apparent width of the thread of mercury. The industrial thermometer can be obtained in many shapes suited to particular applications: some with the thermometer bulb at an angle to the scale, some with the scale rotated to any desired angle.

The temperature range in which the industrial mercury-in-glass thermometer is used is −38 to 950°F. Lower temperatures are limited by the freezing point of mercury, which is −38°F. Consequently, a mercury thermometer cannot be used below this point.

For overrange protection, that is, when the thermometer is subjected to temperatures higher than the highest calibrated point, a small reservoir is formed at the top of the thermometer bore. As the mercury expands past the highest calibrated point, it slowly fills the reservoir, and the thermometer can be subjected to higher temperatures without breakage.

The accuracy of the industrial thermometer, when properly installed and used, is about ±1 per cent of span. To achieve this accuracy, the thermometer should be installed in such a manner that the bulb is at temperature equilibrium with its surroundings. The thermometer bulb should be installed in such a way that the speed of flow of the surrounding medium past the bulb is sufficient to provide rapid heat transfer. It should be immersed to a sufficient length that the heat loss along the thermal well to the surroundings is a minimum; otherwise, what is known as "immersion error" will result. The temperature of the scale and surroundings should be relatively near ordinary room temperature, so that errors due to expansion or contraction of thermometer bore and calibrated scale are small.

The speed of response of the industrial mercury thermometer will vary considerably, depending mainly on the characteristics of the fluid medium in which it is inserted and on the size of the thermometer, particularly the size of the thermal well. The space between the bulb and well may be filled with a conducting liquid such as mercury or oil, in order to increase the rate of heat transfer. The time constant of its response will range from 0.01 min to several minutes, depending on the conditions of use.

The industrial mercury-in-glass thermometer is used in applications such as open tanks containing liquids, cooking kettles, certain molten-metal baths, steam lines, pipe lines for fluid flow, and air ducts. In these cases a simple inexpensive temperature-indicating device, having reasonable accuracy, precision, and speed, is required. It should not be employed when rapidly fluctuating temperatures are to be measured with high accuracy or where fluctuating ambient temperature conditions are encountered. In some instances filling media other than mercury are used, in order to extend the range of the mercury thermometer. Other media such as ethyl alcohol, pentane, toluene, or a hydrocarbon compound are used. (Refer to Table A–3, appendix.)

Bimetallic Thermometer

The bimetallic thermometer is commonly used wherever the industrial mercury-in-glass thermometer is employed. The construction of a bimetallic-strip element provides a rugged yet accurate and simple device for the indication of temperature. A bimetal is composed of two strips of metal welded together, each strip made from a metal having a different coefficient of thermal expansion. This is shown schematically in Fig. 2–3. For a bimetal in the form of a straight cantilever beam, temperature changes cause the free end to deflect, and this deflection can be related quantitatively to the temperature change. The deflection with temperature is nearly linear, depending mainly on the coefficients of linear thermal expansion.

Invar is universally employed as the low-expansion metal. Invar is an iron–nickel alloy containing about 36 per cent nickel. Its temperature coefficient of expansion is about 1/20 that of ordinary metals. As the high-expansion metal, brass is used at lower temperatures and nickel alloys at higher temperatures.

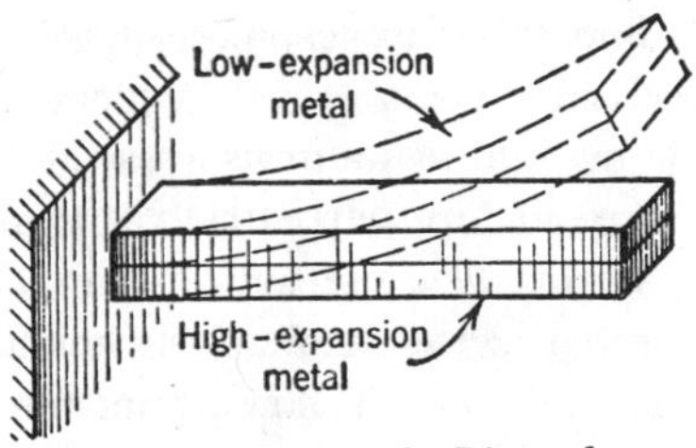

Fig. 2-3. Simple Bimetal

The form of the industrial bimetallic thermometer is shown in Fig. 2-4. The bimetal is wound in the form of a helix, with one end fastened permanently to the outer casing and the other end connected to the pointer stem. A pointer is attached to the upper end of the stem and sweeps over a circular dial to indicate the temperature. When the temperature surrounding the whole stem changes, the bimetal expands and the helical bimetal rotates at

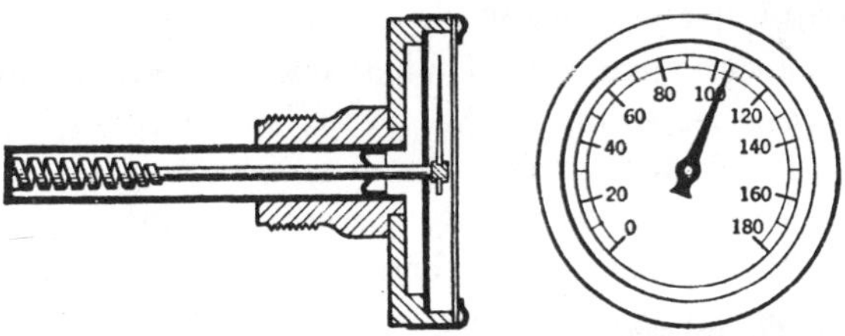

Fig. 2-4. Industrial Bimetallic Thermometer

its free end, thus turning the inside stem and pointer to a new position on the dial.

A thermal well may be used with the thermometer for protection against corrosion and breakage. The material of the thermal well may be brass, steel, stainless steel, or other alloys, depending on the requirements of the installation. Otherwise, the thermometer may be used with a bare stem. The shape of the thermometer can be selected to suit particular needs with the scale and stem at different angles.

The temperature range covered by the industrial bimetallic thermometer is −40 to 800°F. Practical limits to the temperature range are encountered at low and high temperatures, because either the stress of the bimetal during deflection must be limited to reasonable values, or the materials used in constructing the thermometer are subject to creep at high temperatures. A bimetallic thermometer will, in general, withstand about 50 per cent overrange in temperature, depending on the ability of the bimetal to be overstressed. The accuracy of this thermometer, when properly installed, is about ±1 per cent of span.

As with the mercury-in-glass thermometer, the stem should be inserted in the medium far enough to minimize immersion error. The speed of the flowing medium should be high enough to insure rapid heat transfer. Changes in ambient temperature at the head of the instrument do not seriously affect the accuracy, so long as the stem is immersed to a depth that will cover the bimetallic element inside the stem.

The speed of response of a bimetallic thermometer depends greatly on the conditions under which it is installed and is comparable to a mercury-in-glass thermometer.

Pressure-Spring Thermometer

The construction and mechanical operation of all industrial types of pressure-spring thermometers are virtually identical. Briefly, the liquid-expansion thermometer utilizes the cubical expansion of a liquid, generally mercury, to indicate the temperature. The gas-expansion thermometer operates at substantially constant volume, the pressure of the gas being proportional to temperature. The vapor-actuated thermometer is operated by the vapor pressure of a liquid.

The thermal system for all pressure-spring thermometers is very similar to that shown in Fig. 2–5. A metal bulb contains the thermometer fluid, a liquid, a gas, or a liquid–vapor, and is inserted at the point at which the temperature is to be measured. The bulb comes to temperature equilibrium with its surroundings, thereby developing a given pressure or displacement of fluid. A metal capillary is connected to the bulb and transmits the pressure at the bulb to the receiving element at the instrument. The receiving element is a form of bourdon tube or pressure spring and is used to convert the pressure or displacement of fluid in the thermometer bulb into a motion. This motion is multiplied by a linkage to operate a pen arm over a moving chart for recording purposes, or to operate a pointer over a

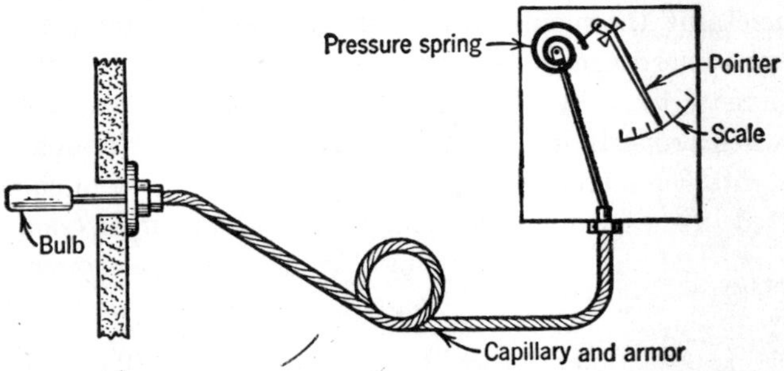

FIG. 2–5. Industrial Pressure-Spring Thermometer

stationary scale for indication of temperature. The thermal system, consisting of bulb, capillary, and receiving element, is a hermetically sealed unit.

The construction of the thermometer bulb and thermal well is shown in Fig. 2–6. The bulb is composed of a cylindrical piece of metal tubing, closed at one end and with the capillary and

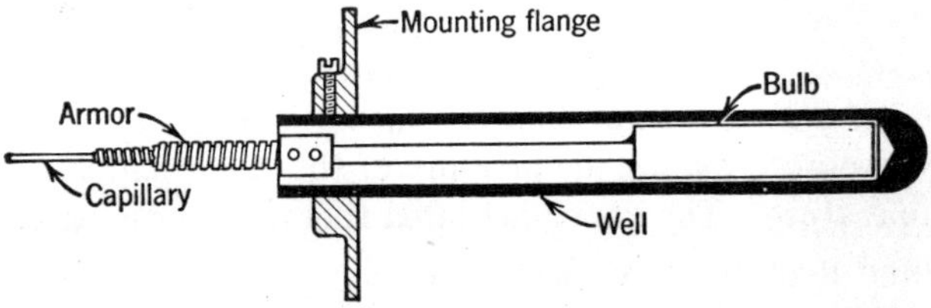

FIG. 2–6. Thermometer Bulb and Well

extension neck inserted at the other end. The size of the bulb varies considerably, depending on the type of filling medium, the temperature span of the instrument, and the length of capillary tubing with which it is used. For comparable temperature ranges and capillary lengths the dimensions of the bulb are given in Table 2–1. It should be noted that the dimensions given in

TABLE 2–1. PRESSURE THERMOMETER BULB SIZES

Type	Diameter, in.	Length, in.	Overall Volume, cu. in.
Vapor actuated	⅜	5	0.56
	½	3½	0.60
	⅜	12	1.3
Mercury	⅜	3¾	0.42
	½	6½	1.3
	25/32	7 1/16	2.9
Gas	1	5	3.9
	¾	10	4.4
	1	10	7.9

the table are by way of example only. They will vary considerably, depending on the manufacturer since there are literally hundreds of combinations of type, range, and capillary length.

The averaging (capillary) thermometer bulb is shown in Fig. 2–7. It is composed of a long bulb of approximately ¼-in.

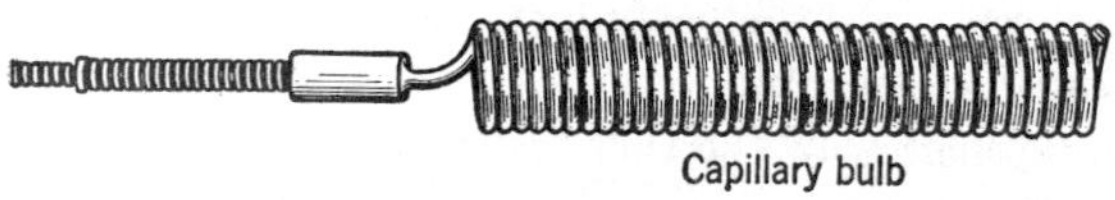

FIG. 2–7. Averaging (capillary) Thermometer Bulb

tubing wound in spiral form and is useful in obtaining average temperature over an extended region, as in a duct or oven.

The materials of the bulb are chosen according to the requirements of the particular application as to corrosion, strength, wear, and so on. Some of the materials in common use for other than mercury-expansion thermometer bulbs are copper, steel, stainless steel, and Monel. For mercury-filled thermometer bulbs, stainless steel is universally used because mercury tends to amalgamate with copper and its alloys. (See Tables A–6 and A–7, appendix.)

The extension neck is made of small-size tubing, and its purpose is to protect the capillary from damage when the bulb is being installed or removed. The extension neck is fastened directly to the bulb, with the capillary inside also fastened to the bulb, and thus both may be bent or shaped to facilitate installation.

A thermal well may be used with the thermometer bulb in applications where the temperature of fluids under pressure is to be measured, where extra corrosion protection is desired, or where extra mechanical strength is required. The mechanical fit between thermometer bulb and thermal well is made as reasonably close as possible. Since heat transfer takes place mainly by conduction, the space between the bulb and well is sometimes filled with oil, mercury, powdered metal, or thin metal inserts in order to increase the contact area between bulb and well. The thermal well may be made from such materials as copper, brass, cast iron, steel, stainless steel, Monel, and nickel, to name a few. (See Tables A-6 and A-7, appendix.)

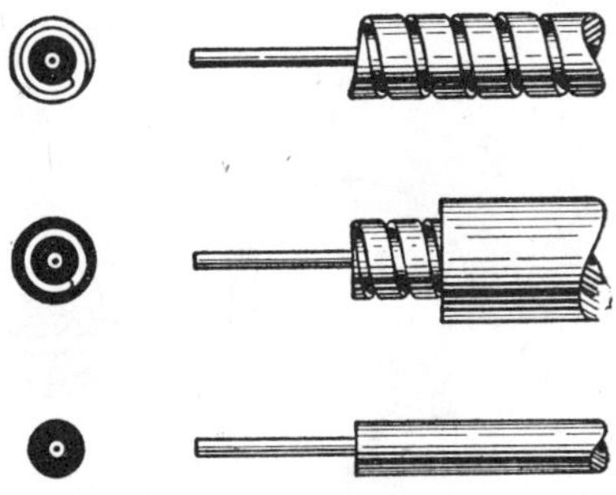

FIG. 2–8. Thermometer Capillary and Armor

The capillary connects the thermometer bulb to the receiving element. The distance between bulb and instrument is generally restricted to 200 ft or less because of the high cost of systems with greater lengths of capillary. The capillary is made of copper or steel for all but mercury-filled systems and is about $\frac{1}{16}$ in. outside diameter and about 0.015 in. inside diameter. Mercury systems employ a stainless-steel capillary with thicker walls. For protection against damage or against corrosion the capillary is enclosed in armor tubing or other protective covering, as indicated in Fig. 2–8. The armor may be spiral-wound bronze, steel, or stainless steel, or it is sometimes covered with rubber or other plastic material. For some applications the capillary is covered with plain copper tubing or plain lead tubing. The capillary tubing is the part of a thermal system most vulnerable to damage. Consequently, the capillary not only should be protected with armor or other covering but also must be installed carefully to avoid mechanical damage.

The purpose of the receiving element is to convert the fluid pressure or the volumetric expansion in the thermal system into a motion which can be amplified by mechanical linkage to operate a pen or pointer. The receiving element in a recording thermometer is always one of the bourdon-tube forms shown in Fig. 2–9. The bourdon tube is a metal tube with a relatively thin wall bent in the form of a circle and extending roughly 250 or 300 geometrical degrees. The cross section of the tube is elliptical. When a pressure is applied to the inside of the tube, the cross

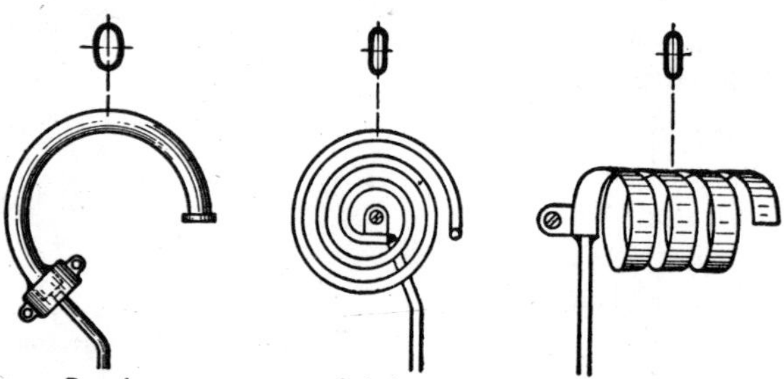

FIG. 2–9. Pressure Springs

section deflects toward a more circular form, and this causes the tube to straighten, thus moving the tip of the tube along an arc For most ordinary bourdon tubes this motion is about ⅛ in. and is nearly linearly proportional to the applied pressure. In order to increase this motion, the bourdon tube is often made with several turns, wound in a spiral shape with all turns in the same plane, or wound in a helical form with all turns at the same diameter as shown in Fig. 2–9. The cross section of the spiral or helix is made quite flat in order to obtain the maximum motion possible for a given pressure change. All three forms of the bourdon tube are used in recording thermometers and in pressure gages. The materials commonly used in the thermometer spiral or helix are phosphor–bronze, beryllium–copper, steel, and stainless steel. The mercury-filled thermometer uses only steel or stainless steel, of course.

The industrial pressure thermometer is either an indicating or recording instrument. It may indicate the temperature by means of a pointer sweeping over a circular dial, or it may record the temperature on a chart of circular form. The details of the mechanism vary considerably, according to the manufacturer, but one typical arrangement of a recorder is shown in Fig. 2–10. The spiral is fastened cantilever fashion at its center, where the capillary tubing enters. The free end of the spiral is connected by a metal link to an extension on the pen arm. The pen arm is thus made to move in an arc over the chart in response to movement of the spiral tip. A zero shift for calibration is generally accomplished by rotating the whole spiral and by moving

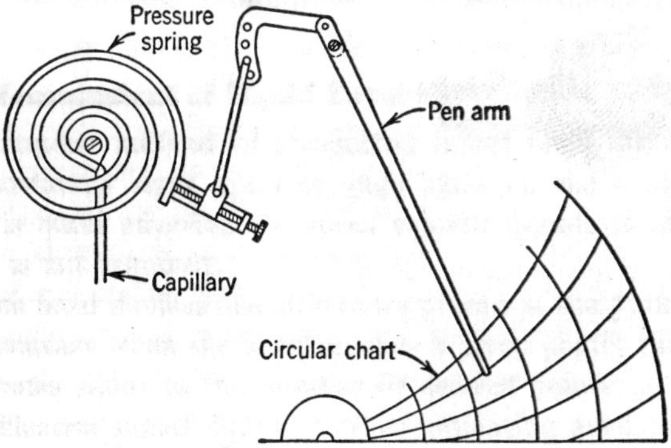

FIG. 2–10. Thermometer Recording Mechanism

the pen arm with respect to its shaft. A range adjustment is made by setting the connecting link so as to vary the ratio of spiral movement to pen movement.

Theory of Volumetric and Pressure Thermometers

A *liquid-expansion* pressure-spring thermometer utilizes the volumetric expansion of a liquid caused by temperature changes to operate the spiral and thus indicate the temperature. The thermal system is solidly filled with the thermal liquid at a high pressure, care being exercised to eliminate all entrapped air. A temperature increase at the bulb results in an expansion of the liquid, which is forced up the capillary tubing, causing the spiral to expand. Thus, a liquid-expansion thermometer is not, strictly speaking, a pressure thermometer but a volumetric thermometer, since pressure changes in the thermal liquid due to temperature expansion are incidental. The fluid is considered incompressible. The stiffness or spring gradient of the bourdon tube or pressure spring therefore has relatively little effect on the calibration of the thermometer.

The relation between volume of a liquid and its temperature is given by the law of cubical expansion:

$$V_1 = V_0(1 + BT) \tag{2-2}$$

where V_1 = final volume
V_0 = initial volume
B = mean coefficient of volumetric expansion (Table A–3, appendix)
T = temperature, degrees centigrade

This equation indicates a linear relation which is not quite true since the coefficient of volumetric expansion B varies slightly with temperature. When a simple equation such as equation 2–2 is used, a mean coefficient B is selected for use in a restricted range of temperature. For comparison its value is given in Table A–3 (appendix).

The more complete relation is

$$V_1 = V_0(1 + \alpha T + \beta T^2 + \gamma T^3) \tag{2-3}$$

where V_1 = final volume
V_0 = initial volume
T = temperature, degrees centigrade

and α, β, γ are the coefficients of volumetric expansion indicated in Table A–3 (appendix).

The volume of liquid required in the bulb is therefore determined by the expansion coefficients of the liquid and the volume change required to operate the receiving element over the required range. With a given receiving element, a wider temperature span requires a greater volume of liquid in the bulb.

Mercury is most widely used as a thermal liquid because of the wide temperature range between its freezing and boiling points. Infrequently ethyl alcohol or toluene may be used. The temperature expansion of toluene and ethyl alcohol is about six times as great as that of mercury, and from that standpoint their use is advantageous.

The scale shape of most liquid-expansion thermometers is nearly linear, especially when mercury is used, because the second-order and third-order terms in equation 2–3 are small. (See Table A–3, appendix.)

The bulb is subject to some expansion on temperature increase while the liquid inside the bulb also expands. The expansion of the bulb is small in comparison to the expansion of the liquid, since the average volumetric expansion coefficient of most metals is small (from 18 to 30×10^{-6} per degree Fahrenheit) compared to most liquids (about 101×10^{-6} per degree Fahrenheit for mercury). However, if the volumetric-expansion coefficients of both metals and liquids vary in more or less the same manner with temperature, then the only effect of bulb thermal expansion is to decrease the net expansion for a given temperature range. In general, however, slight variations in the volumetric coefficients of metals and liquids do occur, and the scale shape of the liquid-expansion thermometer is changed to a very slight degree.

The temperature limits of the mercury-filled pressure thermometer are about −35°F to 1000°F. Although the boiling point of mercury at standard pressure is 675°F, the boiling point is elevated to over 1000°F by filling and sealing the thermal system under high pressure. The mercury thermometer can thereby operate up to 1000°F without encountering vapor-pressure effects of any appreciable magnitude.

A *gas-expansion* thermometer operates on essentially constant volume, the gas expanding according to its temperature. The pressure created by the expanding gas is measured by means of a pressure spring, and the instrument is calibrated for temperature. The physical law covering expansion of a gas is the ideal-gas law,

$$pv = RT \tag{2–4}$$

where p = absolute pressure, pounds per square foot
v = specific volume, cubic feet per pound
T = absolute temperature, degrees Rankine
R = gas constant, for a given gas

The volume of the pressure-thermometer bulb capillary and tubing is substantially constant, and so we may write

$$p = \left(\frac{R}{v}\right) T \tag{2-5}$$

and we see that the relationship between pressure and temperature is ideally linear. For a pressure thermometer of given volume,

$$\frac{p_2 - p_1}{p_2} = \frac{T_2 - T_1}{T_2} \tag{2-6}$$

where subscript 1 denotes the condition at the lowest point on the scale and subscript 2 indicates the condition at the highest point on the scale. When the pressure span of the spiral and the temperature span of the thermometer are known, the pressure range can be determined.

The volume of gas required in the bulb is determined by the gas expansion and by the temperature range of the instrument, since

$$v = \frac{RT_1}{p_1} = R\left(\frac{T_2 - T_1}{p_2 - p_1}\right)$$

A wider temperature range requires a larger bulb if the same receiving element is used. Since the volumetric expansion of a gas is relatively small, the bulb of a gas thermometer is normally larger than that of either a liquid-expansion or vapor-actuated thermometer.

The two assumptions of a perfect gas and constant volume are not completely valid, however. Most gases are not perfect, especially at very low temperatures and very high pressures, in that they do not quite obey the perfect-gas law. Van der Waals' variation of the ideal-gas law is sometimes used,

$$\left(p + \frac{a}{v^2}\right)(v - b) = RT \tag{2-7}$$

where a and b are van der Waals' constants for a gas. In general, however, it is better to use the compressibility charts where the compressibility factor y in the following equation is given for various temperatures and pressures,

$$pv = yRT \tag{2-8}$$

Whenever possible, actual data on temperature, pressure, and volume should be used.

The volume of the thermometer is made up of the bulb, capillary, and receiving element. The volume change of the bulb due to thermal expansion is quite small, the volume change of the capillary is theoretically zero, but the volume change of the receiving element is appreciable, depending on its size. If, now, the bulb is made quite large so that the ratio of bulb volume to capillary and receiving element volume is 100 to 1, for example, then the volume change of the thermal system is negligible, compared to its total volume. We can therefore say that the gas-expansion thermometer operates with constant volume and can use the simple form of gas law.

Nitrogen is nearly always used in the gas-expansion pressure thermometer because it covers a wide range of temperatures and because it is readily available in both quantity and quality. (See Table A–4, appendix.) The scale shape of the nitrogen thermometer is very nearly linear. Calibration is accomplished either by printing the scale or chart to conform to the nonlinearity of the gas expansion or by correcting the nonlinearity with suitable linkage between instrument pointer and receiving element, thereby allowing the use of linear charts and scales.

The temperature limits of the nitrogen gas thermometer are approximately −200 to 800°F. The higher temperature limit is defined only by the ability to keep the gas confined. The metal of the bulb becomes permeable at high temperatures, and, if the bulb is maintained at a high temperature for a long period, a part of the nitrogen in the system becomes lost, thus destroying the calibration of the instrument.

A *vapor-actuated thermometer* operates from the vapor pressure of a liquid that partially fills the thermal system. The vapor pressure is measured by a pressure spring, and the instrument is calibrated in terms of temperature.

Since the vapor pressure depends solely on the temperature at the free surface of the liquid, the vapor-actuated thermometer indicates only the temperature existing at the free surface, wherever it may be in the system. Thus the free surface must always exist at the bulb and not in the capillary or pressure spring (see

Fig. 2–20). The bulb must be large enough to contain all the liquid when the bulb temperature is below the temperature of the capillary and pressure spring. Further, the liquid volume must be large enough to fill completely the capillary and pressure spring when the bulb temperature is higher than the temperature of the capillary and pressure spring. The volume of liquid required in the vapor-actuated thermometer bulb is further deter-

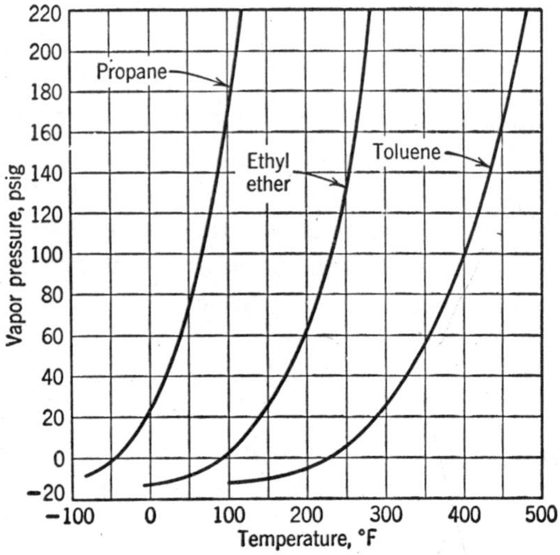

Fig. 2–11. Vapor-Pressure–Temperature Curves

mined by the necessity that the thermal system does not become solidly filled when all the vapor condenses at the lowest temperature of the instrument range and that the liquid does not entirely vaporize at the highest temperature of the scale. The vapor-actuated thermometer bulb may therefore be made relatively small.

There are a number of fluids used for vapor-actuated thermometers. Some of these fluids are shown in Table A–5 (appendix). The most common fluids in use are methyl chloride, sulfur dioxide, ether, toluene, butane, propane, and hexane. The vapor-pressure-versus-temperature curves for a few of these fluids are given in Fig. 2–11. One will immediately note that vapor-pressure and temperature curves are not linear. Therefore, the scale of the vapor-actuated thermometer is not linear but has wider graduations at the higher temperatures, as is shown by the scale reproduced in Fig. 2–12. This results in the disadvantage that temperatures near the lower part of the scale are more difficult to read. However, the vapor-actuated thermometer can be made in relatively narrow temperature spans, so that almost any measured temperature can be placed on the upper portion of the scale.

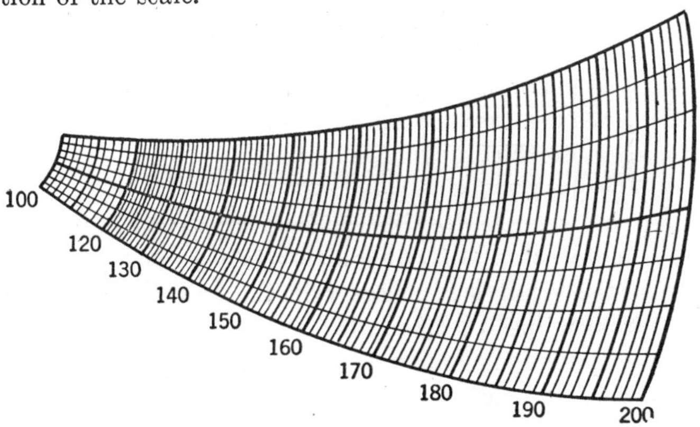

FIG. 2–12. Portion of Vapor-Actuated Thermometer Chart

The range of a vapor-actuated pressure thermometer depends entirely on the filling medium. However, temperatures as low as −50°F and as high as 600°F may be measured. With special filling media the range may be extended somewhat.

Static Accuracy of Thermometers

The calibrated accuracy of most industrial thermometers is ±0.5 per cent of full instrument range. This calibration is made with the bare bulb installed in a well-agitated liquid and with the capillary tubing and instrument case at a fairly steady ambient temperature.

The sources of static error in the industrial pressure thermometer are the ambient-temperature effect, the head effect, the barometric effect, the immersion effect, and the radiation effect.

The effect of ambient-temperature changes at the capillary and at the receiving element of a liquid-expansion thermometer is to cause the instrument to be in error. The ambient-temperature effect for the liquid-expansion thermometer can be calculated from

$$E_T = \frac{V_cT_c + V_eT_e}{V_B} \tag{2-9}$$

where E_T = total ambient error, degrees Fahrenheit
V_c = volume of liquid in capillary, cubic inches
V_e = volume of liquid in pressure spring, cubic inches
V_B = volume of liquid in bulb, cubic inches
T_c = temperature change at capillary, degrees Fahrenheit
T_e = temperature change at pressure spring, degrees Fahrenheit

Suppose, for example, that the temperature at the bulb, capillary, and spiral of a 0–100°F-range mercury thermometer is 60°F. Let the volumes of mercury in the bulb, capillary, and receiving element be 0.8 cu in., 0.2 cu in., and 0.2 cu in., respectively. If the temperature along the capillary *or* at the receiving element changes to 61°F and the bulb remains at 60°F, the instrument will have an ambient error of 0.25°F. It therefore becomes necessary to compensate a liquid-expansion thermometer for temperature changes along the capillary and at the receiving element if the ambient-temperature changes are large.

Ambient-temperature compensation is accomplished in several different ways. First, if the ratio of volume of liquid in the bulb to volume of liquid in the capillary and receiving element is large, say 1000 to 1, ambient errors would be small. Generally, however, this cannot be accomplished without making the thermometer bulb excessively large. Compensation of a liquid-expansion thermometer for ambient-temperature changes at the receiving element is afforded by using either a bimetal strip or a second receiving element, arranged as indicated in Fig. 2–13. It may be seen that, if this is properly constructed, ambient-temperature changes at the receiving element will result in no change of indication at the thermometer. Compensation of a liquid-expansion thermometer at the receiving element *and* at the capillary is accomplished by using a volumetric compensation or a second receiving element and capillary, as indicated in

Fig. 2–14. Volumetric compensation (Taylor Instrument Companies) consists of using a small wire inside the capillary tubing. The wire and capillary tubing are made of different materials (diagram at the left in Fig. 2–14). A temperature increase along the capillary causes the capillary, the mercury, and the

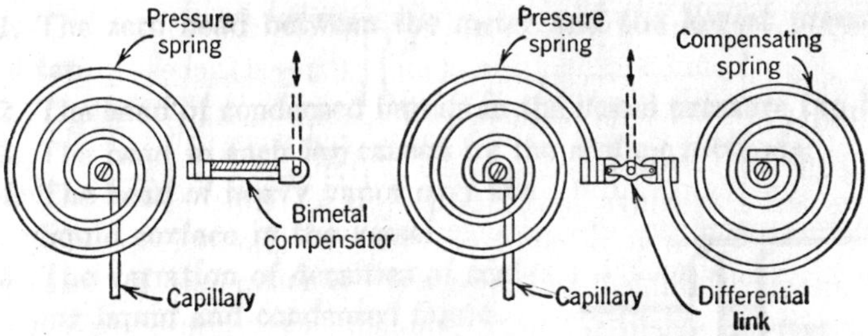

Fig. 2–13. Methods for Ambient-Temperature Compensation at Pressure Spring

wire to expand. If the capillary expansion is greater than that of the wire, the expansion of the mercury can be compensated, so that the net volumetric expansion of the system is negligible. Ambient-temperature changes at the capillary therefore do not affect the operation of the thermometer. The second method,

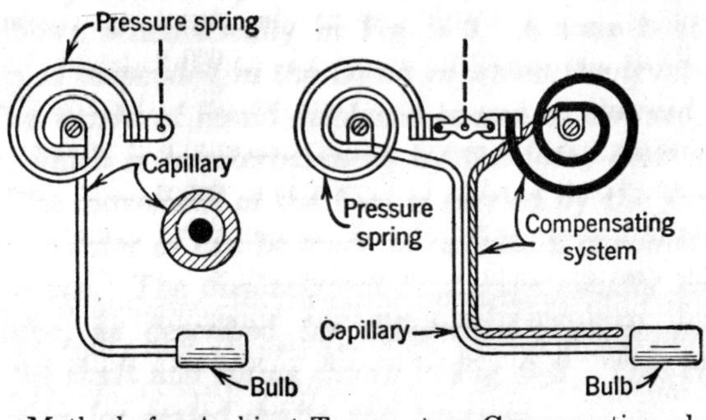

Fig. 2–14. Methods for Ambient-Temperature Compensation along Capillary

shown on the right in Fig. 2–14, consists of using a second thermal system, either mercury-filled or gas-filled, with the capillary tubing of both the compensating system and the measuring system running adjacent to each other. Only the measuring system has a bulb; the compensating system ends with the capillary tubing. Changes of ambient temperature along the capillary or at the receiving element are canceled by mechanically connecting the receiving elements in opposition.

Ambient-temperature compensation may be required with the gas-expansion thermometer as well as with the liquid-expansion thermometer. The degree of ambient error depends on the temperature range of the instrument and the relative volume of the bulb, capillary, and receiving element. Its magnitude can be calculated by the same method used for the mercury thermometer. Ambient error is reduced by making the volume of the bulb as large as conveniently possible and reducing the volume of the capillary and receiving element to a minimum. For a small range of ambient temperatures the volume of the bulb is usually large enough to make ambient errors negligible. The temperature range of the gas-expansion thermometer is normally large, and this further reduces the ambient error. Generally, then, the nitrogen pressure thermometer does not require compensation. When these conditions are not met, compensation is employed. This may take the form of compensation at the receiving element only, by the use of either a bimetal strip or a second gas-filled spiral or helix, as was employed for the mercury-expansion thermometer. Complete compensation is provided by using a second gas-filled element and capillary, arranged as previously described for the mercury thermometer.

The vapor-actuated pressure thermometer does not require ambient-temperature compensation, since the pressure in the thermal system is determined only by the temperature at the free surface of the liquid. Changes of temperature elsewhere in the system cause an expansion or contraction of volume, but this is automatically compensated by establishing a new vapor-pressure equilibrium at the liquid surface, thereby maintaining a constant pressure. The thermometer is arranged so that the free liquid surface is always at the bulb for any relative temperatures of bulb, capillary, and receiving element.

The head effect in a pressure thermometer is caused by placing the thermometer bulb at a considerably higher or lower position (elevation) than the receiving element. The head of liquid between bulb and receiving element will cause a greater or lesser pressure at the receiving element. If the bulb is located 10 ft above the receiving element, as indicated in Fig. 2–15 by distance d, the equivalent pressure at the receiving element of a mercury thermometer is 10 ft of mercury or 59 psi. Since the mercury system is filled at, say, 1000 psi pressure, the error caused in the reading of the instrument is small. If necessary, the mercury thermometer can be calibrated with the bulb in position so as to correct for the head effect. When the bulb is

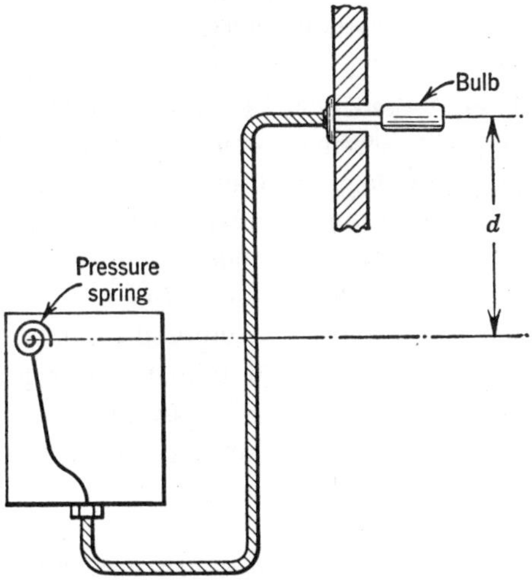

Fig. 2–15. Vapor-Actuated Thermometer Installation and the Head Effect

below instrument level, the head effect is reversed. The head effect in the gas-expansion thermometer is negligible, because, first, the pressure created by a column of gas is small, and, second, the nitrogen thermometer is normally filled with an initial pressure of several hundred pounds per square inch.

The head effect in vapor-actuated thermometers is sometimes appreciable, since the pressure in the thermal system is low and the pressure effect of a head of liquid is of appreciable magnitude. Suppose that the fluid in a vapor-actuated thermometer is water, the temperature range is 300 to 500°F, and the bulb is installed 10 ft above the receiving element. Under these temperature conditions the capillary and receiving element are filled with water. The pressure equivalent of this head is 4.33 psi. By calculation from a steam table, this will cause an error of +0.50°F at 400°F or an error of ±1.0°F at 450°F. The head effect is usually accounted for by calibrating the instrument with the bulb at a height above the receiving element that it will have when actually installed. If the elevation of the bulb is changed, then the instrument requires recalibration. When the temperature range of the vapor-actuated thermometer is below ambient temperature, for example -50 to $+50°F$, the capillary and receiving element are filled with vapor, and the head effect is negligible. If the bulb is installed below instrument level, the head effect is reversed.

The barometric effect in a pressure-spring thermometer is the error in instrument reading due to changes in barometric pressure. This error occurs because the pressure-spring deflection is proportional to the difference in internal and external pressures. Normally, a change in barometric pressure of ± 1 psi may occur. With the mercury-expansion thermometer filled at about 1000 psi and operating solidly filled, the effect of barometric-pressure changes is ± 0.1 per cent and is therefore negligible. In the gas-expansion thermometer, the initial pressure is several hundred pounds per square inch, and the operating pressure even higher. The effect of barometric-pressure changes can probably be safely ignored. The vapor-actuated thermometer, however, operates at pressures in the region of 100 to 200 psi, and sometimes lower. Barometric-pressure changes may cause a small error, which should not be neglected. In spite of this, the error is not usually compensated in the vapor-actuated thermometer because of the disproportionate cost of this feature.

The immersion effect results from conduction of heat along the bulb and thermal well, thereby reducing the temperature at the bulb and causing an error, sometimes called a conduction error. Heat is conducted along the bulb and extension neck and along the thermal well to the outside. A temperature drop exists with this flow of heat, and the temperature indicated by the thermometer is lower than the actual temperature. A thermometer bulb and well should be immersed far enough and the exposed head should be sufficiently insulated to keep all parts of the bulb and well at the same temperature.

When the thermometer bulb is only partially immersed, either in the fluid whose temperature is being measured or in a thermal well, the temperature indicated by a mercury thermometer or gas thermometer will be less than the actual temperature, roughly in proportion to the area of the bulb not exposed. Only a part of the filling medium is expanded by heat. In the vapor-actuated thermometer the correct temperature is measured if the whole free surface of liquid in the bulb is in good thermal contact with the fluid of which the temperature is being measured. Generally, the vapor-actuated thermometer bulb is not so dependent on complete immersion for its accuracy as the mercury thermometer and gas thermometer.

Radiation errors often exist in the measurement of gas and air temperatures because of the ability of the thermometer bulb to "see" solid bodies whose temperatures are lower or higher than that of the gas. These errors may be as large as 40 to 50 per cent. In a duct, for example, the thermometer reads low if the walls are at a lower temperature than the gas. Conversely, if the duct walls are at a higher temperature than the gas, the thermometer may read high. A radiation shield constructed around the thermometer bulb is effective in preventing radiation error. The shield must be arranged so that the hot gases flow between the thermometer bulb and the shield and between the shield and the duct walls. The bulb gains heat by convection and conduction from the gas but loses heat by radiation to the walls. When a shield is used, the heat lost by radiation is reduced, because the shield is more nearly at gas temperature.

The precision or reproducibility of pressure thermometers is related to the questions of ambient-temperature compensation and calibration drift. Ambient-temperature compensation can be made quite accurate, sufficient to maintain the guaranteed calibration accuracy over a wide range of applications. Calibration drift depends on the construction of the thermometer and on the stability of the filling media. Some of the factors are:

- Leakage in the thermal system.
- Mechanical stresses in materials.
- Thermal stresses in materials.
- Fatigue and creep of metals.

All industrial pressure thermometers of good quality are aged and sometimes given thermal massage in order to detect leakage, relieve residual stresses, and check for distortion and permanent set. The calibration of the thermometer is thereafter checked. After installation, the calibration may drift after a period of time, the length of which depends wholly on the details of installation and the amount of maintenance. The calibration of any thermometer should be checked periodically.

Ordinarily the filling media such as mercury, nitrogen, and the lower-temperature liquids for vapor thermometers are used in quite pure forms and are stable over long periods. At higher temperatures it is possible for contamination and decomposition to take place, and these are causes for calibration drift.

The dead zone of industrial pressure thermometers depends on the starting friction and lost motion existing in the linkage and in pen or pointer bearings, and particularly on the friction between pen and chart paper in recording thermometers. The dead zone created by this friction is extremely variable. The mercury thermometer develops a large force in the thermal system by virtue of its liquid expansion. Therefore, it has the smallest dead zone, generally in the neighborhood of 0.05 to 0.10 per cent of full scale. The gas thermometer and vapor thermometer operate from the pressure developed in the thermal system, producing forces generally smaller in magnitude than those of the mercury thermometer. The dead zone in these types may be as large as 0.25 per cent of scale for narrow temperature spans.

Response of Thermometers

The response of the pressure thermometer is determined almost altogether by the heat-transfer characteristics of the thermometer bulb and well and by the characteristics of the fluid in which the element is installed. The characteristics of filling media, bulb, and thermal well that influence response are:

Thermal capacitance.
Thermal conductivity.
Surface area per unit mass.

The characteristics of the fluid surrounding the bulb also affect the response:

Film coefficients of heat transfer.
Mass velocity.

Thermal capacitance.
Thermal conductivity.

A thermometer bulb should have a large area, a small mass, a small specific heat, and a high thermal conductivity. The thermal characteristics of several thermometer materials and filling media are given in Table A–6 (appendix).

As shown by the table, there is little to choose among the various metals for thermometer bulbs. The response of the thermometer therefore depends almost wholly on the size of the bulb, its area, and its method of installation.

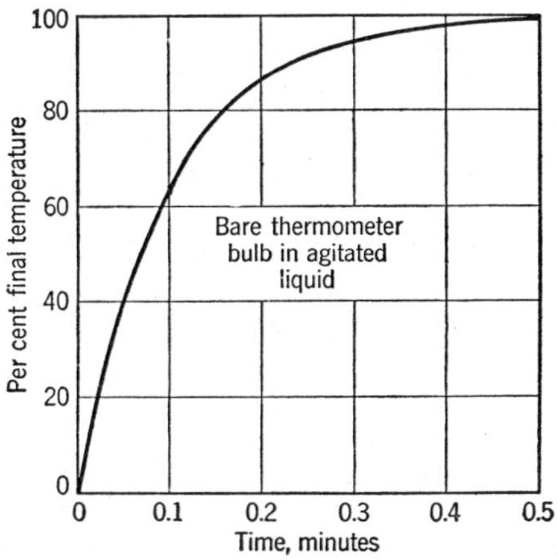

Fig. 2–16. Average Response of Mercury, Gas, and Vapor Thermometers

For a bare thermometer bulb in a well-agitated liquid, the response to a sudden temperature change is shown in Fig. 2–16. The time constant is about 0.1 min, which is normal for a pressure thermometer, either gas-expansion, or vapor-actuated. Notice that the response is essentially first order. The response in a flowing liquid is proportional to the thermal diffusivity, because the heat-transfer film resistance is small, whereas the resistance of the bulb material is a large portion of the total. The response is therefore faster when the thermal diffusivity is greater.

The dip effect is shown in Fig. 2–17. This effect is caused by the sudden expansion of the metal thermometer bulb, which takes place before expansion of the fluid because heat arrives first at the thermometer bulb. Temporarily this causes a net contraction of fluid volume, causing the instrument to indicate a reversed direction of temperature change. The extent of the dip effect is most noticeable in mercury thermometers because of the relatively large quantity of heat that must be transferred through the bulb to the mercury. The lag caused by the dip effect may be of 0.01 min magnitude. The dip effect is present in gas-expansion and vapor-actuated thermometers, but its extent is negligible.

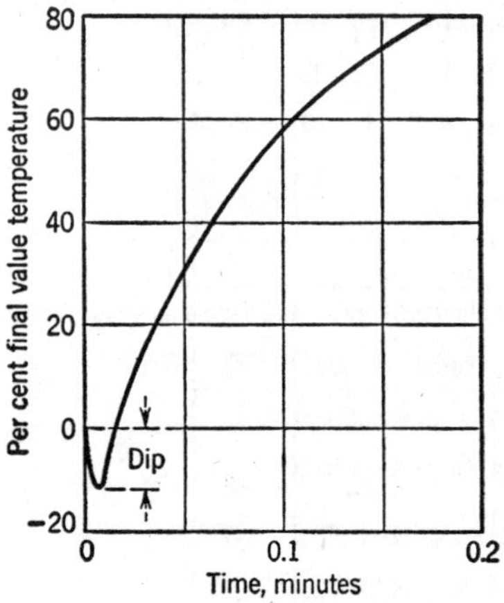

FIG. 2–17. Dip Effect in Mercury Thermometer

For a bare thermometer bulb in moving air, the time constant of response is normally larger than in a moving liquid, sometimes 5 to 10 times as large at low temperatures. The response in a flowing gas is proportional to the inverse of thermal capacity. The effect of thermal resistance of the bulb is small compared to the effect of the film coefficients of heat transfer in a gas. The response is therefore faster when the thermal capacity is small. At higher temperatures, 400°F or higher, the time constant is reduced, because thermal radiation begins to add appreciably to the heat transfer. Radiation depends on the fourth power of absolute temperature and therefore has much less effect at lower temperatures. The comparison of thermal lags in moving air and in moving liquids is illustrated in Figs. 2–18 and 2–19, which show that the lag in moving air is not only larger but also more dependent on the speed of the fluid past the bulb. If the speed of the fluid is slow, a blanket of stagnant fluid develops around the bulb, effectively insulating the bulb from contact with the fluid.

The cross-ambient effect causes an additional lag in response and exists only in the vapor-actuated pressure thermometer. Suppose, first of all, that the capillary and receiving element are at 70°F. As shown in Fig. 2–20, the bulb may be at a higher

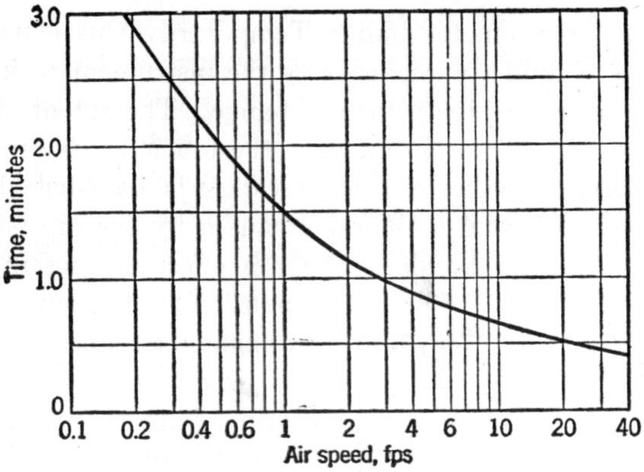

Fig. 2–18. Variation of Thermometer Lag with Air Speed

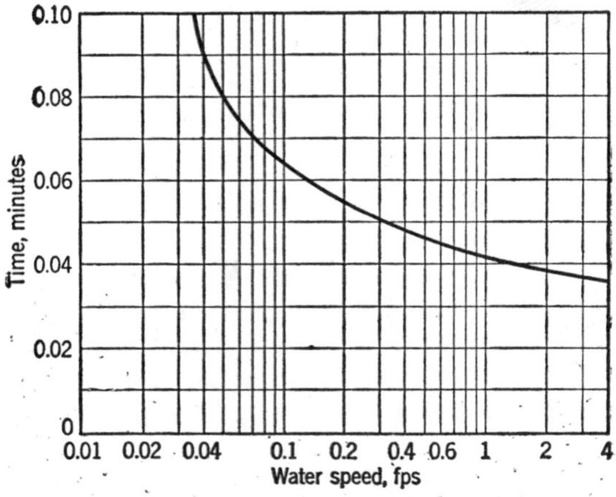

Fig. 2–19. Variation of Thermometer Lag in Moving Water

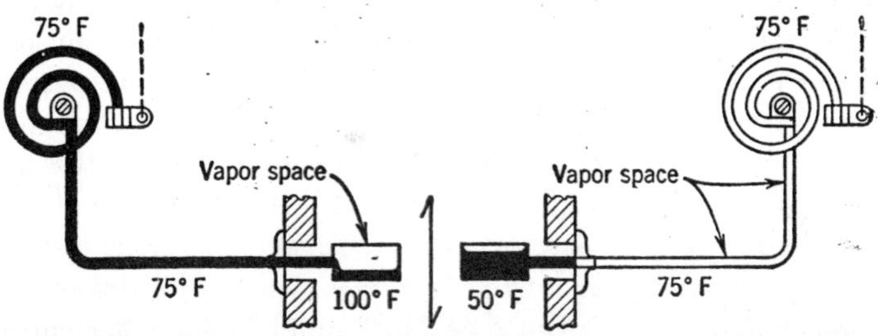

Fig. 2–20. Location of Vapor Space in Vapor-Actuated Thermometer

temperature such as 100°F, or at a lower temperature such as 50°F. When the bulb is at a higher-than-ambient temperature, the receiving element and capillary are solid-filled and the vapor space is in the bulb. Liquid, of course, condenses in the cooler part of the system and vaporizes in the hotter part of the system. When the bulb is at a lower temperature than ambient, the vapor space is in the receiving element and capillary. Now suppose that the temperature range of the instrument is across the am-

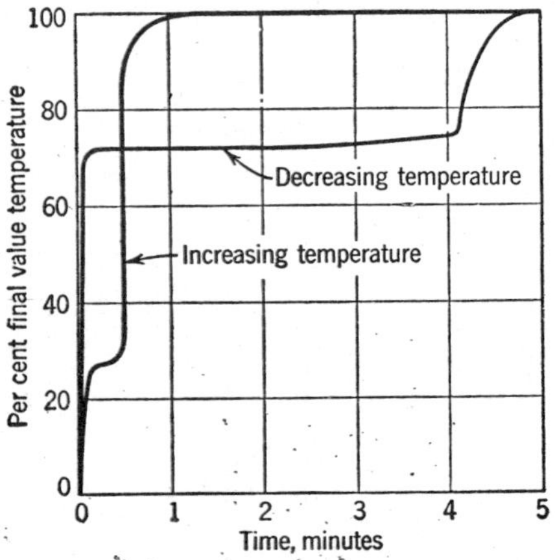

FIG. 2–21. Cross-Ambient Effect in Vapor Thermometers (from *Brown Instrument Co. Tech. Bull.*, "Response Speeds of Pressure Thermometers")

bient temperature, say 0 to 100°F. Then, when the temperature at the bulb crosses from below to above the ambient temperature, liquid must migrate from the bulb and fill the capillary and receiving element. When the bulb temperature drops below the ambient temperature, the vapor condenses into liquid at the bulb. This change of position of vapor space and liquid space takes place automatically and naturally requires time to occur. Thereby, a considerable lag in response is caused, as shown in Fig. 2–21.

The vapor-actuated pressure thermometer should therefore not be used in cross-ambient ranges if a high speed of response is desired. Sometimes to avoid the cross-ambient effect, a vapor-actuated thermometer may have a dual fill: that is, the receiving element and capillary may be filled solidly with a nonvaporizing liquid, which acts as a hydraulic transmission means for the pressure developed in the bulb by the actual thermometric filling medium. The vapor–liquid thermal fluid is thereby confined to the bulb, and the cross-ambient effect is avoided.

The pressure-thermometer bulb is often used with a thermal well. The effect of the thermal well is nearly always to increase considerably the response lag of the thermometer. Response with and without a thermal well is compared in Fig. 2–22. Notice that two effects result: first, the response is much slower in-

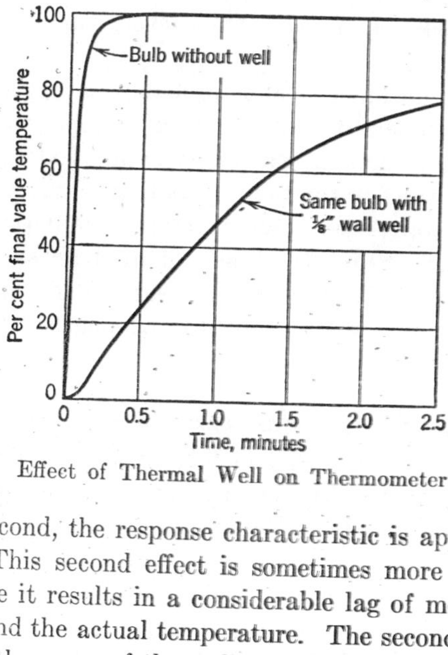

FIG. 2–22. Effect of Thermal Well on Thermometer Response

deed; and, second, the response characteristic is apparently second order. This second effect is sometimes more serious than the first, since it results in a considerable lag of measured temperature behind the actual temperature. The second-order effect is caused by the mass of the well, which is separated thermally from the mass of the bulb by a relatively high thermal resistance.

Heat is transferred between the bulb and well by conduction, convection, and radiation. The direct conduction of heat between bulb and well is governed almost entirely by the area of contact. Since the clearance space between the bulb and well contains air, the amount of actual metal-to-metal contact between bulb and well is small. This contact is often increased by filling the space between bulb and well with a metal powder, graphite, oil, mercury, or thin sheet-metal inserts. The effect of these measures is to increase the speed of response about 10 per cent, or sometimes more. More important, however, they reduce the thermal resistance between bulb and well so that much of the second-order effect is avoided. Reduction of thermal resistance between bulb and well is naturally more effective when the temperatures of flowing liquids are being measured, since the film resistances are small.

Convection transfer of heat between bulb and well is small because the air film between bulb and well is normally very thin and no large mass of air can come into play. When the space is filled with a solid as mentioned previously, no heat convection is allowed.

Radiation transfer of heat is probably an appreciable factor, even at low temperatures, if the space between bulb and well is not filled. At temperatures approaching 1000°F, most of the heat is transferred to the bulb by radiation, very little by conduction and convection. Where thermal-radiation transfer of heat occurs, the surface conditions of the bulb and well are important. A rough and well-oxidized surface is a much better receiver of radiation than a smooth and polished surface. (See Table A–16, appendix.)

Pneumatic-Balance Pressure Thermometer

The pressure thermometer may be arranged for pneumatic operation as indicated in Fig. 2–23 in schematic fashion. The pneumatic system shown is somewhat simpler than those in actual use. The gas-filled bulb is connected by a short length of tubing to the inside of a bellows. The balance pressure on the outside of the bellows is determined by the position of the bellows acting against the control nozzle. Air is supplied to the system through a restriction. Suppose that the temperature at the thermometer bulb increases and the gas expands. The increased pressure of the gas in the bellows pushes the bellows to the right, thereby opening the control nozzle. This allows a greater flow of air into the chamber. The balance pressure is now higher, acts against the bellows, and balances the effect of the higher bulb pressure. The transmitted pressure is higher, and the pressure receiver (an ordinary pressure gage) indicates a change in temperature. The pressure receiver can therefore be calibrated in terms of temperature. This system can be used with either the gas thermal system or a vapor-actuated thermal system, but not with a liquid-filled thermal system.

The advantages of this system are readily apparent. First, the costly capillary and armor are replaced by the ordinary copper tubing connecting the transmitter to the receiver. The capillary between the bulb and the transmitter is usually short. Ambient-temperature effects do not require compensation if the volume of gas on the bellows and tubing is small with respect to bulb volume. The response speed of the thermometer may be improved or at least be no less than in the ordinary pressure-

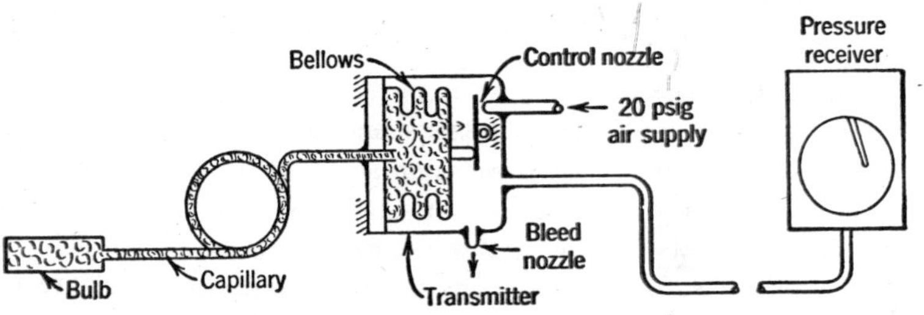

FIG. 2–23. Pneumatic-Balance Pressure Thermometer

spring thermometer, in spite of any lag in the pneumatic system, because expanded gas in the bulb is not required to pass through a long capillary. Also, in industrial forms, an amplifying pilot or relay is used at the transmitter, thereby reducing the pneumatic lag and serving to increase the sensitivity of the control nozzle so that bellows motion is reduced to a minimum. A further advantage is that the dead zone of the thermometer is decreased since the pressure receiver is usually made a bellows-type gage, providing large force for operating the indicating mechanism.

The main disadvantages of this system are the maintenance or service required by the pneumatic system and the fact that, with lower pressures in the gas thermal system, the barometric effect may not be negligible.

Comparison of Pressure-Spring Thermometers

Each type of industrial pressure or transmitting thermometer has its useful application. Table 2–2 attempts to summarize their characteristics.

TABLE 2-2. CHARACTERISTICS OF PRESSURE-SPRING THERMOMETERS

	Mercury	Gas	Vapor
Scale shape (as used)	Linear	Linear	Nonlinear
Temperature limits, °F	−35 to 1000	−200 to 800	−50 to 600
Smallest span, °F approx.	100	160	80
Largest span, °F approx.	1000	800	350
Ambient effect	Yes	Yes	No
Cross-ambient effect	No	No	Yes
Head effect	No	No	Yes
Barometric effect	No	Little	Yes
Dip effect	Yes	No	No
Immersion effect	Yes	Yes	Less
Response speed	Same	Same	Greater
Sensitivity	Greater	Same	Same

The vapor-actuated pressure thermometer is probably the most widely used, because it is somewhat less costly and simpler to maintain than the mercury thermometer or gas thermometer. The vapor-actuated thermometer requires no compensation and has a good speed of response. Its disadvantages are slight. The gas-expansion thermometer has the advantage of good inherent accuracy, and, in addition, it can reach quite low temperatures. The mercury thermometer has the advantage of greater sensitivity because of the large force developed by the thermal system.

Problems

2-1. Compute the temperature at which Fahrenheit and centigrade scales coincide.

2-2. Develop an algebraic equation relating the Fahrenheit and Kelvin scales.

2-3. The lowest and highest recorded weather temperatures in the world are −90 and 135°F. Convert these temperatures to centigrade and Réaumur.

2-4. A mercury thermometer is made from an Invar tube, 0.050 in. inside diameter. It is evacuated and filled with mercury to a height of 5.0 in. at 68°F. Using the simple law of liquid expansion, find the height of the column when the thermometer is completely immersed in a liquid of temperature 600°F.

2-5. A nitrogen thermometer bulb is made of copper with a 0.75 in. inside diameter. The volumetric coefficient of expansion of copper is about 28×10^{-6} per degree Fahrenheit. The temperature range of the instrument is 100 to 600°F, and the pressure range of the system must be 500 psi. Neglecting the volume of capillary and spiral, determine:

(*a*) The initial pressure in the bulb at 100°F.

(*b*) The required inside length of the bulb.

(*c*) The initial pressure in the bulb if thermal expansion of the bulb is considered and if the length of the bulb computed in part *b* is used.

2–6. Using a steam table, plot a curve of vapor-pressure versus temperature for a vapor-actuated thermometer filled with water and having a range of 220 to 320°F.

2–7. A mercury thermometer has a bulb, capillary, and spiral volumes of 0.70 cu in., 0.20 cu in., 0.10 cu in., respectively. The range of the instrument is (*a*) 0 to 500°F; (*b*) 150 to 250°F. The ambient temperature change at the capillary is ±30°F. What is the ambient error in per cent of span?

2–8. A gas thermometer has a bulb and capillary volumes of 2.50 cu in. and 0.10 cu in., respectively, and a range of 200 to 600°F. If the accuracy desired is ±0.5 per cent of scale span, is ambient-temperature compensation necessary for ambient-temperature changes at the capillary of (*a*) 60 to 80°F? (*b*) 70 to 120°F?

2–9. Why is the temperature at the free surface of the liquid of a vapor-actuated thermometer the only temperature measured?

2–10. The vapor thermometer of problem 2–6 is installed with the bulb in a tower 30 ft above the spiral. What is the error caused in the reading of the instrument when the true temperature is 300°F?

2–11. A liquid-expansion thermometer using mercury has a bulb volume of 1.0 cu in. and a temperature range of 0 to 200°F. The bulb is of steel. If the temperature is steady at 150°F and suddenly drops to 100°F, what is the magnitude of the dip effect? Assume that the mercury in the bulb remains momentarily at 100°F. (Cubical-expansion coefficient is 3 times a linear-expansion coefficient.)

2–12. A gas thermometer employing nitrogen has a calibration 0 to 300°F. At the maximum temperature the pressure is to be (*a*) 600 psia, (*b*) 2000 psia. The barometric pressure changes by 3 in. Hg. What error is caused if the thermometer reads 250°F?

2–13. Two thermometers that are identical except that one has a brass bulb and the other a stainless-steel bulb (no well) are used to measure a temperature of 250°F. If the temperature is measured in a fast-moving fluid, which will be the faster in response? Give reasons. If the temperature is measured in fast-moving air (neglect radiation), which will be the faster in response? Give reasons.

2–14. A thermometer bulb is installed in a duct, and hot gases at temperature T flow through the duct at a velocity V. The walls of the duct are at a lower temperature T_D.

(*a*) If the thermometer bulb is unshielded, write a relation between gas temperature, duct-wall temperature, and indicated temperature T_i.

(*b*) If the thermometer bulb is surrounded with a cylindrical shield, write the relations between gas temperature, duct-wall temperature, and indicated temperature.

(*c*) If the thermometer bulb is oxidized stainless steel, ⅜ in. outside diameter by 4 in. long, the shield of oxidized stainless steel, the gas velocity 10 fps, the gas temperature 600°F, and the duct-wall temperature 300°F, estimate the indicated temperature both with and without the radiation shield.

2–15. If the barometric pressure drops 2.0 in. Hg below normal, what is the error caused in a vapor thermometer filled with water and reading 250°F?

3· Thermoelectric Temperature Measurement

The thermocouple pyrometer has the widest useful range and application of any temperature-measuring device. Although the thermocouple was primarily employed for high-temperature measurements, hence the name pyrometer, its usefulness has been greatly extended in recent years by the introduction of simple, reliable, and responsive instruments for measurement of thermocouple voltage. The thermocouple instrument is now used equally for all temperatures, both low and high.

Temperatures encountered in industrial processing are seldom higher than 3000°F, and temperatures ranging from 2000 to 3000°F are commonly met in the steel, glass, and ceramic industries. A vast majority of these measurements are made in furnaces where air or gas temperatures or furnace-wall or roof temperatures must be determined. Some measurement of temperature of liquids is made, however, in molten metal, molten salts, and liquid glass. Confining our attention for the moment to temperatures between 1000 and 3000°F, it must be remembered that radiation transfer of heat plays a primary role in temperature measurements made in this range. Radiation not only influences the speed with which a thermocouple responds, but it also determines what temperature is measured and where it is measured. The thermocouple is used for low temperatures as well; temperatures of both liquids and gases as low as −300°F are measured.

The thermocouple provides an accurate and reliable indication of temperature for many kinds of industrial applications. Thermocouples can be interchanged since they are supplied to a standard calibration. This is one important reason for their wide adoption for industrial temperature measurement. The reliability of the thermocouple is demonstrated by the fact that the platinum versus platinum–rhodium thermocouple is used to define the international temperature scale between 1220°F (660°C) and 1945°F (1063°C).

Thermoelectricity

A thermocouple is composed of two dissimilar wires joined together so as to produce a thermal emf. A simple thermoelectric

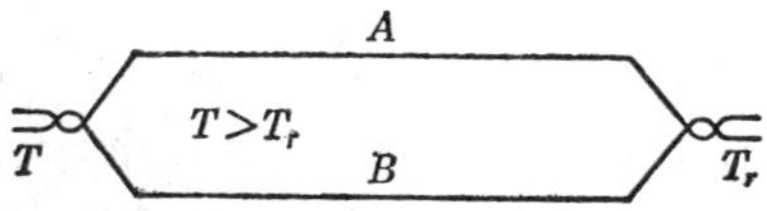

FIG. 3–1. Simple Thermocouple Circuit

circuit composed of two wires of different metals, *A* and *B*, is shown in Fig. 3–1. Assume that the left-hand junction of the wires is the point of measurement. Then the left-hand junction is called the measuring junction. The right-hand junction is called the reference junction, and it is frequently maintained at either 32°F (0°C) or 68°F (20°C)

In 1821 Seebeck discovered that a current flows in a thermoelectric circuit when the temperatures at the junctions are different. There is therefore a thermal emf generated in the circuit which causes the current flow. By the principle of conservation of energy, it is apparent that this energy under static conditions must come from the absorption of heat from external sources.

The *Peltier effect* relates the absorption and evolution of heat at the junctions of a thermocouple to the current flow in the circuit. Under conditions of use, heat is evolved at the reference junction and absorbed at the measuring junction in proportion to the flow of current but independent of the method by which the junction is made. The junction may be welded or soldered or merely an intimate contact. The Peltier effect is proportional to current alone and is different from the Joule heating effect, which is proportional to the square of the current times the resistance I^2R. The Joule heating effect is not relevant to the operation of the thermocouple. Therefore, the junction of the thermocouple produces an emf called the Peltier emf, the magnitude and direction of which depend on the temperature. The Peltier emf differs for different combinations of metals.

The *Thomson effect,* predicted by Thomson (Lord Kelvin), is a relation between the emf generated in a single homogeneous wire and the temperature difference between the ends of the wire. The Thomson emf is proportional to the temperature and the temperature difference in the wire and differs for different metals.

Finally, a thermocouple employs both the Peltier effect and the Thomson effect. The Peltier effect cannot be used alone, since, if two dissimilar metals are joined and a higher temperature exists at the measuring junction, then there must necessarily be a temperature gradient along each wire. This involves the Thomson effect. The separate existence of the Thomson effect has never been demonstrated with homogeneous conductors.

There are, then, four emf's that appear in the thermocouple of Fig. 3–1: the Peltier emf at the measuring junction, the Peltier emf at the reference junction, the Thomson emf along wire A because of the temperature difference, and the Thomson emf along wire B because of the temperature difference.

The Thomson emf along each wire is

$$\text{Thomson emf} = \int_{T}^{T_r} \sigma \, dT \tag{3-1}$$

where σ = rate at which heat is absorbed or evolved per unit temperature difference per ampere current. T is the temperature in degrees Kelvin. The Peltier emf at each junction will be called π in volts.

The total emf of the thermocouple is

$$E = [\pi]_{T_r} - [\pi]_T - \int_{T}^{T_r} \sigma_B \, dT + \int_{T}^{T_r} \sigma_A \, dT \tag{3-2}$$

Simplifying this equation gives

$$E = [\pi]_{T_r} - [\pi]_T + \int_{T}^{T_r} (\sigma_A - \sigma_B) \, dT \tag{3-3}$$

The Peltier and Thomson effects are assumed reversible. According to the second law of thermodynamics for reversible processes, the total entropy change of the system is zero. This leads to the conclusion that the fraction of heat absorbed and converted into electric energy in the thermocouple is equal to the ratio of temperature difference to temperature.[1] Thus,

$$\frac{dE}{\pi} = \frac{dT}{T} \tag{3-4}$$

where the thermocouple has junction temperatures of T and $T + dT$. Substituting from equation 3–3 gives

$$\frac{d\pi + (\sigma_A - \sigma_B)\,dT}{\pi} = \frac{dT}{T} \tag{3-5}$$

and simplifying,

$$\frac{d}{dT}\left(\frac{\pi}{T}\right) + \frac{\sigma_A - \sigma_B}{T} = 0 \tag{3-6}$$

and integrating,

$$\pi = (\sigma_A - \sigma_B)(T_c - T) \tag{3-7}$$

where T_c is a constant of integration.

Experimental evidence indicates that σ is itself nearly proportional to temperature, and so it is assumed that

$$\sigma_A - \sigma_B = kT \tag{3-8}$$

where k is a proportionality constant.

We can now evaluate the thermocouple emf by combining equations 3–3, 3–7, and 3–8:

$$E = [kT(T_c - T)]_{T_r} - [kT(T_c - T)]_T + \int_T^{T_r} kT\,dT \tag{3-9}$$

Performing the indicated operations gives

$$E = kT_c(T - T_r) - \frac{k}{2}(T^2 - T_r^2) \tag{3-10}$$

where the constant of integration is zero since, when the junctions are at the same temperature, the thermocouple emf is zero. The temperature emf relation is therefore quadratic, and most materials follow this equation reasonably well. Values of the constants are given in Table A–8 (appendix), where the themocouple is composed of each material versus lead. The values of the constants in equation 3–10 (Table A–8) vary considerably, depending on the exact composition of the metal and its surface characteristics. Deviations from the quadratic law may be attributed to the manner in which the Thomson emf varies with temperature; the assumption made in equation 3–8 is not always valid. Temperature–emf data are given in Tables A–10 to A–14 (appendix) for standard industrial thermocouples. These data are standardized by the National Bureau of Standards. Data for other metals are given in Table A–9 (appendix).

[1] This equation and equations 3–5 and 3–6 can also be developed by carrying a unit charge of electricity q around the circuit and writing $\Sigma Q/T = 0$, where the heat $Q = qE$.

There are three laws for thermoelectric circuits composed of homogeneous conductors.[2]

Law of Homogeneous Circuit. An electric current cannot be sustained in a circuit of a single homogeneous metal, however varying in section, by the application of heat alone.

This law has not been derived but is generally accepted as an experimental fact. From it we deduce that the thermal emf developed in the thermocouple, when the measuring junction T is at a different temperature from the reference junction T_r, is independent of the temperature gradient and its distribution along the wires. Referring to Fig. 3–1, the only temperatures related to the thermal emf are the measuring-junction temperature T and the reference-junction temperature T_r, and all other intermediate temperatures are of no consequence.

Law of Intermediate Metals. The algebraic sum of the thermal emf's in a circuit composed of any number of dissimilar metals is zero, if all the circuit is at a uniform temperature.

This law follows directly from the second law of thermodynamics. If a net emf did exist, the resulting flow of electric current would cause heating in some part of the circuit, thereby transferring heat from a lower to a higher temperature without application of external work.

By combination with the law of homogeneous circuits, this law makes it possible to insert a measuring device and its connecting wires C into the thermocouple circuit at any point, as shown in Fig. 3–2, without altering the thermal emf, provided that all intermediate junctions are at the same temperature T_i.

Law of Intermediate Temperatures. The thermal emf developed by any thermocouple of homogeneous metals with its junction at any two temperatures T_1 and T_3, respectively, is the algebraic sum of the emf of the thermocouple with its junctions at temperatures T_1 and T_2, respectively, and the emf of the same thermocouple with its junctions at temperatures T_2 and T_3, respectively.

This law deals with the calibration of a thermocouple. If the thermal emf's of several metals versus a reference metal such as platinum or lead are known, the thermal emf of any combination

[2] W. F. Roeser, "Thermoelectric Thermometry," *J. Applied Phys.*, **11**, 388 (1940).

of these metals may be obtained by algebraic difference. (See Tables A–8 and A–9, appendix.) Further, the calibration of the thermocouple can be made by reference only to the difference in temperature of the junctions ($T - T_r$), as denoted in Fig. 3–2. Thereby the reference-junction temperature may be made any convenient value and the measuring-junction temperature found by simple differences. Were it not for this law, the use of a thermocouple in temperature measurement would be made very difficult.

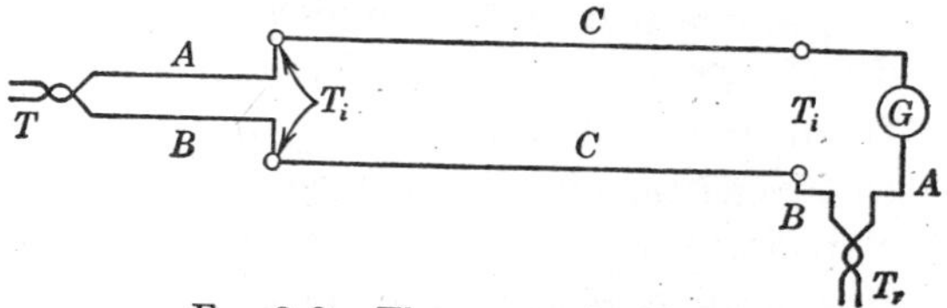

FIG. 3–2. Thermocouple Circuit

When all three laws are combined, it is evident that the algebraic sum of the thermal emf's generated in a thermocouple circuit containing any number of dissimilar homogeneous conductors is a function only of the temperatures of the junctions. It is important that none of the fundamental laws depends on any assumption regarding the exact mechanism of converting heat to electric energy or on any assumption regarding the location of the emf. The only necessary assumption (set forth by W. Thomson, later Lord Kelvin) is that the thermocouple may be regarded as a reversible heat engine.

Industrial Thermocouples

The desirable properties of thermocouples for industrial use are:

Relative large thermal emf.
Precision of calibration.
Resistance to corrosion and oxidation.
Linear relation of emf to temperature.

Thermocouples should have a relatively large thermal emf for a given temperature range in order to use a simple and rugged measuring instrument for indication of temperature. The emf of most thermocouples is in the neighborhood of 10 to 50 millivolts for the desired temperature ranges. The measurement of such small emf's requires an instrument of high quality and one capable of accurately measuring emf's as small as 10 microvolts. Yet the instrument must be sufficiently rugged and simple that it will stand the abuse of long, continuous, industrial service

A thermocouple must be capable of calibration to a standard emf temperature relationship and should maintain that calibration without drift over a long period of time. Thermocouples are made to operate with standard instruments having calibrated charts and scales. The interchangeability of thermocouples and instruments is a principal reason for their wide use and application.

Resistance to corrosion, oxidation, and contamination should be high in order that the thermocouple will have a long life. Frequent replacement greatly increases the cost of performing temperature measurements.

It is desirable to have a linear relation between thermal emf and temperature, not only because the scale is more easily read, but also because the problems of reference-junction compensation are greatly reduced.

The five most commonly used thermocouples are:

Copper–Constantan.
Iron–Constantan.
Chromel–Alumel.
Platinum–platinum, 13% rhodium.
Platinum–platinum, 10% rhodium.

The useful industrial temperature range of common thermocouples is shown in Fig. 3–3. Iron–Constantan thermocouples

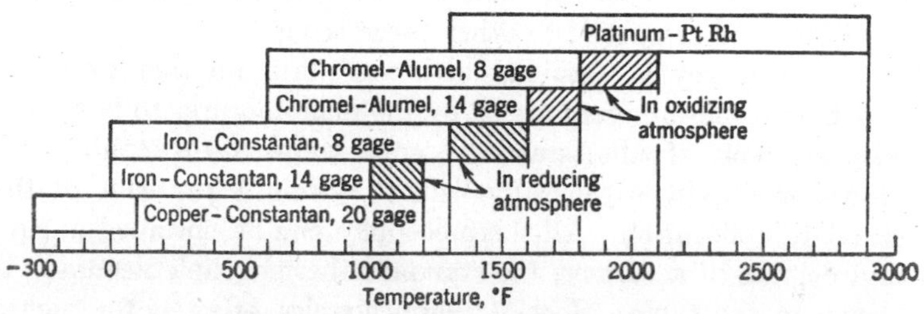

FIG. 3–3. Useful Range of Industrial Thermocouples

may be used at temperatures several hundred degrees higher than the upper limit shown in Fig. 3–3 if there is no oxygen present to attack the iron wires. Chromel–Alumel thermocouples may be employed at higher temperatures than indicated in Fig. 3–3 if excess oxygen is present.

Items that should be noted are that copper–Constantan thermocouples are restricted to low temperatures and platinum thermocouples are used for high temperatures. The copper–Constantan and iron–Constantan thermocouples are used in the same range as the industrial pressure thermometer.

Other characteristics of thermocouples are shown in Table 3–1. The span in degrees Fahrenheit is representative of the

TABLE 3–1. CHARACTERISTICS OF THERMOCOUPLES

Thermocouple, Positive Metal First	Composition		Span, °F for 10 millivolts	Average Accuracy, per cent
	Positive Metal	Negative Metal		
Copper-Constantan	100 Cu	55 Cu, 44 Ni[1]	450 at 100	±¾
Iron-Constantan	100 Fe	55 Cu, 44 Ni[1]	320 at 600	±1
Chromel-Alumel	90 Ni, 10 Cr	94.5 Ni[2]	435 at 1400	±¾
Pt 10% Rh–platinum	90 Pt, 10 Rh	100 Pt	1500 at 2100	±½
Pt 13% Rh–platinum	87 Pt, 13 Rh	100 Pt	1352 at 2100	±½

[1] With approx. 1 Mn and Fe.

[2] With approx. 2.5 Mn, 2 Al, 1 Si.

power of the thermocouple, since the smaller the temperature span for a given change of emf the greater is the change in emf per degree of temperature. A 10-millivolt span was selected for reference since this represents about the smallest span normally used in indicating or recording instruments, although there are a great many exceptions. The platinum thermocouples have a very low emf compared to other thermocouples.

Best accuracy is obtained with the platinum thermocouples which are generally guaranteed by the manufacturer to be within ½ per cent of the standard emf temperature calibration. Iron–Constantan is generally held to within 1 per cent of the standard calibration. All thermocouples can be obtained in several degrees of accuracy; the standard thermocouple accuracy is shown in the table. Special thermocouples selected for higher accuracy may be obtained at premium cost with the platinum couples accurate to within $\pm\frac{1}{4}$ per cent, the iron–Constantan couple accurate to $\pm\frac{1}{2}$ per cent, and the Chromel–Alumel couples accurate to $\pm\frac{3}{8}$ per cent. In addition, thermocouples can be obtained with an error curve giving its deviation from standard calibration. Also a thermocouple can be obtained with a National Bureau of Standards certification when the best accuracy is desired.

The resistance to corrosion, oxidation, and contamination generally depends on one of the metals of the thermocouple. For example, the Chromel of a Chromel–Alumel thermocouple is susceptible to attack in reducing atmospheres (where oxygen is lacking and hydrogen and carbon monoxide are present). The iron of iron–Constantan thermocouples is susceptible to corrosion where oxygen or water vapor is present. Platinum thermocouples are sensitive to contamination at high temperatures by hydrogen, arsenic, phosphorus, silicon, and some metal vapors. Platinum thermocouples are nearly always used with a gas-tight thermal well in which an oxidizing atmosphere should be maintained at all times.

Other special thermocouples have been used at various times mainly for the purpose of resisting corrosion and oxidation in particular applications. These are, Chromel–KA2S, in which KA2S is a metal similar to a stainless steel, iron–Alumel, nickel–nickel molybdenum, Chromel–Constantan, and graphite–silicon carbide.

As we found earlier, the thermal emf–temperature relation of most thermocouples is parabolic. The second-order term is small for all common thermocouples but not negligible. The scale shape of a pyrometer will therefore be nearly linear. Thermocouple emf–temperature data are given in Tables A–10 to A–14 (appendix) for standard thermocouples.

Thermocouples are most commonly made in the form of wires, insulated and welded together at the measuring junction, as indicated in Fig. 3–4. The measuring junction is formed in two different ways: twisted weld and butt weld. The twisted weld is made by twisting the two wires for several turns and welding. The shape of the junction after welding remains more or less unchanged. The twisted weld is made with wires of larger sizes and gives mechanical strength. The butt weld is made by

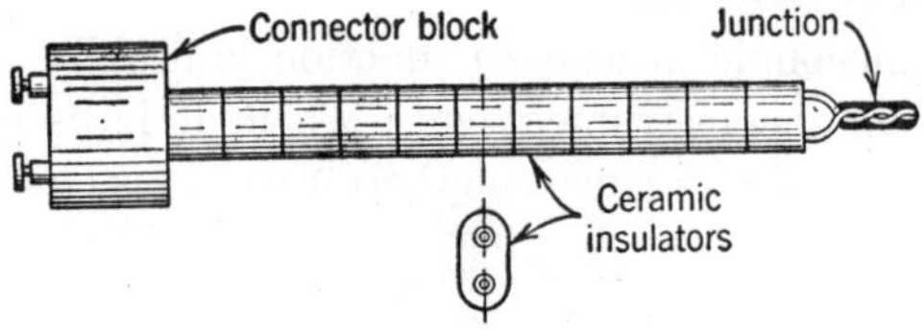

FIG. 3-4. Industrial Thermocouple

fusing the two wires into a round bead. This method is used for small-size wires.

Wire sizes for all but platinum thermocouples are usually one of the following:

B&S Gage	Diam., in.
8	0.1285
14	0.06408
20	0.03196

and, infrequently, 2, 6, 11, 24, 28, 30, and 32 B&S gage are used. Platinum thermocouples are generally 24 B&S gage, 0.0201 in. diameter.

In order to prevent the forming of a second junction, the wires of a thermocouple are insulated from each other by being

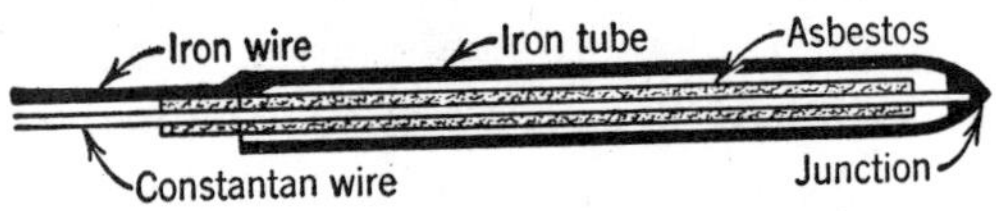

FIG. 3-5. Tube-Type Thermocouple

threaded through porcelain insulators. These insulators will retain their shape up to 2800°F. The ends of the thermocouple wires are connected to a porcelain block having screw terminals for connections.

Tube thermocouples are made of iron–Constantan with the iron in the form of a tube, the Constantan wire running down the center of the tube, as shown in Fig. 3-5. The purpose of this construction is to improve the speed of response by avoiding the use of a thermal well. The iron tube is generally ⅛ in. outside diameter, and the junction is made by swaging and welding the iron tube around the Constantan wire. The Constantan wire is insulated from the iron tube with either asbestos or a ceramic.

Thermocouple Lead Wires

The thermocouple measuring junction is usually located some distance from the instrument with which it is used, sometimes several hundred feet. Since the thermocouple is connected to the instrument by wires, we have two problems: (1) the location of the reference junction, and (2) the errors introduced by the connecting wires.

Variations in temperature of the reference junction cause a change in thermal emf, since the thermal emf depends on the temperature difference of the measuring and reference junctions. If a temperature is to be measured within one degree, the reference-junction temperature should be constant within one degree (unless reference-junction compensation is employed). The reference junction may be located, for example, adjacent to the thermocouple or may be placed near or in the measuring instrument. Ordinarily temperature variations of some magnitude are expected adjacent to the thermocouple because the thermocouple may be installed in a furnace, heat exchanger, or vat. Temperature variations at the measuring instrument are normally more constant, especially when the instrument is installed in a temperature-controlled or air-conditioned room. Therefore, the reference junction is located near and generally inside the measuring instrument.

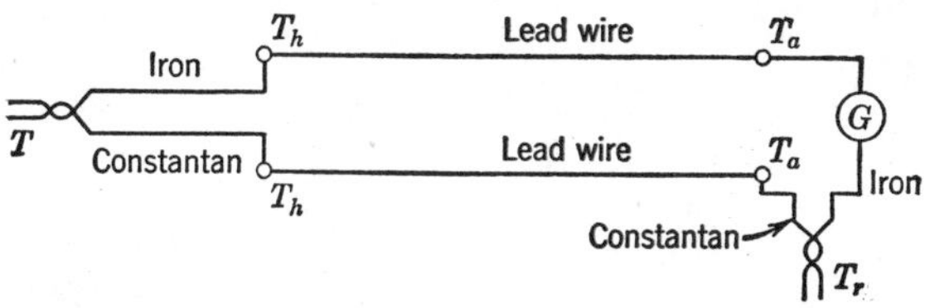

FIG. 3-6. Temperatures in a Thermocouple Circuit

Further, the introduction of connecting wires into the thermocouple circuit has no effect so long as the ends of the wire are at the same temperature. However, the temperature adjacent to the thermocouple may be several hundred degrees while the temperature at the instrument is near ambient temperature. The temperature difference $(Th - Ta)$, shown in Fig. 3-6, may be several hundred degrees. Therefore it is necessary to use connecting wires or lead wires that will avoid error.

There are two methods of selecting lead wires: (1) the thermocouple wires themselves may be extended to the instrument, and (2) the lead wires may be made of thermocouple metals having the same thermoelectric properties as the thermocouple with which they are used.

The first method involves theoretically no error and, as indicated in Table 3–2, is used for copper–Constantan and iron–Constantan thermocouples, and sometimes for Chromel–Alumel thermocouples.

TABLE 3–2. THERMOCOUPLES AND LEAD WIRES

Thermocouple +	Thermocouple −	Lead Wire +	Lead Wire −
Copper	Constantan	Copper	Constantan
Iron	Constantan	Iron	Constantan
Chromel	Alumel	Chromel	Alumel
Chromel	Alumel	Iron	Copper–nickel alloy
Chromel	Alumel	Copper	Constantan (to 125° only)
Platinum–rhodium	Platinum	Copper	Copper–nickel alloy

The cost of long lengths of Chromel and Alumel wire is high, and the cost of platinum wire in even moderate lengths is prohibitive. With Chromel–Alumel and platinum thermocouples less expensive lead wire is used, as indicated in Table 3–2. Iron and a copper–nickel-alloy lead wires, which have a temperature–emf characteristic very close to that of Chromel–Alumel, are widely used. Copper and Constantan lead wire may be used for Chromel–Alumel thermocouples only where high accuracy is not required, since the temperature–emf characteristics of these pairs do not quite match beyond 125°F. For platinum thermocouples, lead wires are made of copper and a copper–nickel alloy, the temperature–emf characteristic of which is the same as for the platinum thermocouple.

The error involved in using lead wire of different material depends on how well the temperature–emf characteristic of the lead wire matches that of the thermocouple. Suppose that, with a Chromel–Alumel thermocouple used in the range 1000 to 2000°F, the thermocouple head temperature is 250°F and the temperature at the reference junction is 32°F. If the lead wires are copper and Constantan, the error caused is the difference between the emf of the different materials at 250 and 32°F. This is 5.28 – 4.97 or 0.31 millivolt. At 1400°F this represents an error of 13.3°F, or 0.95 per cent, which is not an allowable error for most industrial applications. For greater accuracy it becomes necessary to use the iron- and copper–nickel-alloy lead wires.

Naturally it is important that the polarity of lead wires and thermocouples be observed during installation; otherwise, a large error may result.

Lead wires may be solid, or may be stranded for flexibility. Various kinds of insulation may be obtained, such as enamel, cotton covering, asbestos, glass, polyvinyl covering, or lead sheath over an insulating cover.

Thermal Wells

In only a limited number of applications is it practical to expose the bare thermocouple to the fluid in which the temperature is being measured. Any corrosive or oxidizing action is accentuated at higher temperatures, and the resultant action on thermocouple materials makes it advisable to protect them by means of a thermal well or a protecting tube. Even with a thermal well, the corrosion and oxidation may be so rapid that frequent replacements are necessary. In fact, in certain molten-metal baths, it is the practice to use one thermocouple and well for one quick reading of temperature and then discard them.

There are many hundreds of types of thermal wells and similar accessories, but generally the thermal well is made in the form of a tube with a closed end installed, as indicated in Fig. 3–7. The thermocouple assembly on the left is the kind normally used for the measurement of air or gas temperatures, as in a furnace. It is composed of a head, a mounting flange, the thermocouple itself, and the thermal well. The thermocouple assembly on the right is used for pressure vessels and pipe lines, particularly for liquids under pressure. An open-end thermal well, shown in Fig. 3–8, is sometimes used where only slight protection is required, as in furnaces at lower temperatures. An open-end thermal well only protects the thermocouple from mechanical damage and from the erosive effects of fast-moving gases, while supporting the thermocouple from sagging.

Where corrosion may be severe, especially at temperatures over 2000°F or where platinum thermocouples are used, two thermal wells are employed, as shown in Fig. 3–9. A secondary thermal well covers the primary thermal well. The purpose of

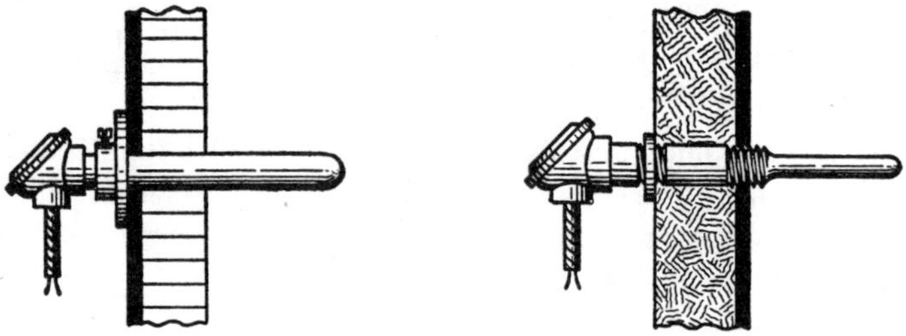

Fig. 3–7. Thermocouple Installations

the secondary well is to prevent sagging of the assembly at high temperature and to protect the surface of the primary well so that it will remain gastight. Secondary wells are made of either metal or ceramic, depending on the requirements.

The materials used for thermal wells are reviewed in Tables A–6 and A–7 (appendix), together with a conservative maximum

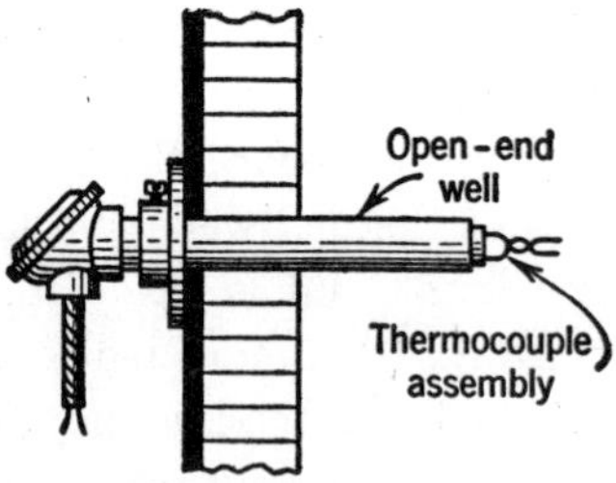

Fig. 3–8. Open-End Thermal Well

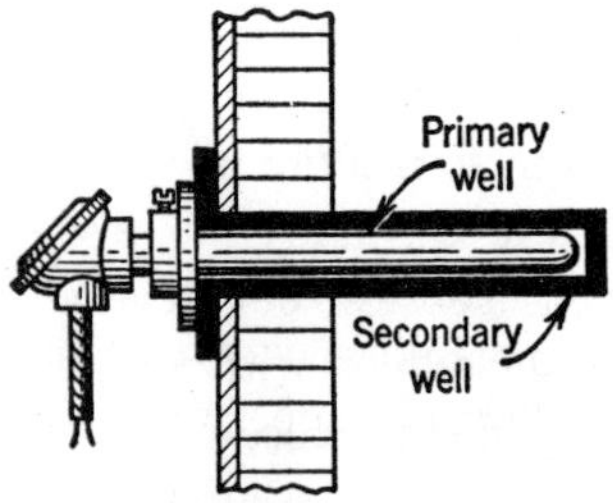

Fig. 3–9. Secondary Thermal Well

temperature and a few remarks on their application. The maximum temperature limit varies with the application. It may sometimes be higher if corrosion conditions are not severe, but, on the other hand, the maximum temperature is seriously reduced when corrosion conditions are severe. Thermal wells of platinum are used only for molten glass since their cost is extremely high.

Almost any of these materials may be used for primary protection; firebrick, Mullite, and silicon carbide are used only for secondary protection.

The mechanical properties which must be considered in the selection of a thermal well are:

Resistance to corrosion and oxidation.
Resistance to mechanical shock.
Resistance to thermal shock.
Resistance to gas leakage.
Mechanical strength.

In some applications the primary purpose of the thermal well is to protect the thermocouple from physical damage. In heat treating, for example, the work is loaded into a small furnace and by accident often comes in contact with the thermocouple, bending or breaking it. For physical protection a metal well is preferred.

When the thermocouple is used for testing in a kettle or ladle filled with liquids or molten metal, it is plunged into the bath and the temperature determined. This imposes a great thermal shock or sudden change of temperature on the thermocouple and its thermal well. Metals generally withstand thermal shock, but some ceramics do not.

The thermal well should be impervious to diffusion of gases through its walls, since some gases attack thermocouple materials. With the platinum thermocouple, special care must be taken to insure no diffusion of gases through the protecting well. A secondary well is sometimes used for prevention of gas leakage. Calorized iron and Nichrome are the metals having lowest permeability, whereas in ceramics, porcelain, sillimanite, and Mullite are superior.

Response of Thermocouples

The speed of response of the thermocouple is determined by its heat-transfer characteristics: the conduction, convection, and radiation of heat between the thermocouple and its surroundings. With a bare thermocouple installed in a flowing liquid most of the heat is transferred by direct conduction between liquid and thermocouple.

In a bare thermocouple installed in a gas at low temperature, say 400°F or less, most heat is transferred by direct conduction, some by convection, and very little by radiation. Direct conduction is greatly increased by increased velocity of the gas past the thermocouple. In the same installation at high temperatures, 1000°F or higher, most heat is probably transferred by radiation, some by conduction, and very little by convection. To illustrate the effect of radiation, Fig. 3–10 shows that a bare thermocouple will reach a higher temperature more quickly than a

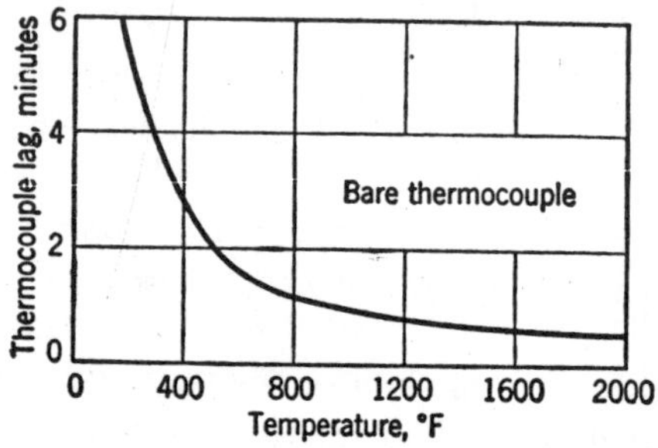

FIG. 3–10. Thermocouple Lag at Various Temperatures

lower temperature. The effect of radiation is noticeable even at temperatures of 800°F. The time constant of the thermocouple is reduced by a factor of three at a temperature of 1700°F. Between 400 and 1000°F heat is transferred by all three means.

When a thermal well is used, the situation differs by virtue of the air space separating the thermocouple measuring junction and the thermal well. This space is not often filled with solid or liquid conducting materials as is the case with an industrial thermometer. Sometimes the thermocouple junction is arranged to make physical contact with the inside wall of the thermal well, but this measure is often ineffective, because the area of contact is necessarily small. Between the well and the thermocouple heat is transferred by conduction and convection through still air at low temperatures, and by radiation and conduction at high temperatures.

Other factors that influence the speed of response of thermocouples are: (1) thermal characteristics of materials, particularly well materials; (2) the size of the thermocouple and well; and (3) the method of application.

With few exceptions the materials of the thermocouple do not affect the speed of response to any great extent. Data are given in Table A–6 (appendix) for many thermal-well materials. Although the thermal resistance of ceramics is higher than that of metals, the thermal capacity of ceramics is somewhat lower. Thermal capacity is more important than any of the other factors at low temperatures, since the film coefficient of resistance is high compared to the thermal resistance of the well material. The film coefficient of heat transfer is greatly influenced by velocity of fluid past the thermocouple. Data on this point were discussed in connection with the speed of response of the industrial thermometer. The speed of air or gas flow should be at least 120 fpm, and the speed of liquid flow should be at least 60 fpm, in order to obtain a reasonable speed of response.

At higher temperatures the total emissivity of the surface of the thermocouple junction and thermal well determines the effectiveness of radiation in transferring heat. A high emissivity means that a greater percentage of the heat available is absorbed. Since the thermal capacity of the thermocouple and well is more or less constant, the greater heat flow brings a faster speed of response. Heat radiators are also good absorbers, and so the effect of radiation works in both directions. For polished metal surfaces the emissivity is quite low, perhaps 0.05, and little heat can be transferred by radiation. For the same metal surface well oxidized, rusted, and blackened, the emissivity may be as high as 0.95 or higher, which allows a large radiation transfer of heat. The quality of the surface also influences the emissivity and heat radiation. A smooth machined surface has a smaller emissivity than a rough surface as in sand-cast wells or rough-machined wells.

In general, ceramic materials have a higher emissivity than most smooth oxidized metals. Metallic materials form an oxide surface, which improves it as a radiation receiver but decreases conduction of heat through the thermal well, because the corrosion scale has about 500 times the thermal resistance of a metal wall.

Some thermal-well materials such as quartz, glass, and Vycor are translucent and transmit rather than absorb radiation of certain wavelengths. In such cases, the thermal well has much less effect on the speed of response, and the thermocouple responds almost as quickly as though the thermal well were not used at all.

The effect of materials of the well is shown in Fig. 3–11, giving response curves for a 14-gage Chromel–Alumel thermocouple with a Vycor well and with a calorized iron well. The high speed of response of the Vycor (a glass) is due to its transmission of radiation.

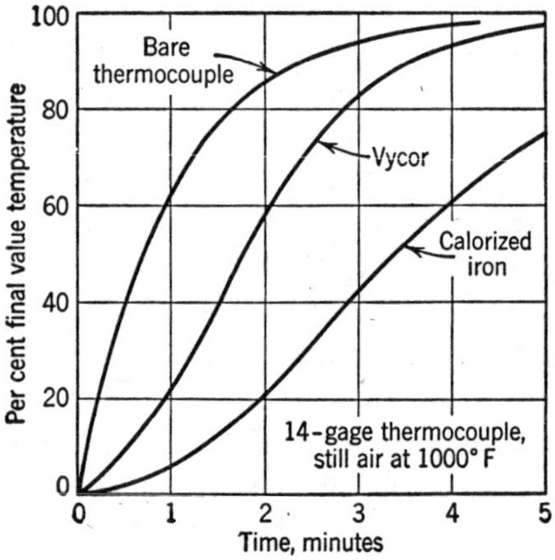

FIG. 3–11. Effect of Well on Thermocouple Response

The size of the thermocouple and well have a greater influence on speed of response than the materials selected. This is caused, in general, by the greater thermal capacity of an increased mass of material. There is a considerable difference in response of bare thermocouples of different-size wire, as shown in Fig. 3–12. These thermocouples were tested at 250°F. The size of the thermal well also influences speed of response.

Thermal wells for the same thermocouple have different diameters and wall thicknesses, depending on their method of manufacture. Cast materials such as Nichrome, Chromel, and iron have thicker walls and slower response than machined wells of Inconel, steel, and iron. Ceramic materials such as Vycor, sillimanite, and porcelain generally are made with walls much less thick than cast metallic materials.

Heavier thermal wells with thick walls are undesirable from the standpoint of response but are sometimes desirable for resistance to corrosion and for higher strength. A careful balance must be made between the lag due to a heavy well and the increased maintenance caused by severe corrosion. Frequent replacement of thermal wells of light construction may be economically more sound than accepting a temperature measurement involving great lag. If the temperatures to be measured are

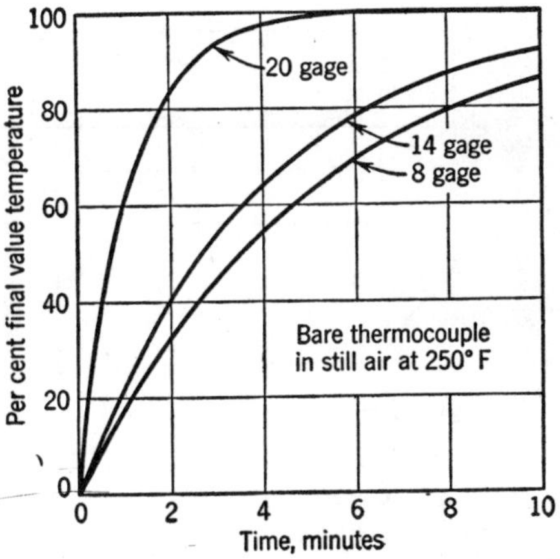

FIG. 3–12. Response of Bare Thermocouples of Different Size

moderately fluctuating, then the dynamic error caused by a large thermal lag will certainly result in measurements that are useless.

The effect of depth of immersion of the thermocouple or well is sometimes to cause an error due to the heat conduction along the thermocouple wires or along the thermal well. Ceramic materials are a little better in this respect, because their higher thermal resistance results in less heat conduction. The 8-gage thermocouple particularly provides a large area for heat flow along the thermocouple wire, and the thermocouple may indicate a lower temperature, even when the thermal well is sufficiently immersed that its heat conduction is negligible.

Radiation errors generally exist to an appreciable degree in measurement of high air or gas temperatures with thermocouples. Radiation effects were previously discussed in Chapter 2 in connection with the thermometer. At temperatures over 1000°F (538°C), the thermocouple receives most of its heat by radiation from the surroundings. These surroundings are indicated in Fig. 3–13 for an electric furnace, and they may be the heating elements, furnace walls, and furnace hearth or floor. The temperature of each of these points is different; the heating elements may be 2500°F, the walls 2400°F, and the hearth 2100°F. The temperature indicated by the thermocouple will depend on the relative amount of radiation received from each of the sources. It is possible that the thermocouple will indicate a temperature

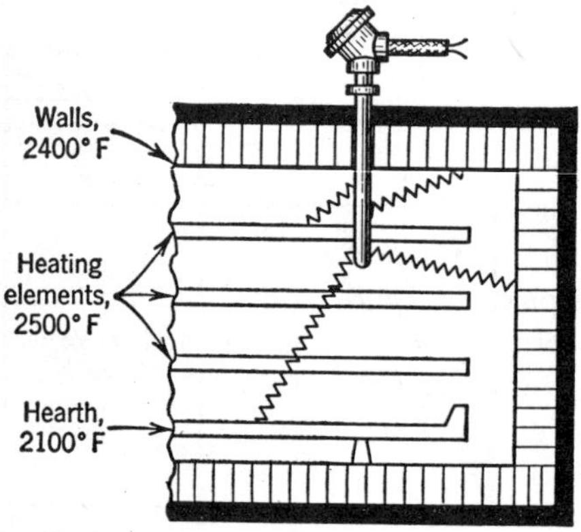

FIG. 3–13. Radiation Effect in Thermocouple Installations

anywhere between 2100 and 2500°F. If the thermocouple is at 2300°F, then it receives radiation from the heating elements and from the walls but loses heat by radiation to the hearth. With still air, which normally exists in small furnaces, the direct conduction of heat to the thermocouple is probably negligible. Such radiation conditions exist in the measurement of furnace-, duct-, and engine-exhaust temperatures. The question of *what* temperature is measured is generally relegated to the background, as long as a stable relation between thermocouple indications and product quality is established.

In the measurement of temperature of flowing gases, there are several methods of employing thermocouples so that true gas temperature may be measured or at least approximated.[3,4] Er-

[3] *Temperature, Its Measurement and Control in Science and Industry,* Reinhold Publishing Corp., 1941, p. 775.

[4] E. M. Moffat, "Methods of Minimizing Errors in the Measurement of High Temperatures in Gases," *Instruments,* **22,** No. 2 (Feb. 1949).

rors using simple thermocouple installation may be as high as 50 per cent. First, a radiation shield may be installed around the hot junction, as with a thermometer. It is essential that the hot gases flow around both the shield and the thermocouple. Two or more shields may be used to increase accuracy. The velocity of the gases should be as high as possible; a venturi arrangement is sometimes used to increase the gas velocity. Figure 3–14 shows a simple radiation shield. Second, the thermocouple may be installed in a cylindrical shield, closed at one end. The gases are pumped into the shield past the thermocouple measuring junction and out of the pipe or duct. The velocity of the gases is thereby increased, and the heat transfer is improved. Obviously this method is worth while only when the gases originally have low velocity. Third, a radiation shield installed around the hot junction and maintained at the same temperature as the surrounding walls prevents heat loss from the

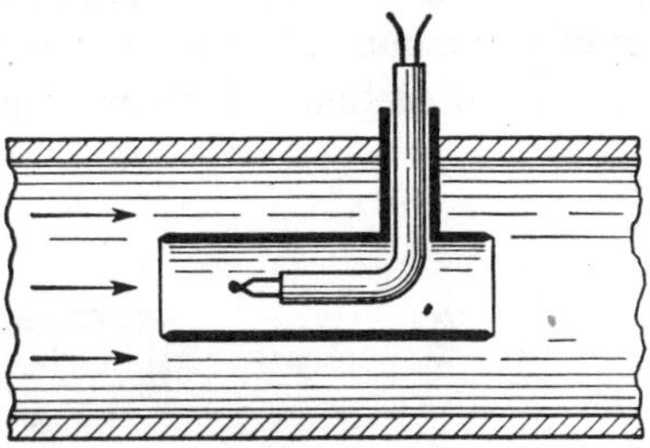

FIG. 3–14. Thermocouple and Single Radiation Shield

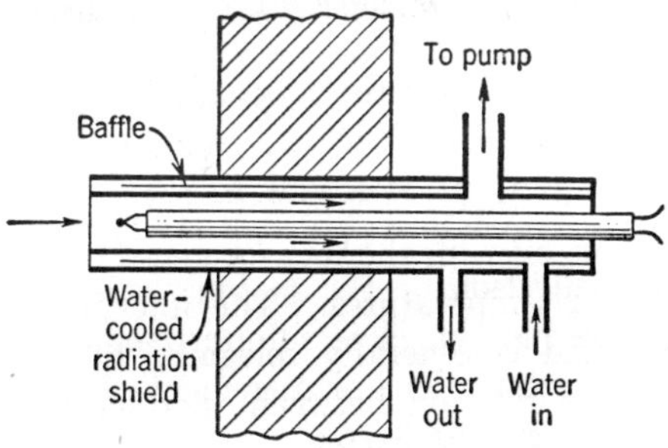

FIG. 3–15. Thermocouple with Cooled Radiation Shield and Forced Convection

thermocouple by radiation. If the surrounding walls are higher in temperature than the gases, water cooling of the shield is required. If the walls are at a lower temperature than the gases, then electric heating is required. Fourth, if methods 2 and 3 are combined, as in Fig. 3–15, the effects of the radiation shield and the increase in gas velocity are to increase further the accuracy of measurement. This method has been used to measure gas temperatures in furnace boilers.

The Millivoltmeter

The industrial millivoltmeter is a calibrated d-c galvanometer especially constructed for the measurement and indication of thermocouple emf. It is the simplest and least expensive of all

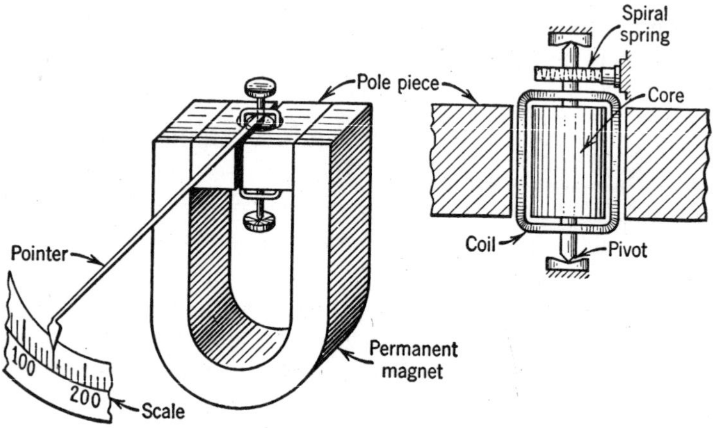

FIG. 3–16. Millivoltmeter Movement

the instruments for use with thermocouples. The millivoltmeter operates on the principle of the D'Arsonval type of galvanometer, wherein the rotation of a coil in a steady magnetic field is proportional to the magnitude of electric current through the coils. The deflection of the moving coil is caused by the electromagnetic torque, which is counterbalanced by the torque of a spiral "hair" spring. The arrangement of the millivoltmeter mechanism is shown in Fig. 3–16.

The permanent magnet is constructed of special alloy steel of carefully controlled quality which insures a high yet constant magnetic field. Metal pieces called pole pieces are situated at the poles of the magnet and serve to concentrate the magnetic field at the moving coil. Generally a soft-iron core is placed inside the moving coil in order to reduce the air gap and increase the field at the coil.

The coil is wound in rectangular form of fine wire and is mounted generally on jewel pivots at the top and bottom. The electric connections to the coil are formed by the hairsprings at the upper and lower ends. Other means such as a suspension of the coil on two fine wires are sometimes used for mounting the millivoltmeter coil. Generally a pivot arrangement is preferred because of its ruggedness. One of the hairsprings is arranged with a movable base so that the zero position of the coil and its pointer can be adjusted.

The millivoltmeter is primarily an indicating instrument, and the pointer operates over a scale of about 5 or 6 in. length. The

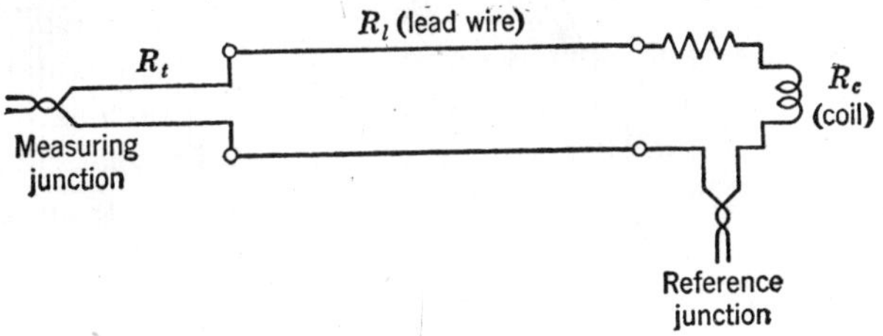

FIG. 3–17. Millivoltmeter and Thermocouple Circuit

range of the millivoltmeter is not normally less than 10 millivolts, which corresponds to a temperature range of 0 to 330°F (0 to 182°C) with an iron–Constantan thermocouple.

The circuit of the millivoltmeter pyrometer is shown in Fig. 3–17. By writing Ohms's law for the circuit, the following equation is obtained:

$$e_i = e\left(\frac{R_s + R_c}{R_t + R_l + R_s + R_c}\right) \tag{3–11}$$

where e_i = voltage at millivoltmeter terminals
e = thermocouple emf
R_s = resistance in millivoltmeter
R_c = resistance of millivoltmeter coil
R_l = resistance of lead wire
R_t = resistance of thermocouple

Apparently then, the emf measured at the millivoltmeter terminals is not the emf of the thermocouple. For example, if R_s,

R_c, R_e, and R_t have typical values of 550, 50, 0.15, and 2.5 ohms, respectively, the relation between indicated and true voltage is

$$e_i = 0.9956e$$

The millivoltmeter therefore indicates less than the true thermocouple voltage. This is readily overcome by simply calibrating the millivoltmeter arbitrarily to read correctly when a given external resistance $(R_t + R_l)$ is assumed. The emf range of the millivoltmeter is adjusted by selecting the gradient or spring rate of the hairsprings and by adjusting resistance R_s, thus controlling the amount of pointer movement for a given change of emf.

The resistance of the millivoltmeter $(R_s + R_c)$ should therefore be as large as possible, in order to reduce the effect of changes in external resistance. Suppose that in the previous example the resistance of the lead wire becomes 5.0 ohms when its length is increased. The relation between indicated and true voltages is now

$$e_i = 0.9915e$$

and the millivoltmeter will be in error by 0.41 per cent.

In order to reduce the error caused by changes in external resistance, the internal resistance of the millivoltmeter should be as high as possible. Usually, the resistance is about 600 ohms. The millivoltmeter is then calibrated with a fixed external resistance, such as 2 ohms, so that it reads directly in thermocouple emf. Changes of external resistance cause an error, as indicated in Table 3–3. In some cases an adjustable resistor is placed in the internal circuit of the millivoltmeter as in the Harrison-Foote millivoltmeter circuit so that changes in external resistance can be compensated by manual adjustment.

Changes of ambient temperature at the millivoltmeter cause two effects. First, the internal resistance of the millivoltmeter may change because of the temperature coefficient of resistance of the internal resistors. This error is generally avoided by making the internal resistor R_s of Manganin, which has a negligible temperature coefficient of resistance, or by selecting a material which has a slight negative coefficient of resistance, so as to compensate for the positive coefficient of resistance of the coil.

Table 3–3. Millivoltmeter Error from Changing External Resistance *

External Resistance, ohms	Error in % Scale
0.5	+0.25
1.0	+0.16
2.0	0.0
3.0	−0.17
4.0	−0.34
5.0	−0.50
6.0	−0.67

* Assuming millivoltmeter internal resistance is 600 ohms.

Second, the greatest error caused by the changes in ambient temperature is the changes in temperature of the reference junction. If reference-junction temperature compensation is not used, then the magnitude of the error is proportional to the change in emf caused by the change of temperature of the reference junction. For example, if the millivoltmeter with a Chromel–Alumel thermocouple reads 1000°F with the reference junction as 32°F, then the instrument will read 960°F when the reference-junction temperature is 75°F and the error is −40°F. Ordinarily the reference junction is not placed in an ice bath to maintain it at 32°F since this procedure would involve too much maintenance for an industrial instrument. Therefore, the millivoltmeter is calibrated for a reference-junction temperature of 75°F, which is approximately room temperature.

Reference-junction temperature compensation is required because the fluctuations in temperature of the millivoltmeter case are usually sufficient to cause an appreciable error in reading. If the changes in reference-junction temperature are more than ±5°F, then reference-junction compensation should be used. Compensation is accomplished simply in a millivoltmeter by arranging a bimetallic strip so that it shifts the position of the fixed end of one of the hairsprings, as indicated by Fig. 3–18. When the reference-junction temperature decreases, the emf of the thermocouple increases. This causes a greater current flow through the millivoltmeter coil and results in a greater torque, tending to move the pointer upscale. On the other hand, the bimetal compensator expands and moves the end of the hairspring sufficiently to cause a torque which is equal and opposite to the torque due to greater emf. The pointer then remains stationary. The accuracy of compensation depends on the exactness with which the torque due to the increase in emf balances the torque due to the shift of the hairspring by the bimetal. The accuracy of compensation is such that errors due to change in reference-junction temperature are a small fraction of 1 per cent.

The response of the millivoltmeter is essentially that of a galvanometer and is governed by the design of the galvanometer.

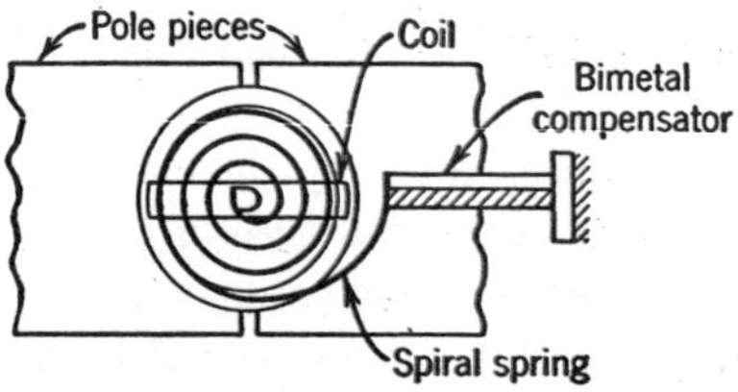

FIG. 3–18. Millivoltmeter Reference-Junction Compensation

The damping of the galvanometer is determined by the external resistance of the circuit. When the external resistance is 2 ohms and the series resistance (R_s in Fig. 3–17) is 550 ohms, the galvanometer is designed to have critical damping with an equivalent time constant of about 0.015 min. Normally this time constant is negligible compared to the time constants of thermocouples, so that little dynamic error exists in the millivoltmeter. When the total resistance of the circuit is smaller than the design value, the galvanometer becomes overdamped and is slower in response, and, when the resistance is greater, the galvanometer is underdamped and will oscillate. The magnitude of the change in external resistance is not usually sufficient to cause a great change in damping action of the galvanometer.

The static accuracy of the millivoltmeter without a thermocouple is generally better than ± 1 per cent of full scale. The mean accuracy of the millivoltmeter and a standard iron–Constantan thermocouple will be ± 1.5 per cent of full scale. The best accuracy that can be achieved is probably ± 1 per cent unless an error curve is used. The dead zone of the millivoltmeter is small but quite variable, because it depends on the friction in the moving-coil system.

The Null-Potentiometer Circuit

The null potentiometer operates by comparing an unknown emf against a known emf. This is accomplished in the fundamental potentiometer circuit of Fig. 3–19. Assume for the moment that the battery supplies a constant voltage over the resistors A, S, and B. The potential drop across the slide-wire S is therefore known, and, if the slide-wire is evenly wound, then the position of each potential interval is known. When an unknown potential at the thermocouple exists, a current will flow through the thermocouple circuit when the known and unknown potentials are unequal. The galvanometer in the thermocouple circuit need not be calibrated; its pointer simply moves one way for a given direction of current flow and in the opposite direction for a reversed flow of current.

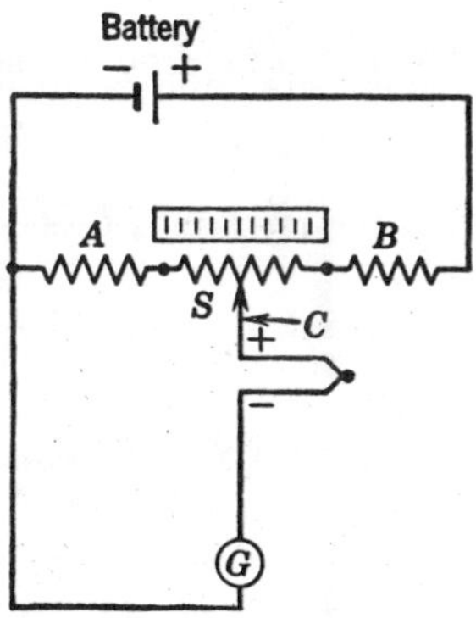

FIG. 3–19. Elementary Potentiometer Circuit

A scale can be attached to the slide-wire S, and the position of the slider C on the scale can be calibrated for emf of the thermocouple. With a thermocouple having a known emf–temperature relationship the scale can be calibrated in terms of temperature.

The calculations of the simple potentiometer circuit are made as follows. Let

S = slide-wire resistance
E = battery voltage
e_0 = thermocouple emf at lowest scale point
e_r = span of thermocouple emf
A, B = circuit resistors (Fig. 3–19)

The potential drop across resistor A is

$$e_0 = iA \tag{3-12}$$

where i is the current through resistor A and the slide-wire. The potential drop across the slide-wire is

$$e_r = iS \tag{3-13}$$

The resistor A must therefore be

$$A = \left(\frac{e_0}{e_r}\right)S \tag{3-14}$$

The potential drop across resistor B is

$$E - (e_0 + e_r) = iB \tag{3-15}$$

and the resistance B is

$$B = \left(\frac{E}{e_r}\right)S - \left(\frac{e_0}{e_r} + 1\right)S \tag{3-16}$$

Apparently then the purpose of resistor A is to adjust the calibration of the "zero" point in the slide-wire, and the purpose of resistor B is to adjust the range of the slide-wire.

The resistance of the thermocouple, lead wire, and galvanometer do not affect the accuracy of the potentiometer circuit if it is assumed that the elements of the circuit are perfect. This is the most important advantage of the null-potentiometer circuit.

For practical industrial use the basic potentiometer circuit of Fig. 3–19 must be modified for the following reasons: first, changes in resistances due to temperature must be avoided; second, the battery voltage must be kept constant; and, third, change in temperature of the reference junction of the thermocouple causes an error unless compensated.

All the resistors of the potentiometer circuit may be made of Manganin, which has a negligible temperature coefficient of resistance. Changes in temperature of the circuit then have negligible effect on the accuracy of the calibrated slide-wire.

The battery voltage is not likely to remain constant since the current drain is of appreciable magnitude. A battery generally does not fulfill the requirements for an emf constant to 1 part in 1000 for a wide range of ambient temperature, a constant internal resistance, and a life of at least several thousand hours. A voltage regulator might be used, but this method is somewhat more expensive. The method universally adopted is to use a good-quality battery in the circuit and either periodically or continuously adjust the current in the battery circuit to a standard value. This procedure is called standardization.

The circuit arrangement of Fig. 3–20 for standardization is common. The standard cell is a primary voltage standard of 1.0190 ± 0.0005 volts at 20°C. The standard cell has almost infinite life if the current drain is zero. Therefore, the standard cell, to be accurate, must be used only under conditions where the current drain is negligible or zero. A standard cell operates best when used at a constant temperature of about 68°F (20°C). The temperature should never go above 140°F (60°C) or below 32°F (0°C). The battery for the current source of the circuit may be a good-quality dry cell, a LeCarbone cell, or an air cell.

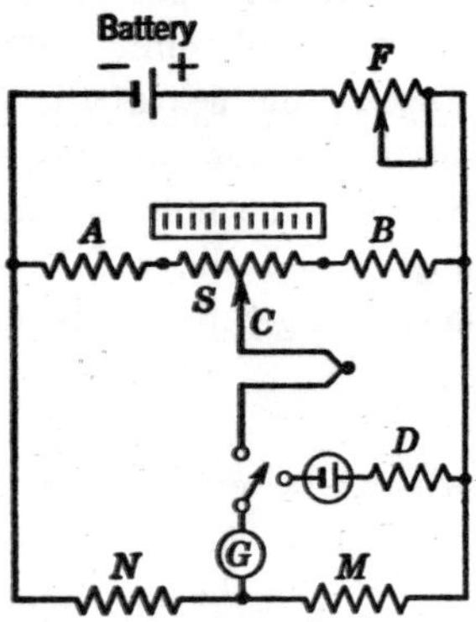

FIG. 3–20. Potentiometer Circuit

The circuit arrangement of Fig. 3–20 is intended for periodic standardization. The switch in the galvanometer circuit is thrown to connect the standard cell into the circuit. The resistor D is merely for the purpose of limiting the current through the standard cell. The requirement is that the current through the resistors A, S, and B is a constant, fixed value. This condition is determined when the galvanometer current is zero. At this condition, the emf of the standard cell is equal to the potential difference across the resistor M, which is determined by the current through the resistors M and N in accordance with the laws of parallel circuits. The current through resistors M and N is proportional to the potential difference across them and is proportional to the battery voltage minus the potential difference across the adjustable resistor F Therefore, standardization is accomplished by throwing the switch to the standard cell and adjusting the resistor F so that zero current flows through the galvanometer. If this operation is performed quickly, the life of the standard cell is prolonged. Usually the operation of standardization is performed about once every hour if automatic and once every day if done manually.

Reference-junction compensation is accomplishd either manually or automatically, generally the latter, through suitable manipulation of resistor N. The analysis is as follows: All that is desired is that, when the reference-junction temperature changes, the slider C should remain stationary. If the reference-junction temperature increases, a decrease in thermocouple emf is experienced. The potential drop across resistors A plus L is now greater than the potential drop across resistor N plus the thermocouple emf. (L is the portion of the slide-wire S resistance to the left of the slider.) If we either manually or automatically increase the magnitude of the resistor N a slight amount, the null balance of the circuit is restored without having moved slider C. The circuit therefore indicates the same measuring-junction temperature as before. This requires

$$i_a(A + L) = i_n N + e \tag{3-17}$$

where i_a = current through A
i_n = current through N
e = thermocouple emf

and where we are assuming that the actual change of resistance of N is small, so that the total effect on the branch currents of the parallel circuit is negligible. The currents are therefore

$$i_a = \frac{E_{ab}}{A + S + B} \tag{3-18}$$

$$i_n = \frac{E_{ab}}{M + N} \tag{3-19}$$

where E_{ab} is the potential drop across resistors A, S, and B. Substituting these equations in equation 3-17, we get

$$\frac{e}{E_{ab}} + \left(\frac{1}{\frac{M}{N} + 1}\right) = \frac{A + L}{A + B + S} \tag{3-20}$$

The right-hand side of equation 3-20 is constant for any given measuring-junction temperature. It is evident that, if the reference-junction temperature increases, the thermocouple emf e decreases, and this may be compensated by increasing the magnitude of the resistor N.

For automatic reference-junction compensation, the resistor N is made of nickel, which has a large positive temperature coefficient of resistance. The resistor N and the thermocouple reference junction are housed in an enclosure so that they will be at the same temperature. The accuracy of compensation depends on the linearity of the thermocouple emf–temperature relationship and how well it matches the reciprocal function of the ratio of M and N in equation 3–20. This is the principal reason for desiring a linear relationship of thermal emf in a thermocouple. The nickel compensation resistor must be matched with the thermocouple emf curve for every type and range of thermocouple. Generally the accuracy can be made better than a small fraction of 1 per cent over quite a wide temperature range.

The thermocouple circuit demands careful workmanship in return for high accuracy. The resistors must be accurate and are usually assembled in compact fashion in an enclosure so that their temperature is uniform. The slide-wire is wound in helical form around a core and has 500 to 2000 convolutions exactly spaced. The slide-wire is made from the largest-size wire practicable in order to reduce the effect of wear.

Industrial Potentiometers

The industrial potentiometer incorporates a null-potentiometer circuit with various refinements, as indicated in the previous section. The main difference in the various potentiometers lies in the means by which the null balance is achieved: that is, the method by which the contactor is moved across the slide-wire so as to bring the current in the thermocouple circuit to zero.

The *manual-balance potentiometer* generally uses a galvanometer for detection of null balance, a slide-wire consisting of both fixed and adjustable resistors, a reference-junction compensating resistor for manual operation, and a current-standardizing resistor for manual operation. The slide-wire is usually circular in form with an indicating scale attached and may be calibrated in either emf or temperature. The emf of the thermo-

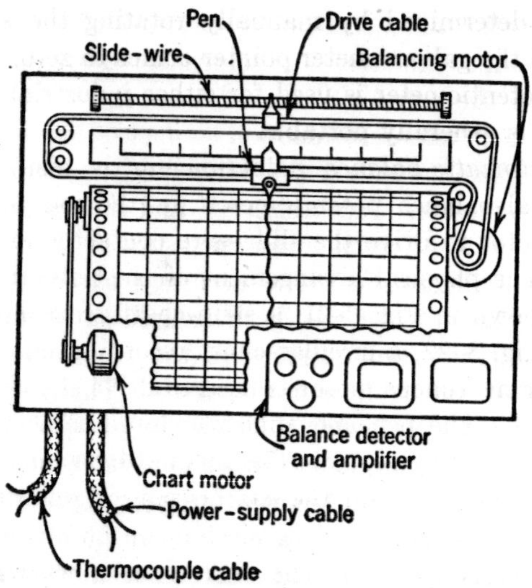

FIG. 3-21. An Automatic-Balance Strip-Chart Potentiometer Recorder

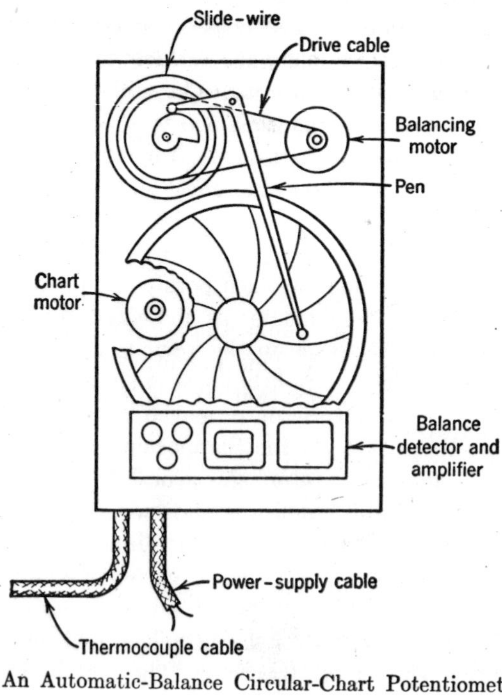

FIG. 3-22. An Automatic-Balance Circular-Chart Potentiometer Recorder

couple is determined by manually rotating the slide-wire and scale until the galvanometer pointer comes to zero. The manual-balance potentiometer is used for either industrial or laboratory work and is generally portable.

The *automatic-balance potentiometer* is distinguished from the manual-balance potentiometer by the use of a reversible electric motor to drive the slide-wire contactor and balance the circuit. The physical arrangement of the potentiometer instrument is shown in Fig. 3–21, a strip-chart recording potentiometer, and Fig. 3–22, a circular-chart recording potentiometer. A separate synchronous motor is used to drive the chart at the desired speed. The pen and slide-wire are simultaneously driven from the balancing motor. The method by which the balancing motor is operated from the null circuit is described briefly in the following sections. The instrument also contains the dry cell, the standard cell, the reference-junction compensation means, and the standardization mechanism.

The mechanical-balance potentiometer employs a galvanometer with an accessible swinging pointer, which is momentarily clamped in its position about once every 3 sec. A ratchet and pawl arrangement then "feels" the position of the pointer, and, if it is not at zero, the slide-wire contactor is moved by means of a mechanically connected continuously rotating electric motor to bring the galvanometer to zero. The mechanism is arranged so that the circuit is brought to balance in several steps. The galvanometer is damped so that it arrives at a stable position between clamping periods. A pointer or pen is made to operate over a scale or chart by connecting through gearing or cables directly to the slide-wire. This potentiometer usually includes reference-junction compensation and either manual or automatic standardization.

The system used by The Brown Instrument Company is shown in Fig. 3–23 and gains its advantage by elimination of the galvanometer altogether. The d-c potential of the thermocouple circuit is applied to a vibrating-reed converter. This converts the d-c potential to a filtered a-c potential at the output of the transformer. The a-c potential passes a stage of voltage amplification and a stage of power amplification and is made to operate a reversible induction motor. The motor drives the slide-wire contactor to restore balance. The direction of motor drive is determined by the polarity of the filtered and amplified a-c potential. The elimination of the galvanometer results in more ruggedness, and the balancing action can be made smooth and fast.

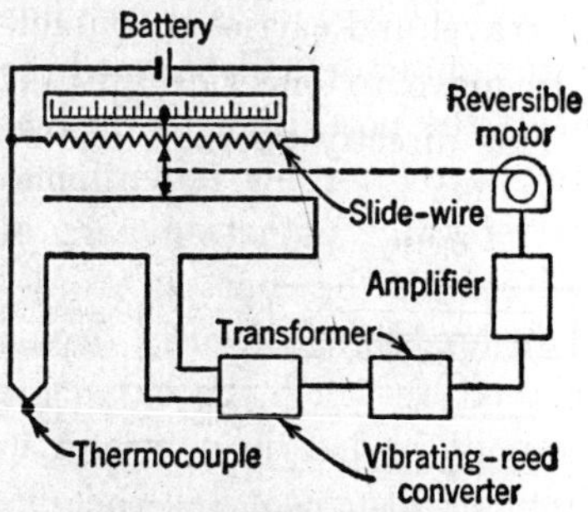

FIG. 3–23. Schematic Arrangement—The Brown Instrument Company Potentiometer

An automatically balancing potentiometer made by the Leeds & Northrup Company can be described in similar terms.

The scheme of the Leeds & Northrup high-speed potentiometer is shown in Fig. 3–24. A carbon-microphone modulator is used to convert the thermocouple emf to an a-c signal. A stage of voltage amplification passes this signal to a Thyratron push–pull drive for operating the motor. The motor drives the slide-wire for balancing the emf in the measuring circuit. A tachometer generator is also operated by the motor for stabilizing purposes. Because of good stability the instrument

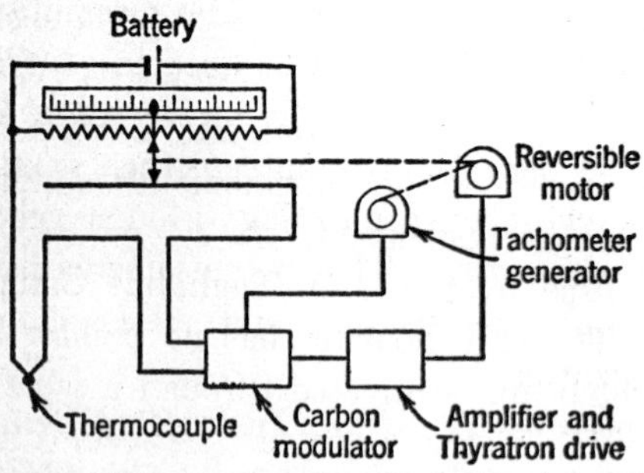

FIG. 3–24. Schematic Arrangement—Leeds & Northrup Company Potentiometer

can operate at very high speeds for recording fast-changing temperatures.

The system of the Bailey Meter Company uses saturable core reactors in order to convert the d-c potential of the thermocouple circuit into an a-c potential. This potential then is amplified to operate a reversible motor, as described previously for corresponding instruments.

The system used by The Bristol Company, shown in Fig. 3–25, employs a galvanometer with a free-swinging pointer. The galvanometer pointer has a very small travel and carries a contact. An unbalance causes the pointer to move to one side until it makes one of the contact points. This directly operates one of

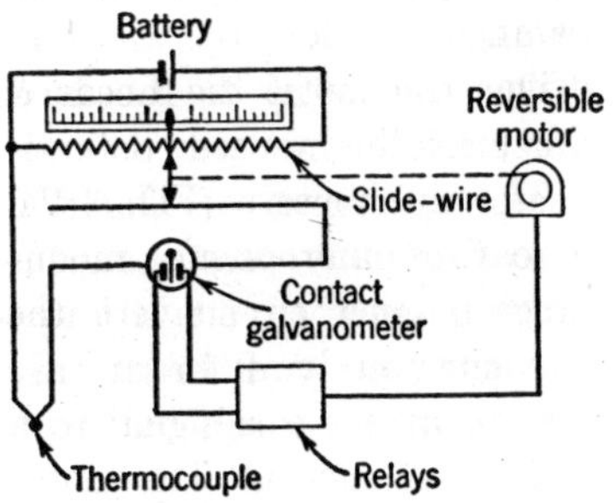

FIG. 3–25. Schematic Arrangement —The Bristol Company Potentiometer

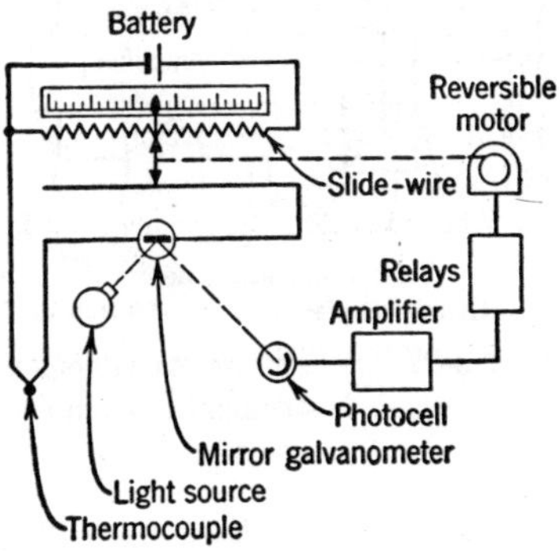

FIG. 3–26. Schematic Arrangement —C. J. Tagliabue Potentiometer

the relays, and a reversible electric motor moves the slide-wire contactor to the balance point. At balance the galvanometer pointer is between the two contacts and the motor remains stationary. The inertia of the galvanometer is small, enabling a fast response of the instrument.

The system used by the C. J. Tagliabue Company is shown in Fig. 3–26. The galvanometer has no pointer but includes a mirror. A light beam is directed through a window in the galvanometer housing and is reflected by the mirror back out the window and onto a photocell. On a circuit unbalance, the mirror rotates and moves the light beam on or off the edge of the photocell. The current from the photocell is electronically amplified to operate relays for driving the motor. When the slide-wire contactor is back at balance, the light beam is again at the neutral edge of the photocell. The galvanometer is totally enclosed, has small inertia and therefore fast action, and is quite rugged in construction.

The null system used by the Wheelco Instruments Company is sketched in Fig. 3–27. The standard potentiometer circuit is employed, and a galvanometer is used to detect unbalance between the thermocouple potential and the slide-wire potential. The vane on the galvanometer pointer operates between two oscillator coils, and, when the vane is out of the balance position, the amplifier drives the motor to rebalance the circuit.

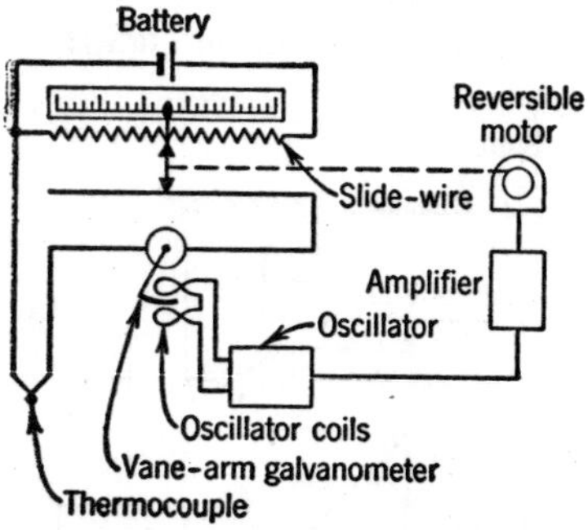

Fig. 3–27. Schematic Arrangement —Wheelco Instruments Company Potentiometer

The pyrometer of The Foxboro Company uses a capacitance balance circuit in a system completely different from the ordinary self-balancing potentiometer. As shown in Fig. 3–28, this method operates by comparing the unknown potential to a standard potential by measuring the relative discharge of a fixed and variable capacitor carrying these respective potentials. C_1 is a fixed capacitor,

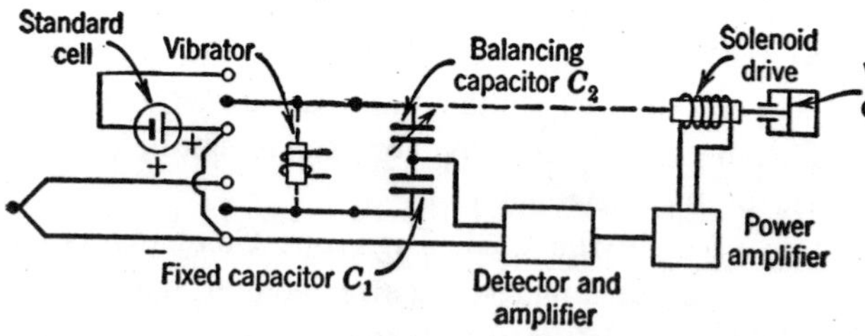

Fig. 3–28. Schematic Arrangement—The Foxboro Company D-C Potential Instrument

and C_2 is a variable capacitor. The contact switches are operated by a vibrator at line frequency. When the vibrator switches are in the upper position, capacitor C_1 is charged by the thermocouple potential and capacitor C_2 is charged by the standard cell potential. When the switches move downward after a small fraction of 1 sec, the capacitors are short-circuited with opposite polarities. If the electrostatic charge at each capacitor is not the same, a current flow appears in the detector circuit.

The d-c unbalance at the detector is converted to an a-c unbalance by means of a second section of the same vibrating-reed converter used for capacitor switching. This potential is then passed through voltage and power amplification to drive a viscous-damped double-solenoid-drive motor. The drive mechanism mechanically positions a calibrated variable capacitor C_2 for rebalancing the circuit. The main advantage of this system is that the circuit requires no battery and no standardization, since direct operation from the standard cell requires very little current drain. Reference-junction compensation is accomplished at a resistor in the standard-cell circuit, and the resistor has a selected temperature coefficient of resistance.

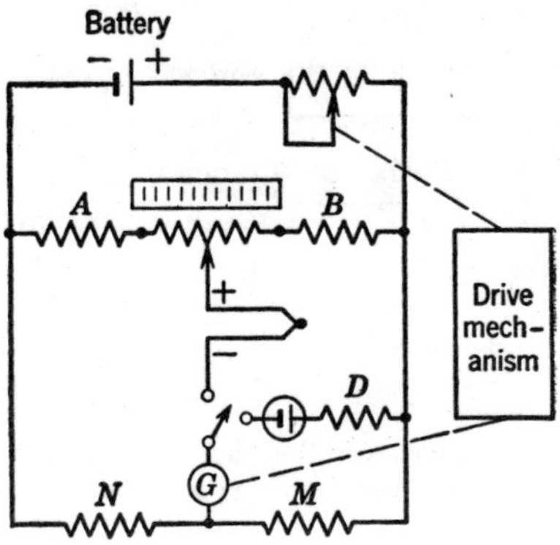

FIG. 3-29. Method for Automatic Standardization

Automatic standardization is in general use with the automatic-balance potentiometer. The null-potentiometer circuit is reproduced in Fig. 3-29. The current source is standardized, by one of the automatic-balancing methods described previously, to adjust the battery resistor F until the detector (a galvanometer or other means as just described) indicates zero current flow in the standard-cell circuit. This operation is accomplished completely automatically about once every hour by a suitable timer and switches. Generally the drive mechanism, either mechanical or electromechanical, is "borrowed" from the measuring circuit to perform standardization. Sometimes a second driving mechanism of smaller proportions is supplied specifically for operating the battery resistor and standardizing the circuit. Also the standardization may be made semiautomatic in that it is instigated only by the manual operation of a push button.

The *accuracy* of *automatic-balance potentiometers* is generally better than ±0.25 per cent of span, based on emf only; that is, without a thermocouple. The root-square accuracy of the potentiometer and of a standard iron–Constantan thermocouple is probably ±1.1 per cent of full scale. The best accuracy that can be achieved without special error-curve calibration is probably ±0.33 per cent. The dead zone of the automatic-balance potentiometer depends on the construction of the slide-wire. For all normal conditions of use, the balancing system positions the slide-wire contactor to the nearest convolution on the slide-wire. The dead zone is generally between 0.05 and 0.15 per cent of full scale. There is probably no other industrial instrument in wide use having an accuracy and sensitivity comparable to the industrial automatic-balance potentiometer.

Problems

3-1. Calculate the thermal emf of a copper–Constantan thermocouple in degrees centigrade with the reference junction at 0°C. Check the emf values at −200, +200, and +400°C with the standard tables. Why is there a difference?

3-2. It is desired to measure the temperature (about 1100°F) in a furnace having a highly oxidizing atmosphere. The temperature changes so rapidly that a bare thermocouple must be used. Which type would you select? Why?

3-3. If we assume that a *selected* Chromel–Alumel thermocouple is in contact with the work in a furnace and a portable precision potentiometer of ±0.05 per cent accuracy is used, what is the root-square accuracy that can be attained?

3-4. In using a copper–Constantan thermocouple at 200°F, the copper and Constantan lead wires were inadvertently interchanged. The reference junction is at 32°F, and the thermocouple head is at 80°F. What is the error in the **reading** of temperature? Under what conditions would no error be encountered when the lead wires are reversed?

3-5. An iron–Constantan thermocouple is used with a carbon-steel thermal well at a temperature of 700°F. It is suggested that the well be filled with mercury to improve response. Should you adopt the suggestion? Give two reasons.

3-6. The supervising engineer would like to have an indicating millivoltmeter of 500 ohms internal resistance installed in his office in order to check a critical process temperature. A 12 B&S gage iron–Constantan thermocouple and lead wires with a resistance of 0.00921 and 0.0406 ohm per foot, respectively, are used. The distance from the hot junction to the millivoltmeter is 210 ft. What static error is involved if the manufacturer supplied the instrument calibrated for 3 ohms external resistance? What about galvanometer damping?

3–7. In the complete null-potentiometer circuit of Fig. 3–20, assuming perfect operation of the galvanometer, which of the following conditions are likely to cause a static error?

(*a*) Contact resistance of 10 ohms between slider and slide-wire.
(*b*) Contact resistance of 10 ohms at battery resistor F.
(*c*) Change in value of standard-cell resistor *D*.
(*d*) Change in potential of standard cell.
(*e*) Change in potential of battery.

3–8. In the complete null-potentiometer circuit of Fig. 3–20, the following values are used: $E_b = 1.2$ volts, $F = 30$ ohms, $S = 10$ ohms, $M = 50$ ohms, $N = 5$ ohms. For a range of 100 to 500°F with an iron–Constantan thermocouple, find the values of resistors *A* and *B*.

3–9. With an iron–Constantan thermocouple at a temperature reading of 400°F, suppose that the battery voltage of the potentiometer declines at the rate of 11.6 per cent per day. The accuracy required is ±4°F. How often should the circuit be standardized? Hint: Compute by ratios.

3–10. An iron–Constantan thermocouple is calibrated with its reference junction at 32°F. What correction must be made to the indicated emf if the cold junction is actually at 75°F? Use tables.

3–11. A Chromel–Alumel thermocouple is calibrated with its cold junction at 32°F. What is the indicated emf if the hot junction is at 1820°F and the reference junction is at 100°F? Use tables.

3–12. Twenty copper–Constantan thermocouples are used with a 20-position selector switch and an automatic-indicating potentiometer. Sketch a circuit for this system, showing particularly what is done with the reference junctions.

3–13. A thermocouple is to be imbedded in the work in a furnace having high-frequency heating. Can you suggest what difficulties might be encountered if an automatic potentiometer of either the galvanometer or electronic type is used?

3–14. A thermocouple has its hot junction at 500°F and its reference junction at 75°F. One of the thermocouple wires becomes heated to 600°F at a point midway between hot and cold junctions. What effect does this have?

3–15. An automatic potentiometer of 5 millivolts span is available for measuring a temperature of 325°F. Select a thermocouple, and give the temperature range of the instrument, the emf range of the instrument, and the root-square accuracy that might be expected at this temperature.

3–16. In the complete potentiometer circuit of Fig. 3–20 what is the purpose of resistors *M* and *D*?

3–17. When a thermocouple and well are placed in an air stream of constant velocity with the walls of the enclosure at a higher temperature than the air, a radiation error exists. Assume that the surface area of the enclosure is large compared to the area of the thermal well. Derive the following equation for true air temperature.

$$T = T_i - \frac{Ke}{h}\left[\left(\frac{T_w}{100}\right)^4 - \left(\frac{T_i}{100}\right)^4\right]$$

where T = true air temperature, degrees Rankine
T_i = indicated temperature, degrees Rankine
e = emissivity
K = radiation constant, 0.172
h = convection coefficient, Btu/hr/ft²/°F
T_w = wall temperature, degrees Rankine

3–18. Describe the differences in Peltier heat and Joule heat. Why is the Joule heat not important in the theory of the thermocouple?

3–19. Derive the relations,

$$\frac{\pi}{T} = \frac{dE}{dT}, \quad \frac{\sigma_A - \sigma_B}{T} = -\frac{d^2E}{dT^2}$$

(*a*) The derivative (dE/dT) is called the thermoelectric power. Why?
(*b*) Does the second derivative give a maximum or a minimum or both?
(*c*) What is the condition for maximum thermoelectric power?

3–20. Two millivoltmeters are to be installed at different locations operating from the same thermocouple. Would a series connection or a parallel connection be more desirable? Derive an expression in each case for the voltage at the millivoltmeter terminals.

3–21. Two millivoltmeters are connected in parallel to a thermocouple. The millivoltmeters have different internal resistances R_1 and R_2. The millivoltmeter coils have the same resistance R_c. The thermocouple and lead-wire resistance is R_{ce}. Derive an expression for the voltage at the millivoltmeter terminals, if a negligible distance is assumed between the meters.

4· Resistance Thermometers

The industrial resistance thermometer is being more widely used in recent years because of its recognized accuracy and simplicity for industrial operations. A further advantage is that resistance temperature measurement allows very small increments of temperature to be detected; temperature changes as small as 0.03°F can be measured in industrial processing without great difficulty In the laboratory the resistance thermometer is a standard for precise temperature measurement. The success of the resistance thermometer is due primarily to Callendar who, in 1886, rescued this method of temperature measurement from difficulties with contamination of the resistance element. The introduction in recent years of the resistance-thermometer instrument employing an a-c detection and balancing mechanism simplified the instrument for industrial use.

The resistance thermometer is, at present, useful within about the same temperature limits as the expansion thermometer, the copper–Constantan thermocouple, and the iron–Constantan thermocouple; the limits are approximately −300 to 1200°F. A platinum resistance thermometer is used to define the international temperature scale between −310°F (−190°C) and 32°F, and between 32°F and 1220°F (660°C).

Thermal Coefficient of Resistance

The resistance thermometer operates from the change of electrical resistance of a substance with temperature. Usually this substance is a metal, but nonmetallic materials called semiconductors have also been used. The industrial resistance thermometer nearly always employs platinum, copper, or nickel. A material suitable for use as a resistance-thermometer element should have, first, a continuous and stable relationship between resistance and temperature, and, second, a relatively high thermal coefficient of resistance.

The resistance of most metals increases with temperature; that is, their conductivity decreases. The variation of the electrical resistance with temperature is assumed to be of the form,

$$R = R_0[1 + a_1t + a_2t^2 + a_3t^3 \cdots a_nt^n] \tag{4-1}$$

where R = resistance, ohms
R_0 = resistance, ohms at 0°C
t = temperature, degrees centigrade
$a_1 \cdots a_n$ = constants

For general purposes an average coefficient of resistance, a_1 in the foregoing equation, is given in most tables where all terms higher than the first power are assumed to be zero. Thereby a linear relationship of resistance and temperature is assumed over a small temperature range. This method is not sufficiently accurate for use in resistance thermometry. For platinum two constants are sufficient up to about 1200°F (649°C) and even higher. Its equation is therefore quadratic and varies from a linear relationship by about 7.0 per cent at the upper limit. Three constants are generally used for copper, although the relationship is very nearly linear up to 250°F (121°C). Three constants are necessary for nickel in the common temperature range, and its relationship is appreciably nonlinear. Resistance-temperature data on some metals are given in Table A–15 (appendix).

The coefficients of resistance may vary considerably with the purity of the metal and with its heat treatment. In fact, the temperature coefficient of resistivity has been used as a test for impurities. This is considered a disadvantage in the use of the resistance thermometer: namely, that the primary element is sensitive to contamination and oxidation and requires protection. With careful use the resistance thermometer may be employed considerably above the limits given, provided limitations due to physical construction are not exceeded.

The resistance of copper is considerably lower than that of either nickel or platinum, and, although this is a disadvantage because accurate resistance measurements are required, its stability and linearity of resistance versus temperature make it especially suitable for the measurement of temperatures near ambient.

TABLE 4–1. RESISTANCE PROPERTIES OF METALS

	Resistivity, ohms/mil-ft at 68°F	Melting Limit, °F	Temperature Limit, °F
Platinum	59.1	3224	1200
Copper	9.4	1981	250
Nickel	38	2651	600

Resistance-temperature measurements below 32°F may be made, but, although platinum has been used as low as −400°F, the industrial resistance thermometer is not often used below −150°F. Probably this is due to the fact that temperatures as low as −150°F are not often encountered in industrial operations. Equation 4–1 with the constants of Table A–15 (appendix) is not valid much below 32°F. The resistance-versus-temperature relationship is radically altered at very low absolute temperatures by the phenomenon of supraconductivity. At a temperature of about 1 to about 50°K, depending on the metal, the resistance drops suddenly to almost zero. Naturally a resistance thermometer cannot be used below this point.

In precision resistance thermometry a platinum resistance element is used. The resistance–temperature relation for platinum elements is given by the famous Callendar equation:[1]

$$T = \left[\frac{R_t - R_0}{R_{100} - R_0}\right] 100 + \delta \left[\frac{T}{100} - 1\right] \frac{T}{100} \tag{4–2}$$

where T = temperature at resistance element, degrees centigrade
R_t = resistance at temperature T, ohms
R_0 = resistance at 0°C, ohms
R_{100} = resistance at 100°C, ohms
δ = constant, characteristic of each element (about 1.49 to 1.50 per degree centigrade)

[1] See E. F. Mueller, "Precision Resistance Thermometry" in *Temperature, Its Measurement and Control in Science and Industry,* Reinhold Publishing Corp., 1941, p. 162.

Since the resistance of platinum is given by using two constants in equation 4–1,

$$\frac{\delta}{100} = \frac{1}{1 - \left(\frac{A_1}{A_2}\right)\frac{1}{100}} \tag{4–3}$$

The accuracy of the Callendar equation is in excellent agreement with the thermodynamic scale, and correction is not required in the region 32 to 1832°F. The three constants, R_o, R_{100}, and δ are determined by calibration of the thermometer at 0, 100, and 444.6°C. The last is the sulfur point.

Industrial Resistance-Thermometer Bulbs

The industrial resistance-thermometer bulb is made in various forms, but essentially it consists of a coil of fine wire wound on or in a frame of insulating material, as indicated in Fig. 4–1. First of all, the resistance wire is arranged so that good thermal conductivity is provided and a high rate of heat transfer obtained. Second, the windings should be made in such a manner that physical strain is negligible as the wire expands and contracts with temperature changes. Electrical resistance also changes with strain (or with mechanical stress), and this effect is employed in the electrical strain gage for measurement of small displacements. Third, inductive effects due to the position of the wire in a coil must be minimized in order to make the measurement one of resistance only. The resistance bulb is annealed and aged after manufacture in order to stabilize its electrical properties. Needless to say, the resistance wire itself must be homogeneous in order to avoid any localized emf, because of the thermoelectric effect when the bulb is not at uniform temperature and in order to avoid variation in temperature coefficient of resistance.

The platinum resistance-thermometer bulb usually has a resistance at 32°F between 10 and 35 ohms, although values as high as 125 ohms are in use The wire is wound on a notched or perforated mica frame so that the shape is more or less that of a round coil, as indicated in the diagram on the left in Fig. 4–1. Another form, indicated on the right in Fig. 4–1, is made with the platinum wire clamped between two mica plates. This latter form makes the thermometer bulb more compact. Improvement in heat-transfer characteristics is obtained by using a metal frame having spring qualities so as to insure physical contact between all parts of the bulb The platinum bulb is used industrially between the limits of −300 and 1200°F.

The copper resistance bulb usually has 10 ohms resistance at 32°F, less than other bulbs because of the low resistivity of copper. Since it is used only at moderate temperatures, the frame may be composed of a plastic or of glass instead of mica. The copper bulb is used industrially between the limits of −40 and 250°F.

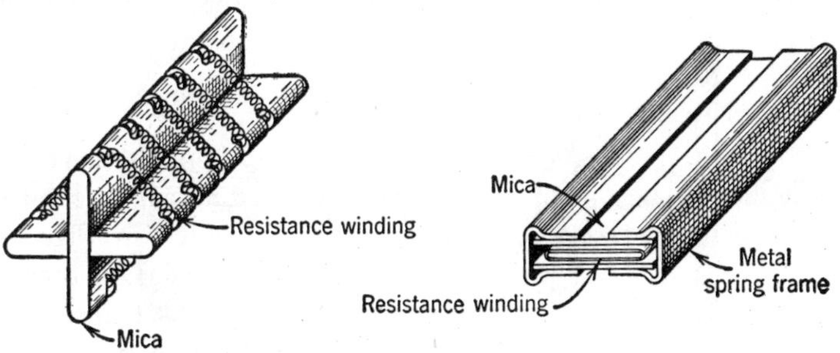

FIG. 4-1. Resistance-Thermometer Elements

The nickel resistance bulb has a resistance at 32°F of 100 to 300 ohms, depending on the circuit in which it is used. A high resistance is employed because the resistivity of nickel is high, it is a relatively inexpensive material, and high resistance makes resistance measurements easier. Since the nickel bulb is used in the moderate-temperature range, the wire may be wound with plastic or glass insulation. The nickel bulb is used industrially between −300 and 600°F.

A resistance-thermometer element is sometimes made of a strip of very thin nickel foil with a resistance of about 10 ohms. As a bolometer it is used to receive radiant energy. The bolometer is discussed more fully under radiation-receiving elements.

Another type of resistance-thermometer element is made in the form of a woven-wire mesh cloth. This type is particularly suitable for some surface-temperature measurements.

The connections to the wires of the resistance bulb are made carefully in order to avoid contact resistance and thermoelectric effects. Contact resistance is avoided by soldering, fusing, or welding joints. The thermoelectric power of copper and platinum or of nickel and copper junctions is not negligible. Consequently thermoelectric emf's can be avoided only by making certain that all such junctions in the circuit are at the same temperature.

Resistance-thermometer bulbs are generally used with a thermal well, except when the temperature measurement is being made in dry air. Platinum bulbs may be fitted with a well of porcelain, brass, or stainless steel which is sometimes sealed. The well is in the form of a tube, as shown in Fig. 4–2, or the sensitive end of the well may be flattened to provide better heat conduction between well and resistance bulb. Often a secondary well, made of one of various metals, is used to protect further the resistance element from contamination.

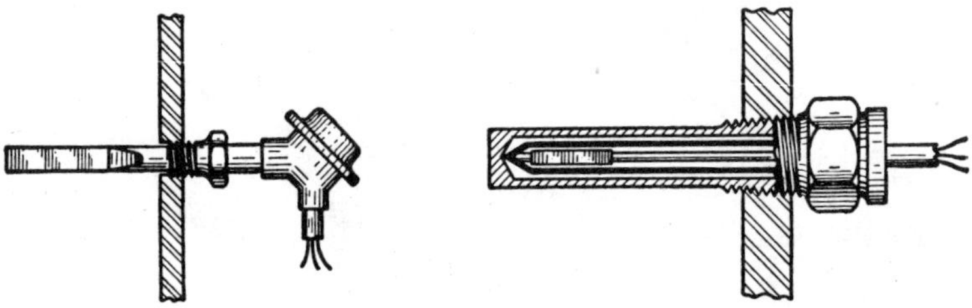

FIG. 4–2. Industrial Resistance-Thermometer Bulbs

The speed of response of modern resistance-thermometer bulbs is approximately equal to the speed of response of thermocouples when they are used under the same conditions. The typical response of a resistance-thermometer bulb is shown in Figs. 4–3 and 4–4. Here again we see the same effects as were noted for pressure thermometers and thermocouples. The resistance thermometer responds much more quickly in a fast-flowing liquid than in air. A greater part of the lag is caused by the necessity for using a well with the inevitable air space separating the resistance element and the well. The thermal resistance of the air space is more influential in causing lag when the thermometer is installed in liquid than in air, because the resistance of the heat-transfer film around the well is small compared to the resistance of the air space inside the well. In air, radiation trans-

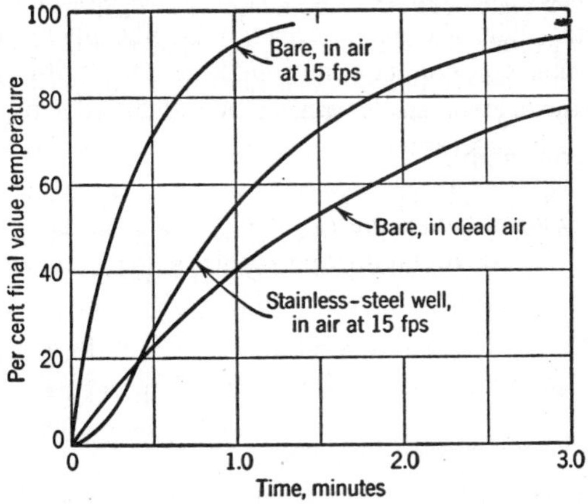

FIG. 4–3. Response of Resistance Thermometer in Air at 200°F

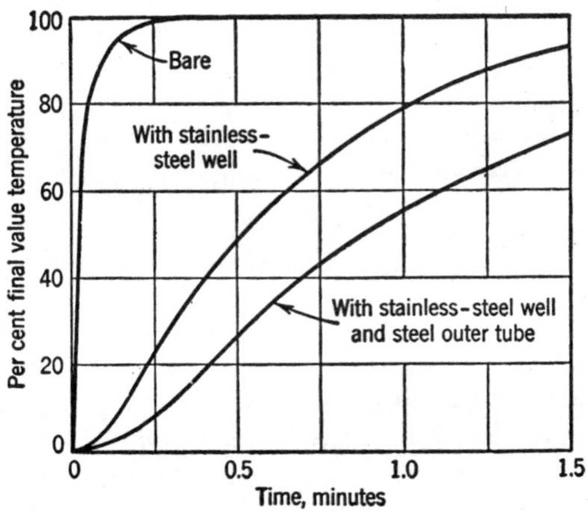

FIG. 4–4. Response of Resistance Thermometer in Agitated Water at 150°F

fer of heat occurs, and the condition of the surface of the well is important. As a rule, rough and darkened surfaces are better receivers of radiation. The secondary well nearly doubles the lag, as would be expected.

The accuracy of resistance thermometers is better than that of either thermocouples or expansion thermometers. Standard industrial resistance-thermometer bulbs are generally accurate to ±0.25 per cent, varying somewhat for different temperature ranges. To reduce the static error, selected resistance bulbs may be used. Thus the *standard* resistance-thermometer bulb has better accuracy than a *selected* iron–Constantan or copper–Constantan thermocouple.

The precision and reproducibility of resistance thermometer bulbs are, in general, better than those of thermocouples. This is primarily due to the absence in the resistance thermometer of anything resembling a reference junction. The temperature is measured by the change of resistance of the bulb, which is altogether located in the fluid in which the temperature is measured.

Resistance-Thermometer Circuits

Measurement of temperature with a resistance thermometer reduces to a measurement of electrical resistance. The techniques of resistance measurement are well advanced, and high accuracy and precision are not difficult to obtain.

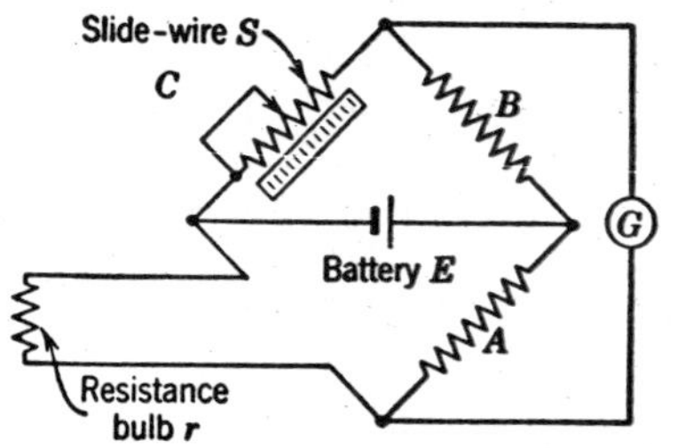

Fig. 4–5. Simple Wheatstone Bridge

The Wheatstone bridge with various modifications required for use with a resistance-thermometer bulb is the most common method of measurement. Whether the resistance bridge is operated with alternating or direct current is immaterial in the following discussion, and both are in use. An a-c capacitance bridge may also be used.

The *Wheatstone bridge,* shown schematically in Fig. 4–5, is the basic system for measurement of resistance. A battery applies a potential across two of the junction points, and a galvanometer connects the other two points. A and B are fixed resistors, while r is the resistance-thermometer bulb, and S is an adjustable slide-wire. All resistors and slide-wires are made of Manganin in order to avoid any effect due to ambient-temperature changes. The Wheatstone bridge is a null method of measurement, because at the balance or null point the potential across resistors A and B must be equal to the potential across r and S. Thus, if we ignore the resistance of the lead wires,

$$i_a A = i_b B \tag{4-4}$$

and

$$i_a r = i_b S$$

where i_a and i_b are the currents through resistors A and B, respectively. Combining gives

$$r = \left(\frac{A}{B}\right) S \tag{4-5}$$

Therefore, adjustment of slider C on slide-wire S balances the bridge. The bridge may be calibrated for temperature with slider C reading on a calibrated scale.

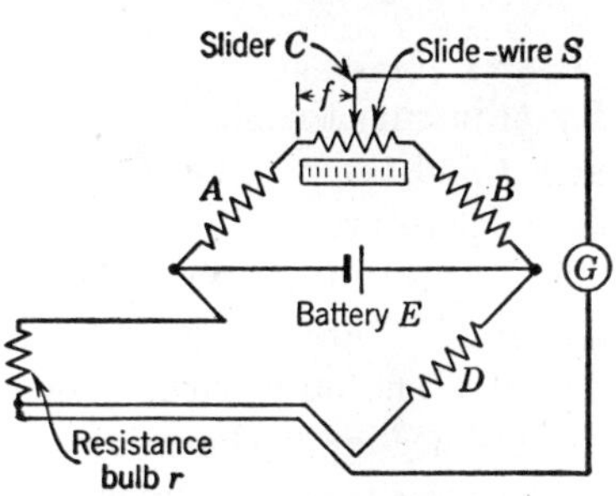

FIG. 4-6. Callendar–Griffiths Bridge

Three improvements must be made in the simple Wheatstone bridge. First, contact resistance between slider C and slide-wire S is directly in the bridge circuit. Contact resistance may amount to a fraction of an ohm but is quite variable, depending on the condition of the slide-wire. Dirt, dust, and wear may increase contact resistance. Second, the thermometer bulb r is connected into the bridge by lead wires of some length, perhaps 1000 ft or even more. Variation in temperature of the lead wires causes a variable resistance which is directly in the bridge circuit. Third, the current through the bulb causes a Joule heating effect $i_a^2 r$, which is variable, depending on the magnitude of the current. The heat thus generated raises the temperature of the thermometer bulb an amount depending on the thermal capacity of the bulb and the rate of transfer of heat from the bulb to its surroundings.

The *Callendar–Griffiths bridge* of Fig. 4-6 is an improved circuit in common use. The slide-wire S is placed at the top of the bridge where it may lie in both arms of the bridge. Balance in the bridge is accomplished by adjusting the slider C so that current through the galvanometer is zero. As before,

$$i_d D = i_b(B + S - fS) \tag{4-6}$$

$$i_d r = i_b(A + fS) \tag{4-7}$$

where f is the fraction of slide-wire S lying in the A arm of the bridge, and i_b and i_d are the currents through the corresponding arms of the bridge. Dividing gives

$$r = D \frac{\dfrac{A}{S} + f}{\dfrac{B}{S} + 1 - f} \tag{4-8}$$

By adjusting the slider C along slide-wire S the bridge is balanced and the position of the slider may be calibrated in terms of temperature.

The effect of contact resistance is avoided because the contact and slider C are not in the bridge circuit but in the galvanometer circuit. The only effect of variation in contact resistance is to cause a negligible change in operation of the galvanometer while the accuracy of the bridge is unimpaired.

The Siemens three-lead method is used in the circuit of Fig. 4–6 to compensate for variable lead-wire resistance. The galvanometer connection is now made at the resistance bulb and forms the third lead wire. Each lead wire is made of copper wire of the same diameter and length, so that the resistance of each is the same. The three lead wires are run in the same cable so as to pass through the same temperature conditions. If the resistance of each lead wire is R_l, then one lead wire is in the r arm of the bridge and one in the D arm of the bridge. The resistance of the galvanometer lead wire has negligible effect. Rewriting the foregoing equations and including lead-wire resistance gives

$$i_d(D + R_l) = i_b(B + S - fS) \tag{4-9}$$

$$i_d(r + R_l) = i_b(A + fS) \tag{4-10}$$

Dividing results in

$$\frac{r + R_1}{D + R_1} = \frac{\dfrac{A}{S} + f}{\dfrac{B}{S} + 1 - f} \tag{4-11}$$

If the right side of equation 4–11 is made unity,

$$r = D \tag{4-12}$$

Therefore, the balance of the bridge is independent of lead-wire resistance only at one particular setting of the slide-wire S; that is, the right-hand side of equation 4–11 can be made unity only for a given setting of the slide-wire contactor. At one temperature of the scale, say 50 per cent, the compensation is exact, and resistances A and B must be equal in magnitude. At other temperatures there is a slight and almost negligible static error when there are changes of ambient temperature along the lead wires. With this circuit the resistance of the lead wire should be reduced as much as possible by using large-diameter copper lead wire, and the lead-wire cable should be installed where minimum change of temperature is encountered.

Heating effect is proportional to the current,

$$H = \left(\frac{E}{r + D}\right)^2 r \tag{4-13}$$

But in order to achieve compensation resistor D is equal to the mid-value of r. The heating effect is therefore

$$H \cong \frac{E^2}{4r} \tag{4-14}$$

The change in bulb current for full-scale temperature change is therefore small when the battery voltage (or bridge voltage) is small and when the resistance of the thermometer bulb is large.

The *double-slide-wire bridge* of Fig. 4–7 (Leeds & Northrup Company) is designed to improve lead-wire resistance compensation. Two slide-wires S_1 and S_2 are so mechanically connected

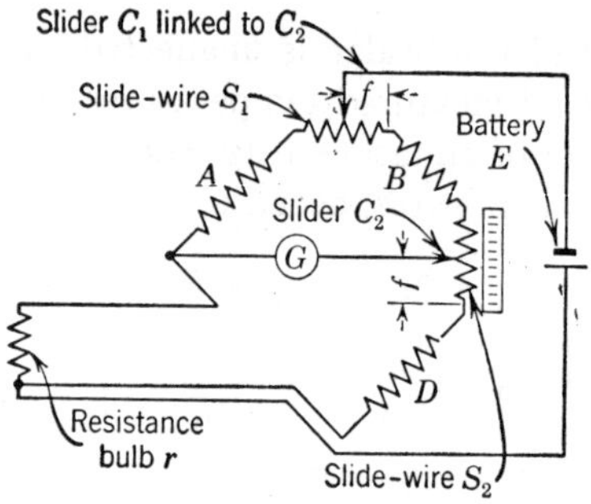

FIG. 4–7. Double-Slide-Wire Bridge—Leeds & Northrup Company

that their contactors always move in unison. Balance of the bridge is expressed by

$$i_a(A + S_1 - fS_1) = i_b(fS_1 + B + S_2 - fS_2) \quad (4\text{–}15)$$

$$i_a r = i_b(fS_2 + D) \quad (4\text{–}16)$$

Dividing as before gives

$$r = \frac{(A + S_1 - fS_1)(fS_2 + D)}{fS_1 + B + S_2 - fS_2} \quad (4\text{–}17)$$

A little analysis of this equation reveals that simultaneous adjustment of both slide-wire contactors balances the bridge, and it may be calibrated in terms of temperature.

Contact resistance in this circuit has almost no effect on accuracy, since the contacts are in the battery ciruit and galvanometer circuit.

Lead-wire compensation is accomplished by the Siemens three-wire method as in the previous circuits. Rewriting equation 4–16 and including lead-wire resistance, we have

$$i_a(r + R_1) = i_b(fS_2 + D + R_1) \quad (4\text{–}18)$$

Combining gives

$$\frac{r + R_1}{fS_2 + D + R_1} = \frac{A + S_1 - fS_1}{fS_1 + B + S_2 - fS_2} \quad (4\text{–}19)$$

If the right-hand side of equation 4–19 is unity, then

$$r = fS_2 + D \quad (4\text{–}20)$$

and the accuracy of the bridge is unaffected by changes in lead-wire resistance at all measured temperatures of the scale; that is, the right-hand side of equation 4–19 can be made unity for any setting of the two slide-wires, because

$$A + S_1 - fS_1 = fS_1 + B + S_2 - fS_2 \tag{4–21}$$

reduces an identity if $S_2 = 2S_1$ and if $A - B = S_1$. Therefore, with a properly constructed bridge and with equal resistances of lead wires, compensation is exact.

Heating effect is proportional to the current through the bulb,

$$H = \left(\frac{E}{r + A + fS_1}\right)^2 r \tag{4–22}$$

Substituting from the above equalities gives

$$H = \left(\frac{E}{3r + A - 2D}\right)^2 r \tag{4–23}$$

The current through the bulb is therefore not quite constant. However, by selecting a large value for resistor A and using a small battery voltage, the change in heating effect is reduced to negligible proportions.

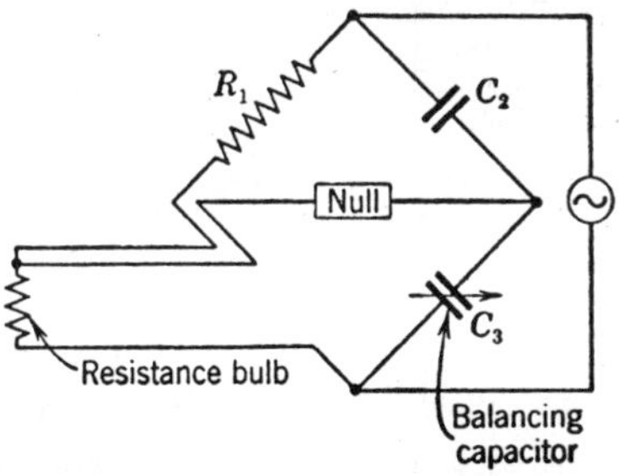

FIG. 4–8. Capacitance Bridge—The Foxboro Company

The *capacitance bridge* (The Foxboro Company) is shown in schematic form in Fig 4–8. An a-c supply of 1000 cps is used, and the balance of the bridge is detected by electronic means. The variable capacitor C_3 is used for balancing the bridge by adjusting the impedance of its branch. Thus,

$$\frac{Z_r}{Z_3} = \frac{Z_1}{Z_2} \tag{4–24}$$

where Z is the impedance of the corresponding branch of the bridge. Substituting for the complex impedance of each branch,

$$\frac{\dfrac{r}{\dfrac{i_3}{\omega C_3}}}{} = \frac{R_1}{\dfrac{i_2}{\omega C_2}} \tag{4-25}$$

where i is the current through capacitors and where ω is the frequency of alternating current. For balance of the bridge the current through each capacitor is the same and,

$$r = R_1 C_2 \left(\frac{1}{C_3}\right) \tag{4-26}$$

Balance of the bridge is therefore accomplished by adjustment of capacitor C_3 against variation in resistance r of the thermometer bulb.

Lead-wire resistance compensation is accomplished by the use of the Siemens three-wire method. The compensation is not exact, since a one-to-one ratio of bridge impedances cannot be maintained at all times, as it is in the double-slide-wire bridge. Lead-wire compensation is therefore made exact at one point of the bridge setting while at other points the error is nearly negligible. The effect of contact resistance is completely eliminated because there is no contact or slide-wire. Heating effect at the resistance-thermometer bulb is reduced by using high-resistance values in the bridge, although they are not usually so high as those of the Callendar–Griffiths or double-slide-wire bridge, and by employing a low potential supply across the bridge.

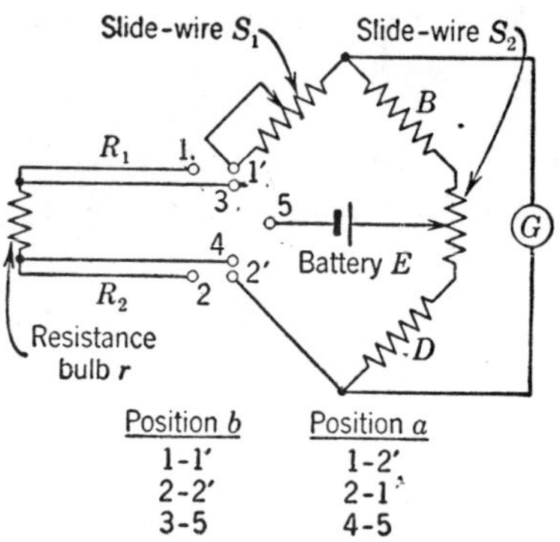

FIG. 4-9. The Mueller Bridge for Laboratory Use

The *Mueller bridge* commonly employed in laboratory measurements of high precision is shown in Fig. 4-9. Slide-wire S_2 is used only to make the resistances of the right-hand arms of the bridge equal. The method involves switching the resistance-bulb leads from one arm of the bridge to the other arm of the bridge. The four-wire resistance bulb is used instead of the three-wire resistance bulb as in the Siemens method of lead compensation. In position b,

$$S_b + R_1 = r + R_2 \tag{4-27}$$

and in position a,

$$S_a + R_2 = r + R_1 \tag{4-28}$$

and by adding, we get

$$r = \frac{S_a + S_b}{2} \tag{4-29}$$

Therefore, the measurement of resistance, and thereby of temperature, is independent of the resistance of the lead wires, whether they are equal or not.

The switching is accomplished by a commutator with amalgamated contacts; the commutator construction is equivalent to a three-pole double-throw switch. The bridge resistors are precision type and the slide-wire is made up of six decades from 10 to 0.001 ohms. It is important that the temperature of the bridge coils or resistors be maintained constant at about 35°C. This is done by immersing the coils in a stirred bath or by placing them in a thermostatic housing.

Null-Bridge Resistance Thermometers

The industrial resistance thermometer is a combination of a thermometer bulb and measuring circuit, such as those previously described, and an instrument of the automatic-balance potentiometer type. It is apparent that an automatic-balance potentiometer instrument composes most of the parts required for making an automatic-balance resistance thermometer. The instrument for operation with one of the null bridges measures the unbalance potential and drives either a slide-wire or a variable capacitor for rebalancing the bridge. Naturally the instrument for a resistance-thermometer system requires no standard cell, no standardization, and no reference-junction compensation.

The system employed by the Leeds & Northrup Company is shown schematically in Fig. 4–10. The double-slide-wire d-c bridge is used in which both slide-wire contactors are mechanically linked together and operated simultaneously by the bal-

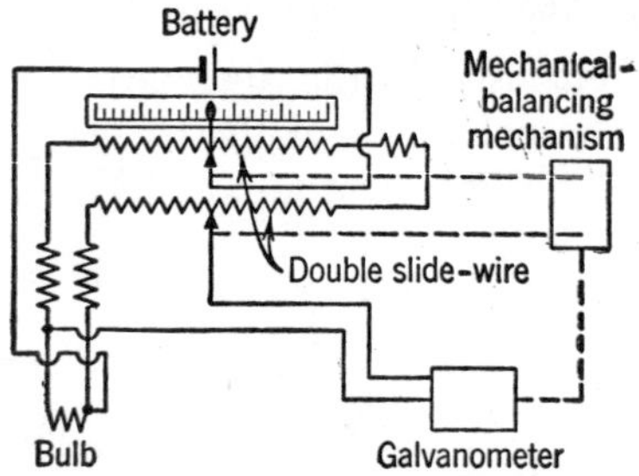

FIG. 4–10. Schematic Arrangement—Leeds & Northrup Company Resistance Thermometer

ancing mechanism. The galvanometer detects the unbalance of the circuit, and the galvanometer deflection is sensed by the mechanical-type periodic-balancing mechanism. Both slide-wire contactors are then moved until the balance of the bridge is restored.

The system used in The Brown Instrument Company resistance thermometer is indicated in Fig. 4–11. The Callendar–Griffiths d-c bridge with Siemens three-wire compensation is used. Unbalance is detected and converted to an a-c signal, amplified, and then used to drive the reversible motor. The motor moves the slide-wire contactor to restore the balance of the bridge.

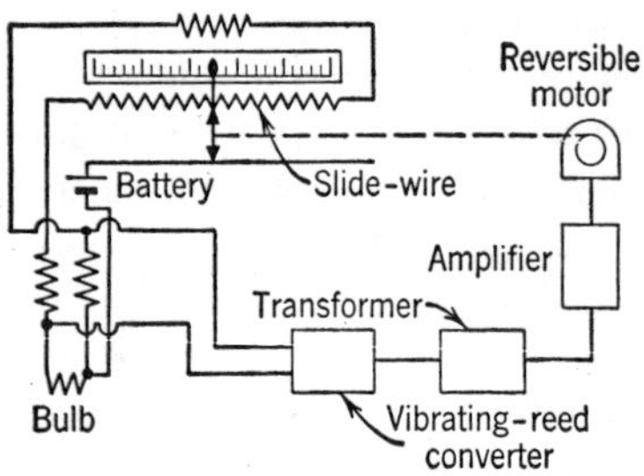

FIG. 4–11. Schematic Arrangement—The Brown Instrument Company Resistance Thermometer

The system used by The Foxboro Company is shown schematically in Fig 4–12. The a-c capacitance bridge is used in order to obtain the advantages of a variable capacitor for balancing the bridge against resistance variations of the thermometer bulb. This instrument uses an a-c potential of 1000 cps frequency as the bridge supply. The balancing mecha-

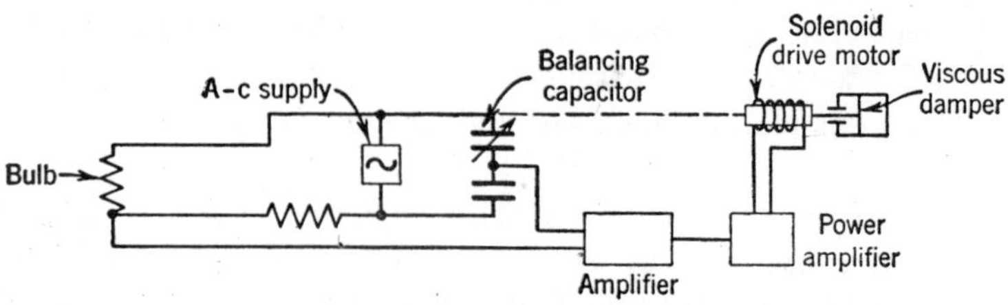

FIG. 4–12. Schematic Arrangement—The Foxboro Company Resistance Thermometer

nism is almost identical to the system used in the automatic-balance d-c potential instrument. The unbalance of the bridge is an a-c signal to the amplifier which, by means of an amplifier and the solenoid motor, moves the calibrated air capacitor to balance the bridge.

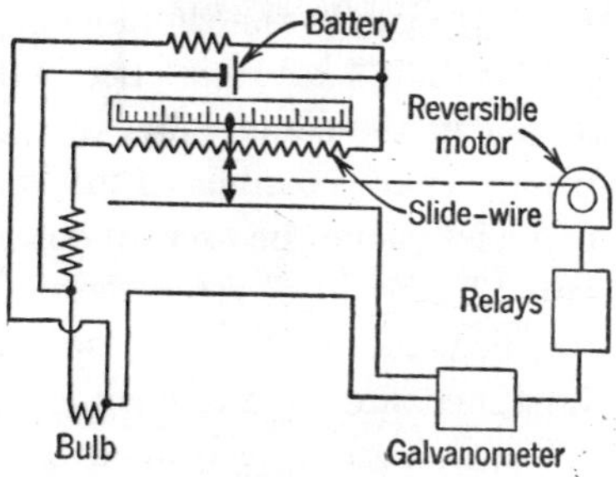

Fig. 4–13. Schematic Arrangement—C. J. Tagliabue Resistance Thermometer

The arrangement of the C. J. Tagliabue resistance thermometer is shown in Fig. 4–13. The Wheatstone d-c bridge is employed with the Siemens three-wire compensation. The galvanometer, a light mirror with a photocell, detects an unbalance of the bridge and operates electric relays which drive a reversible motor for moving the slide-wire contactor and restoring balance in the bridge.

The system used in The Bristol Company and the Bailey Meter Company resistance thermometers is indicated by Fig. 4–14. The use of Callendar–Griffiths a-c bridge allows the a-c signal resulting from an unbalance of the bridge to be amplified and used to drive the reversible motor and reposition the slide-wire

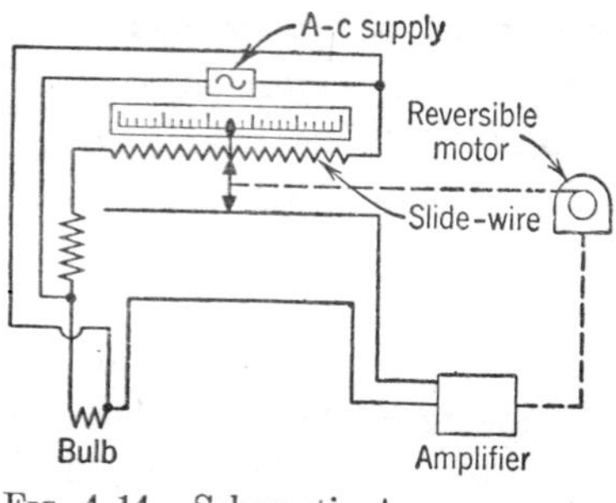

Fig. 4–14. Schematic Arrangement—Bailey Meter Company and The Bristol Company Resistance Thermometers

The potential supply for the bridge, either a-c or d-c, is one of the important factors to consider in the use of any resistance thermometer. A variation in this potential is not serious but does cause very slight static errors and may be the cause of a slight calibration drift. The d-c bridge usually employs one of the better types of batteries which is replaced periodically. A battery eliminator is sometimes preferred because it reduces maintenance. The a-c bridge derives its potential supply from line power. If variations in line voltage of more than 10 per cent are present, a constant-voltage transformer of adequate power capacity should be used.

Each lead wire to the resistance-thermometer bulb should be made of exactly the same-length and of the same-diameter wire so that their total resistance is the same. Then, too, they should be run in the same cable through as uniform temperature conditions as are possible. Finally, the largest practical-size copper lead wire should be used to obtain a small total resistance of lead wire. It should be remembered that with the Siemens three-wire system compensation in the null-bridge is not effective if the resistance variation of each lead wire is not the same.

The industrial resistance thermometer is inherently an accurate and sensitive means of temperature measurement, and the points concerning potential supply and lead-wire resistance are stressed for that reason only. In most applications these points are the source of no difficulty. Ordinarily the temperature span of an automatic-balance resistance thermometer may be made much smaller than that of the thermocouple potentiometer although there are some exceptions. Temperature spans as low as 25 or 50°F allow very small temperature changes to be detected, and the question of accuracy therefore becomes very important.

The accuracy of the null-bridge automatic resistance thermometer is generally better than ±0.25 per cent, based on resistance measurement only, that is, without a resistance-thermometer bulb. The root-square accuracy of resistance thermometer and thermometer bulb is ±0.35 per cent.

The dead zone of the null-bridge resistance thermometer depends on the construction of the balancing mechanism as in the automatic potentiometer since they are nearly the same. The dead zone is generally between 0.03 and 0.15 per cent of span and corresponds to the smallest practical increment of resistance that may be measured.

The manual resistance thermometer is composed of a resistance-thermometer bulb and a manually balanced Wheatstone bridge. The instrument includes a d-c galvanometer and therefore employs a d-c bridge with a battery supply. In construction it is very similar to the manual potentiometer. It may be portable for use in checking automatic instruments, or it may be permanently employed for the manual measurement of temperature. Sometimes a large number of resistance-thermometer bulbs are used with one manual instrument for determining temperatures at a number of locations, as in an air-conditioned building. High-precision Wheatstone bridges such as the Mueller bridge may be obtained if more precise work is desired.

Deflectional Resistance Thermometer

Simple indicators such as ambient-temperature thermometers and aircraft thermometers employ the deflectional-type circuit indicated in Fig. 4–15. The circuit is similar to a d-c Wheatstone bridge, except that there is no adjustable resistor for balancing the bridge. The unbalance potential of the bridge is used as an indication of the change of resistance. Therefore, the name deflectional is used to distinguish the instrument from the null-bridge type. A potential is applied to the bridge by the battery and is adjusted by the resistor C. The instrument is a type of indicating millivoltmeter.

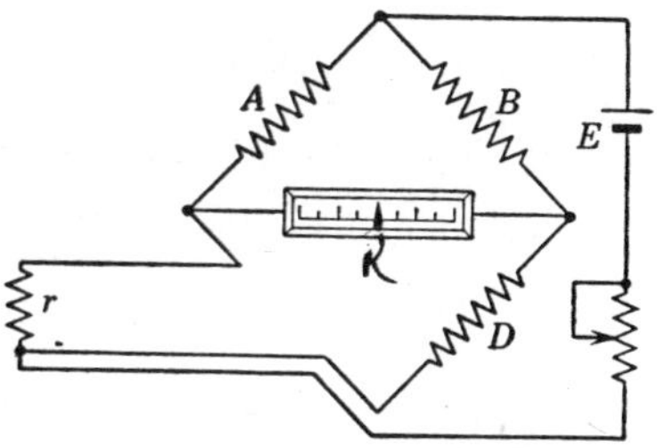

FIG. 4-15. Deflectional Resistance Thermometer

As the resistance of the thermometer bulb r increases, a potential drop appears at the instrument which is calibrated in terms of temperature. The bridge is balanced at some temperature, usually the temperature at 50 or 75 per cent of span. Assume that the bridge is balanced at a temperature corresponding to a bulb resistance r_b. The potential drops across A and B are equal,

$$i_a A = i_b B \tag{4-30}$$

and

$$i_a r_b = i_b D \tag{4-31}$$

where i_a is the current through resistor A, and i_b is the current through resistor B. Therefore,

$$r_b = \frac{D}{B} A \tag{4-32}$$

When the resistance of the bulb r increases, there will be a potential drop across the instrument, which can be found through Kirchhoff's circuit laws,

$$e = E\left[\frac{r - r_b}{xr + y}\right] \tag{4-33}$$

where e = emf at instrument

$$x = \frac{r_b}{R} + \frac{D}{R} + \frac{A}{R} + \frac{r_b}{A} + 1$$

$$y = \frac{AD}{R} + A + r_b$$

R = internal resistance of instrument

By inspection of the previous equation it may be seen that, when the temperature is at the balance value r_b, the emf at the instrument is zero. The mechanical zero of the millivoltmeter instrument is therefore adjusted to the value of temperature corresponding to r_b.

The Siemens three-wire compensation is used, and the arms of the bridge are made approximately of equal resistance. The effect of changing lead-wire resistance is therefore reduced to a minimum. The fixed resistors of the circuit are made of Manganin in order to avoid ambient-temperature effects. The heating effect at the resistance-thermometer bulb is reduced by selecting high resistances in the bridge circuit and employing a small battery voltage.

Variations of battery voltage produce corresponding variations in emf at the instrument and result in corresponding errors. It is necessary therefore to readjust the battery voltage from time to time to maintain a constant potential at the bridge. For this reason a good battery producing a relatively constant voltage is desirable. Often a battery eliminator is used to supply the circuit, or in aircraft a regulated d-c supply is employed.

The deflectional resistance thermometer is nearly always used in the low-temperature range between about 0 and 150°F with a nickel resistance-thermometer bulb.

The root-square accuracy of the deflectional resistance thermometer using the industrial millivoltmeter and an industrial resistance-thermometer bulb is probably ±1.1 per cent of full scale. The dead zone is generally variable, since it depends on the friction in the moving-coil system of the millivoltmeter.

Problems

4–1. A copper resistance-thermometer bulb has a resistance of 12 ohms in ice. Find its resistance at 0, 100, and 200°F.

4–2. In the simple Wheatstone bridge of Fig. 4–5 the resistances A and B are 1000 ohms each. The copper bulb of problem 4–1 is used, and the temperature being measured is 100°F. The contact resistance at slide-wire S suddenly increases to 0.1 ohm. Find the error in temperature reading.

4–3. In the Callendar–Griffiths bridge of Fig. 4–6 the resistor A is 1000 ohms and the slide-wire S is 500 ohms resistance. A nickel resistance bulb has 300 ohms resistance at 0°C. When the resistance thermometer reads 50 per cent of full scale, the compensation of lead resistance is exact. If the thermometer calibration is to start at 0°C, what is its span?

4–4. A null-bridge resistance thermometer of 100 to 300°F range is used to measure a temperature of 250°F. With a standard resistance-thermometer bulb, what is the mean accuracy in reading?

4–5. Could the field winding of an electric generator be used to determine the temperature inside the generator? If so, explain how it could be done.

4–6. A platinum resistance-thermometer bulb is to have a resistance of 30 ohms in ice. What length of wire (diameter 0.004 in.) should be used? What is its resistance change from the ice point to the steam point?

4–7. A deflectional resistance thermometer must have a constant potential across the bridge. Suggest a means for manually "standardizing" the battery current.

4–8. An automatic resistance thermometer has a range of 0 to 300°F with a platinum resistance bulb of 100 ohms in ice. It is desired to use a nickel resistance bulb of 100 ohms in ice. What is the new range of the instrument?

4–9. A deflectional resistance thermometer is arranged to operate with a bulb and 10 ft of lead wire. The bulb is relocated at a 1000-ft distance from the instrument. What changes in the millivoltmeter itself might be required? Why?

4–10. Prove that for a platinum resistance element

$$\frac{\delta}{100} = \frac{1}{1 - \left(\frac{a_1}{a_2}\right)\frac{1}{100}}$$

4–11. Derive equation 4–33.

4–12. The Callendar four-wire compensation method may be arranged in the Wheatstone bridge of Fig. 4–5 by connecting a loop of wire in series with the slide-wire arm and carrying the loop out to the resistance bulb and back without electric connection at the bulb. Derive the equations of balance, and determine whether this method of compensation is exact at all points of the scale.

4–13. Refer to Table A–15 (appendix), and derive for silver the value of the constants a_1, a_2, and a_3 in equation 4–1. Give the maximum per cent error of the equation in the range for which data are available.

5· Radiation Temperature Measurement

Temperature measurement made by means of radiation from a heated body possesses so many advantages that the radiation pyrometer and optical pyrometer have come into wide use in recent years. The most important advantages to industrial-process operations are that no direct contact is necessary with the body at which the temperature is measured and that temperature measurements are relatively independent of the distance between measuring element and heated body. Furthermore the body may be either moving or stationary, a fact that makes the radiation-temperature-measuring device particularly adaptable to continuous industrial processing.

Generally speaking there is no upper temperature to which radiation means of measurement are limited. As long as the radiation of the heated body is known or can be deduced mathematically, a temperature can theoretically be assigned. The international temperature scale is defined above 1945°F (1063°C) by means of an equation for radiant energy for a black body. Radiation-temperature-measuring devices are used primarily for temperatures above 2000°F in the same range as the platinum thermocouple.

Radiation means of temperature measurement may be used at low temperatures as long as a sufficiently sensitive element is employed to detect the small amounts of radiation. Industrial temperature measurements to as low as 200°F can be made by radiation methods—in the range of the expansion thermometer, resistance thermometer, and iron–Constantan thermocouple pyrometer.

Laws of Radiation

Radiant energy is the energy transmitted in the form of electromagnetic waves and is exemplified by radio and radar radiation; infrared, light, and ultraviolet radiation; and X-ray and gamma-ray radiation. To locate temperature-radiation phenomena in the electromagnetic spectrum, Fig. 5–1 should be studied. Radiation transfer of heat takes place in the ultraviolet, visible, and infrared regions from about 1/100 to 100 microns.

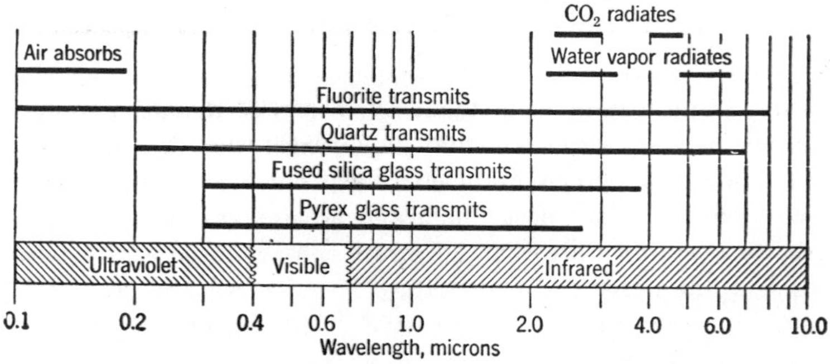

FIG. 5–1. Transmission Band of Materials (see also Fig. 6–2)

Beyond the ultraviolet, that is, at shorter wavelengths, X rays and gamma rays are to be found. Below infrared, at longer wavelengths, are the familiar Hertzian or radio waves and alternating electric waves. The visible part of the spectrum extends from about 0.38 micron (extreme violet) to 0.78 micron (extreme red).

Wavelength of radiation is specified in centimeters. The micron (abbreviated μ) is 10^{-4} cm and the angstrom (abbreviated A) is 10^{-8} cm.

A black body is a body that absorbs all radiation falling on it without transmitting or reflecting any. In other words, a black body radiates energy at all spectral wavelengths at a maximum rate corresponding to its temperature. The black-body concept is important because it is the basis on which radiation laws are founded.

Kirchhoff's law relates the radiated energy to absorbed energy: any body in thermal equilibrium emits as much heat radiation as it receives at any given wavelength and temperature.

The amount of radiant energy from a black body is given by the Stefan–Boltzmann law,

$$\Phi_b = \sigma A T^4 \tag{5-1}$$

where Φ_b = radiant flux (radiant energy per unit time), ergs per second
σ = Stefan–Boltzmann constant 5.77×10^{-5}, erg/(sec)(cm^2)(deg^4)
A = area of black body, square centimeters
T = temperature, degrees Kelvin

This law agrees well with experimental data, thereby providing a means for determining quantitatively the amount of radiant energy from a black body. This law is often expressed for two bodies exchanging radiation,

$$\Phi_b = \sigma A (T_2^4 - T_1^4) \tag{5-2}$$

where one body is at a higher temperature T_2 than the other T_1. In more convenient form,

$$Q = KA \left[\left(\frac{T_2}{100} \right)^4 - \left(\frac{T_1}{100} \right)^4 \right] \tag{5-3}$$

where Q = radiant energy, Btu per hour
K = radiation constant, 0.172
A = area, square feet
T = temperature, degrees Rankine

The emissive power of a black body at a given wavelength is found by Planck's radiation law,

$$W_{b\lambda} = \frac{C_1}{\lambda^5} \frac{1}{e^{C_2/\lambda T} - 1} \tag{5-4}$$

where $W_{b\lambda}$ = spectral radiant flux density, ergs/sec/cm^2/0.01 micron zone
λ = wavelength, centimeters
e = base natural logarithms
C_1 = 3.7403×10^9 ergs/sec/cm^2/0.01 micron zone
C_2 = 1.4384 centimeter degree
T = temperature, degrees Kelvin

Thus the distribution of energy at various wavelengths of radiation is known, and experimental agreement is good. A plot of this equation for several temperatures is given in Fig. 5–2. It is apparent that the intensity of radiation varies appreciably with wavelength. Also it may be seen that the point of maximum radiant intensity shifts to the shorter wavelengths as the tem-

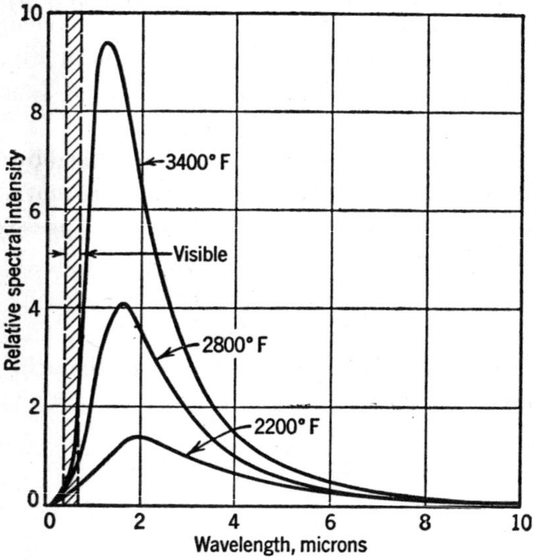

FIG. 5–2. Spectral Distribution of Radiant Energy

perature increases. The point of maximum radiant intensity is given by Wien's displacement law,

$$\lambda_m T = 2900 \tag{5–5}$$

where λ_m is in microns and T is in degrees Kelvin. Therefore, as the temperature of the radiating body increases, the point of maximum radiant intensity λ_m decreases proportionately. This is a common phenomenon observed in the change of color of a body being heated. A metal gradually heated changes its color from red (a long wavelength) to yellow and white as the intensity of radiation increases at the shorter wavelengths of the visible spectrum.

Two principles are used for the construction of radiation-temperature-measuring devices.

First, we may measure the total energy of radiation from a heated body and thereby obtain a *radiation pyrometer*. This energy is represented by the area under the curves of Fig. 5–2 and is given by the Stefan–Boltzmann law. The radiation pyrometer, then, is intended to receive the maximum amount of radiant energy at the widest range of wavelengths possible.

Second, we may measure the spectral radiant intensity of the radiated energy from a heated body at a given wavelength. For example, if a vertical line is drawn on the graph of Fig. 5–2, the variation of intensity with temperature can be found. The *optical pyrometer* is founded on this principle.

Actually it is difficult to build a practical optical system that is responsive either to radiation of all wavelengths or to spectral radiation of exactly one wavelength. Therefore a practical radiation pyrometer responds to a wide band of radiation of approximately 0.1 to 8.0 microns width in the visible and infrared, and the actual width of this band depends entirely on the physical construction of the radiation receiver. An optical pyrometer responds to a very narrow band of radiation of 0.01 micron width at about 0.65 micron in the red portion of the visible spectrum.

Black-Body Conditions

The realization of good black-body conditions in practice is not difficult; otherwise, radiation methods of temperature measurement would never have become practical. One reason for this is that the radiation from many types of surfaces and from certain cavities is almost indistinguishable from black-body radiation.

Black-body radiation is more generally represented by saying that its emissivity is 1.0. Total emissivity e is defined as the ratio of the total radiation from a nonblack body to that from a geometrically similar black body at the same temperature. When the radiation from a body is less than it would be if it were a black body, the total emissivity is less than one. Spectral emissivity e_λ is similar: that is, the ratio of the monochromatic radiation of wavelength λ from a nonblack body to that from a geometrically similar black body at the same temperature.

The total emissivity of tungsten at 2430°K, for example, is about 0.25, which means that tungsten at this temperature radiates only one-quarter the radiation of a black body. The total emissivities for many open surfaces is given in Table A–16 (appendix). The important feature of these data is that open oxidized surfaces and open rough surfaces generally have a high total emissivity, whereas clean and polished surfaces generally have a low total emissivity. Spectral emissivities are also given in Table A–16 (appendix). For example, clean steel has a spectral emissivity of 0.35, but oxidized steel has 0.80.

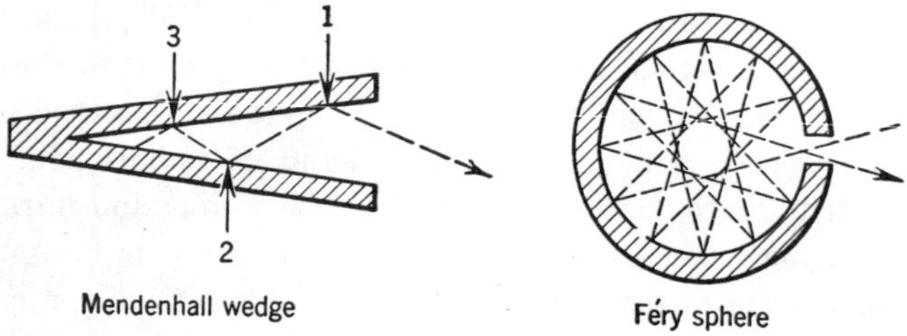

Fig. 5–3. Black-Body Devices

The Mendenhall wedge and Féry sphere shown in Fig. 5–3 represent black-body cavities, no matter of what material the shapes are constructed so long as they are opaque and at uniform temperature. In the wedge, for example, it is seen by the geometry of the figure that radiation "rays" streaming from the cavity must be reflected many times. Radiation may originate at point 1 and issues from the cavity together with radiation reflected at point 1. This latter comes from point 2, where a part originates and a part is reflected. The reflected part comes from point 3, and so on. After as few as ten reflections, the radiation is practically black-body radiation, even if the total emissivity of the material is substantially less than one. The Féry sphere of Fig. 5–3 is also a black-body cavity, by virtue of the multiple reflection of radiation on the inner surfaces. The augmented radiation issues from the sphere as black-body radiation, even if the total emissivity of the material is low.

Rough and oxidized open surfaces are nearly perfect black bodies. The rough-machined or cast surface can be conceived as having a large number of microscopic cavities. Therefore, even if the material was originally of bright clean metal, its emissivity is increased by transforming the surface to a rough one. Oxidized surfaces are represented in the same manner wherein crystals or fine flakes of metal oxides lie on the surface in random directions. The coating of surfaces with finely divided metal such as carbon, platinum, or gold performs the same function in increasing emissivity. Nearly perfect black-body surfaces are manufactured by depositing on the surface what are known as platinum black and gold black, even though the emissivity of smooth platinum or gold is very low indeed.

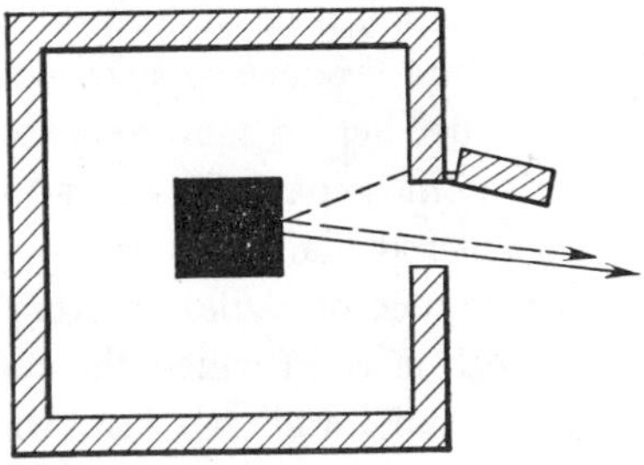

FIG. 5-4. Heated Body in a Black-Body Cavity

An industrial furnace also represents a black-body cavity similar to the Féry sphere, and radiation from a heated furnace is generally taken as that from a black body, so long as there are relatively few openings in the furnace walls, there is no open flame of large proportions, and temperatures in the furnace are relatively uniform. We may construct a black-body cavity for use with a furnace or other heating device by inserting in the wall of the furnace a metal or ceramic tube with a closed end. The tube will become heated to an average furnace temperature, and radiation issuing from the tube will closely approximate that of a black body.

Heated bodies inside a black-body cavity at the same temperature also become black bodies, whether their emissivity is one or less than one. The body and cavity of Fig. 5-4 are at uniform steady temperature. The body absorbs a part of the radiation from the walls of the cavity and reflects the rest. That is,

$$\alpha + \rho + \tau = 1 \qquad (5\text{-}6)$$

where α = absorption factor
ρ = reflection factor
τ = transmission factor

When the heated body is viewed from the outside of the cavity, the reflected radiant energy is, of course, seen. By Kirchhoff's law that the total emissivity of a radiating body is equal to its absorption coefficient for black-body radiation, the heated body must radiate the same amount of energy that it absorbs; otherwise, temperature equilibrium would not be established. Therefore, when the heated body is viewed from outside the cavity, the absorbed part of the energy is received as well as the reflected part. Since these two parts together are black-body radiation, then a heated body inside a black-body cavity at the same temperature produces black-body radiation. By this means metals or other heated materials inside an industrial furnace, which is for the most part closed and has little or no open flame, may be considered black bodies.

Appearance of radiation from a black body is independent of the direction from which the radiant surface is observed. This follows from Lambert's law, which states that the intensity of reflected radiation varies approximately with the cosine of the angle made with the normal to the surface; for perfectly diffusing surfaces the brightness is therefore constant under all angles of observation. This law holds approximately for nonblack-body radiation.

Radiation-Receiving Elements

The purpose of a suitable radiation-temperature-measuring device is to convert the radiant energy from the heated body into a sensible indication of temperature. For this conversion almost any of the temperature-measuring elements mentioned in previous chapters may be used. The thermocouple in the special form of a thermopile or a vacuum thermocouple and the resistance thermometer in the special form of a bolometer are in common use, and even the bimetallic spiral has been used.

The thermopile used as a radiation receiver is shown in Fig. 5–5. It consists of a number of thermocouples connected in series, the measuring junctions of which have been flattened and blackened to provide better absorption of radiation. The thermopile may possess 8 to 16 or even more such measuring junctions. The reference junction of the thermopile is located at the outer ring where the individual thermocouples are connected. The thermopile operates from the radiant energy focused on it. The actual rise in temperature of the measuring junction for steady-state conditions depends on:

1. The amount of radiant energy received.
2. The heat loss by radiation and conduction from the thermopile.

The thermopile may be calibrated for radiant energy absorbed versus emf output, and it may be used with an automatic-balance potentiometer in the same manner as a thermocouple.

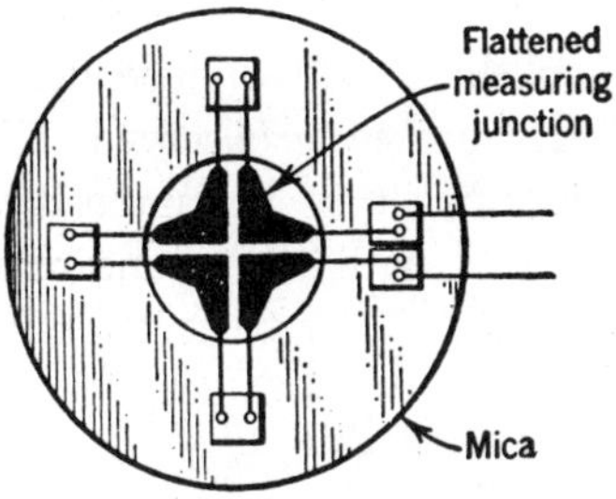

FIG. 5-5. Thermopile

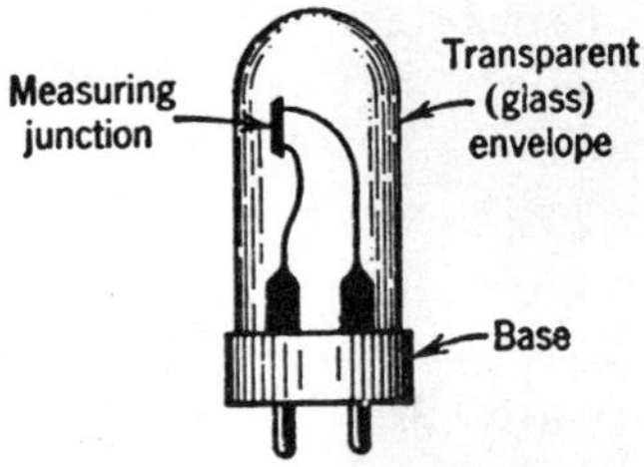

FIG. 5-6. A Vacuum Thermocouple

The vacuum thermocouple is shown in Fig. 5-6 and consists of an evacuated glass envelope in which is supported a single thermocouple made from fine wire. The measuring junction of the thermocouple carries a receiver of blackened platinum foil and is positioned in the center of the glass envelope. The reference junction is generally located in the base of the unit. When radiant energy is directed through the glass envelope and strikes the thermocouple junction, the temperature of the measuring junction is increased over that of the reference junction, and a thermal emf appears. The heat loss from the thermocouple by conduction and convection is greatly reduced by the use of an evacuated envelope and by employing small-diameter thermocouple wire. By these means an appreciable temperature rise of the measuring junction and therefore a measurable emf can be obtained from a relatively small amount of radiant energy received at the measuring junction.

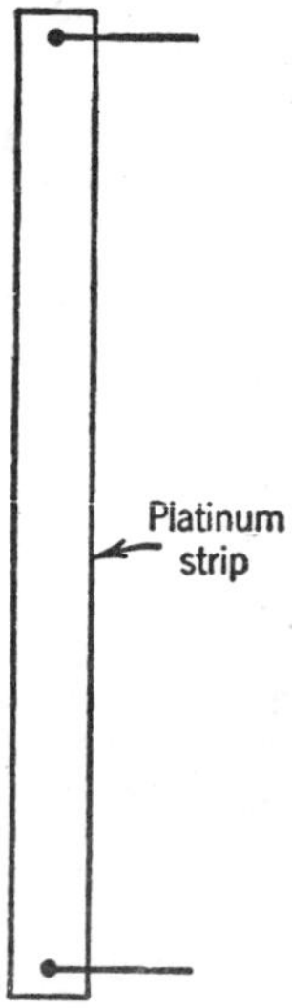

Fig. 5-7. A Bolometer (10 times size)

The bolometer, shown in Fig. 5–7, is a resistance-thermometer element, and one form is made from a strip of nickel foil about 0.1 micron in thickness, about ½ in. long, and of sufficient width to give about 10 ohms resistance. The strip is then coated with gold black or platinum black to increase its absorptivity. The rise in temperature of the strip depends on the amount of radiant energy absorbed and the heat loss by radiation and conduction. The resulting change in resistance can be determined by a Wheatstone bridge and indicated by a resistance thermometer.

The photoelectric tube and photoelectric cell are also used as radiation-receiving elements. In general, these are sensitive only to given portions of the spectrum and for this reason are classed with optical or partial-radiation pyrometers. Cesium oxide photocells sensitive mainly in the short-infrared band are often used, the current output of the cell varying with the amount of radiant energy received. (See photoelectric pyrometers.)

Radiation Pyrometers

The radiation pyrometer, consisting of a radiation-receiving element and a potentiometer instrument, has come into relatively wide use, because it is able to measure directly the temperature of work surfaces in a furnace. It can measure the temperature of moving objects, it is less affected by corrosive atmospheres than a thermocouple, and it can measure higher temperatures than a thermocouple. Probably most important in many applications, however, is its very fast speed of response.

The most common type of radiation-receiving unit is shown schematically in Fig. 5–8. A lens is used to concentrate the radiant energy from the source on the diaphragm and on the thermopile (or vacuum thermocouple). Sighting glasses are shown in the unit to enable the proper line of sight and proper focus to be established. The position of the lens or the position of the diaphragm is usually made adjustable for calibrating purposes.

The mirror-type radiation receiver is shown in Fig. 5–9. Here a diaphragm system together with a mirror is used to focus the radiation on a thermopile (or vacuum thermocouple). The lens-to-thermopile distance is adjustable for proper focus. This unit

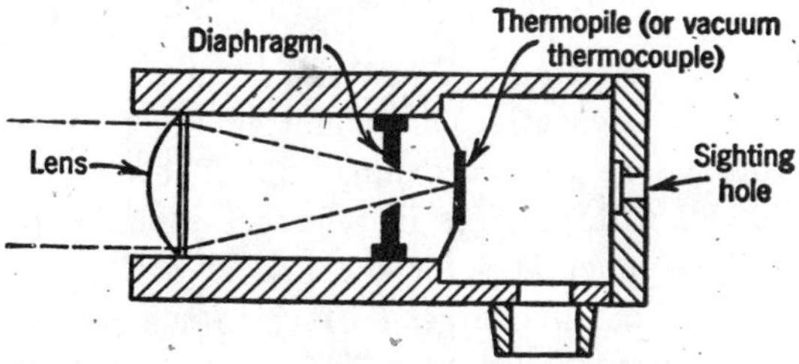

FIG. 5–8. Lens-Type Thermal Radiation Receiver

has an advantage in having no lens, since lens absorption and reflection are avoided.

Several effects in the use of radiation receivers should be considered. These are the reference-junction temperature effect, the distance effect, the absorbing-media effect, and the effect of emissivity.

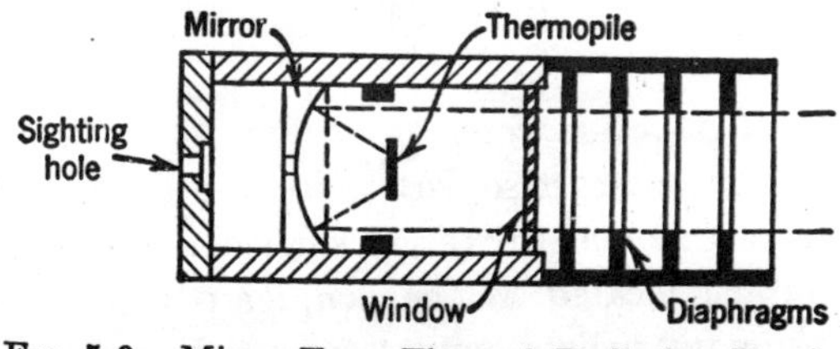

FIG. 5–9. Mirror-Type Thermal Radiation Receiver

The temperature of the reference junction of the thermocouple should be kept substantially constant. The emf generated by the radiation receiver depends on the difference in temperature of the measuring and reference junctions. The reference junction is arranged so that it is in close thermal contact with the housing of the unit. If the housing temperature rises, the temperature rise of the measuring junction and of reference junction is approximately the same. Thus, if the thermoelectric power (dE/dT) of the thermopile is substantially constant, the error is small. In measuring low temperatures, the emf of the thermocouple is small, and the error is appreciable. In measuring high temperatures, the thermocouple emf is large, and the error is therefore negligible.

When the temperature of the housing might rise higher than 150°F, air or water cooling of the radiation-receiver housing should be supplied. This serves to maintain the reference-junction temperature constant. Most radiation units are arranged with a nickel compensating coil shunting the two wires of the thermocouple at the reference junction. The change of resistance of the coil with changes in reference-junction temperature approximately compensates for the loss of emf.

The distance from the target surface to the receiver and the size of the target area must be considered in installing the radiation unit. With all types of radiation receivers the target area must be large enough to fill the field of view; otherwise, insufficient radiation is received, and a negative static error results. The "angle" of the lens or mirror is usually selected so that the target diameter is a given fraction of the distance from target to receiver. This target-area factor is generally between 10:1 and 20:1; that is, the distance from target to receiver should not be greater than 10 or 20 times the maximum useful diameter of the target.

The lens-type radiation receiver of Fig. 5–8 is arranged so that the image of the target surface is focused on the diaphragm opening. This means that the target surface and the diaphragm opening should be located at the conjugate foci of the lens; otherwise, the image will be out of focus. For this reason an adjustment of the position of either the diaphragm or the lens is provided. The error is less than ±10°F when the distance from target to receiver is equal to the true focal distance plus or minus one third. The mirror-type radiation receiver of Fig. 5–9 is arranged so that the image of the front diaphragm is focused on the thermopile by the mirror. Therefore, the measured temperature is substantially independent of target distance if the field of view is filled.

Absorbing media at ambient temperature between the target and the thermopile reduce the amount of radiation received by the thermopile. Such substances as smoke, dirt, and gases absorb radiation and cause the instrument to read low. On the other hand, hot gases, flame, and high-temperature carbon particles may add to the radiation and cause the instrument to read too high.

Gases such as CO_2, SO_2, water vapor, and ammonia have absorption and radiation bands in the wavelengths useful to the radiation receiver. If these media are prevalent, then one recourse is to use a target tube. Sometimes an open tube which is purged with air will keep the line of view free from absorbing media. Hot gases and flames in the line of sight should be avoided.

Reradiation by the lens, window, or mirror of the radiation receiver may cause a slight error unless the lens, window, or mirror is maintained at the same temperature as the housing. This error is reduced by employing air or water cooling at the housing of the radiation receiver so as to stabilize temperature conditions within the housing. Undirected radiation falling on the walls and other parts of the receiver may disturb radiation conditions between walls and thermopile. This effect is reduced by blackening the walls and arranging shields and diaphragms to reduce intraradiation. Absorption of radiation by lenses or windows does not cause an error, providing the radiation receiver is calibrated with these in place. The effect is merely to limit the response of the receiver to a restricted wavelength band of radiation as shown in Fig. 5–1.

The total emissivity of the target surface is an important factor in the operation of the radiation receiver. Emissivity of the target surface presents two problems. First, black-body conditions are closely approximated in closed furnaces, and the measurement of temperature can usually be made without corrections for emissivity. Second, temperatures of open surfaces such as molten metals and hot steel plate cannot be measured accurately unless corrections are made for the low emissivity of such surfaces. A target tube is nearly always employed in the measurement of temperature in molten metals or other hot fluids, simply because one can never be sure of the emissivity value of these surfaces. Slag and oxides form over the surface of most molten metals, and the surface may be at a considerably lower temperature than the metal underneath. For open surfaces of sheet material such as steel plate and calendaring rolls it is necessary to know the total emissivity before accurate measurements of temperature can be made. When the total emissivity of the surface is known, the following method may be used to correct the temperature readings. The radiant energy actually received by the radiation unit is assumed to be given by the Stefan–Boltzmann law and the emissivity of the surface,

$$\Phi = e\sigma A T^4 \tag{5-7}$$

where e is the total emissivity of the surface. The energy is related to the apparent temperature by the same law,

$$\Phi = \sigma A T_a^{\,4} \tag{5-8}$$

where T_a is the apparent temperature and is the temperature indicated by the radiation unit sighted on the surface in question. The true temperature is then

$$T = e^{-1/4} T_a \tag{5-9}$$

For the radiation pyrometers discussed, the correction procedure given by equation 5–9 is somewhat in error. These pyrometers are not total-radiation pyrometers, because the use of the lens restricts the response to a given band of wavelengths. Therefore, the Stefan–Boltzmann law does not strictly apply. The effect of nonblack-body conditions depends on the emissivity applying to the given wavelength band. The total emissivity applies to the full wavelength band, and these two emissivities are different. Actually the correction is within the limits of error with which total emissivities are known. Correction tables for each make of radiation pyrometer can be obtained, however.

The effect of a gray absorbing medium, that is, a medium in which the per cent absorption of radiation is more or less equal at all wavelengths, can be calculated. In this case the transmission factor R merely replaces the total emissivity in equation 5–9.

When the total emissivity of the surface is below about 0.8, accurate temperature measurement is difficult because of the uncertainty in the value of total emissivity. Under these conditions it is preferred to calibrate the radiation unit under actual operating conditions. A further reason is that, with low total emissivity, reflections on the target surface from higher-temperature surfaces are possible, making the temperature measurement more uncertain.

The static accuracy of the radiation unit calibrated under black-body conditions is better than ±0.5 per cent. Under operating conditions, the errors incurred by ambient-temperature fluctuations, the distance and focusing arrangement, the effect of absorbing media, and the uncertain total emissivities may increase the static error.

The response of the radiation unit is extremely advantageous and is one reason for its wide use. Generally speaking, the time constant of response is approximately a first-order type and about 0.01 to 0.02 min in magnitude. The reason for this fast response is the small thermal capacitance of the thermopile or vacuum thermocouple.

The target tube for use with the radiation unit is made of either a ceramic or a metal such as silicon carbide, sillimanite, Inconel, nickel, chrome–nickel, wrought iron, or steel. The tube forms a black-body target, and the length and diameter are proportioned so that multiple reflection brings the total emissivity close to one, as in the Mendenhall wedge. The use of a target tube introduces a slower speed of response, since the tube itself must be heated and cooled by the furnace or heated enclosure surrounding the tube.

The industrial millivoltmeter or automatically balanced potentiometer instrument is ordinarily used to measure the emf developed by the radiation receiver. The receiver may be connected to the instrument with ordinary copper lead wire, because both the measuring and reference junctions of the thermocouple or thermopile are located in the radiation receiver.

Photoelectric Pyrometers

The radiation-temperature-measuring instrument operated by photoelectric means is not so widely used as the radiation pyrometer, but its simplicity and convenience are valuable in many industrial operations. Usually it is employed in the temperature range from 1500 to 3000°F. Photoelectric methods possess great advantages in speed, since the response of a photosensitive cell to radiation is immediate.

The relative responses of a photovoltaic (blocking-layer) cell and a photoemissive tube are shown in Fig. 5–10. Normally the response is limited to the visible red and near infrared by employing a red filter with the tube or cell. The photovoltaic cell directly produces an emf proportional to the amount of radia-

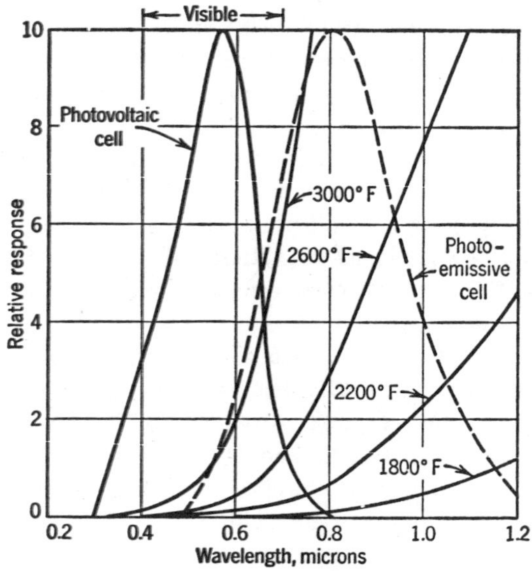

FIG. 5–10. Response of Photoelectric Cells and Black-Body-Radiation Intensity

tion received. The photoemissive tube produces an electron current proportional to the amount of radiation received. The response of the photovoltaic cell shows a slight hysteresis of little importance in temperature measurement and is reasonably stable. Since the phototube and photocell are used for infrared radiation, it is necessary to insure that the tube or cell does not become overheated.

The receiver shown in Fig. 5–11[1] is for radiation measurement of temperature. The cone of radiation passing ultimately

[1] B. M. Larsen and W. E. Shenk, "Temperature Measurement with Blocking Layer Photocells," *Temperature, Its Measurement and Control in Science and Industry,* Reinhold Publishing Corp., 1941, p. 1150.

to the photocell is defined by the area of the first diaphragm. The protecting window is made of thin glass and serves to protect the filter and cell from physical damage. The filter is used for temperature measurements in the range of 1800 to 2200°F in order to reduce the infrared radiation passed to the photocell. This helps prevent overheating of the cell. The response of the photovoltaic cell when the cell itself is above a temperature of about 120°F is unstable, and air or water cooling may be required. By shunting the cell with an external resistance of 50 to 300 ohms a potentiometer-type instrument can be used to

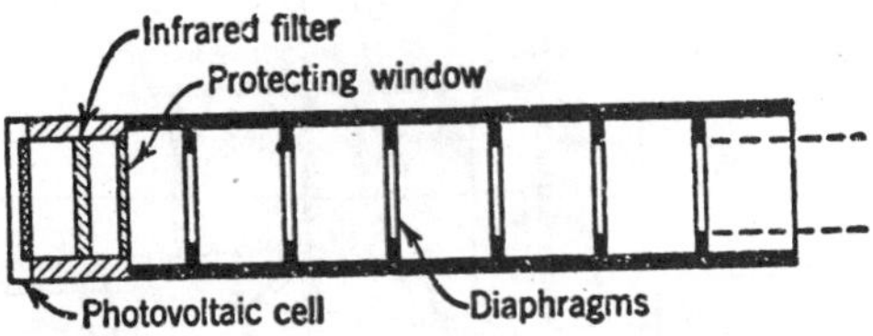

FIG. 5-11. Photoelectric Radiation Receiver

record the temperature at the target. At temperatures of about 2000°F and below, an amplifier may be required, because the cell output is too small to operate directly a potentiometer instrument. The system is calibrated by comparison to an optical pyrometer. Its calibration should be checked frequently, owing to gradual changes in characteristics of the cell.

The advantages of using the photocell receiver are that its response is extremely rapid, the unit is simple and rugged, and no reference-junction and compensation problems exist. Its limitations lie in the necessity for maintaining the cell at low temperature and in the inaccuracy caused by changes and drift in cell characteristic. Nevertheless, this method is employed in measurement of temperatures in open-hearth furnaces and in steel-rolling and welding temperatures.

The photoelectric pyrometer shown in Fig. 5-12 is a product of The Brown Instrument Company. It employs the principle of automatically matching the radiation from a standard source and the radiation from the target surface. The photoemissive vacuum tube produces an electron current which is proportional to radiation intensity. The lens concentrates radiation from the target surface onto a phototube in the sighting tube. The response from the phototube is amplified through a bridge-type circuit to regulate the current through the standard lamp until the current from each phototube matches. A milliammeter (a more or less standard industrial millivoltmeter) indicates the current through the standard lamp. The use of two identical phototubes tends to cancel any changes in their characteristics, and the accuracy of this type of photoelectric pyrometer is therefore reasonably good.

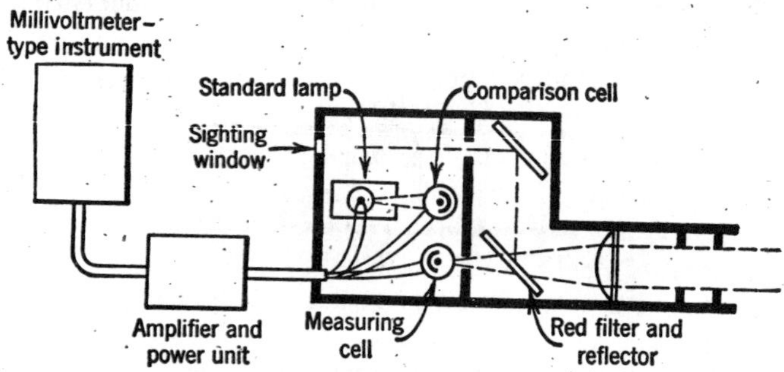

FIG. 5-12. Photoelectric Pyrometer—The Brown Instrument Company

The response of any photoelectric pyrometer is as fast as its instrument can be made to operate, since the response of a phototube is instantaneous for purposes of temperature measurement. Therefore, this type of pyrometer is especially useful in measuring the temperature of fast-moving objects such as are found in steel-rolling mills.

The photoelectric pyrometer operates relatively independent of ambient-temperature changes and is not greatly influenced by distance of target surface, so long as the field of view is completely filled. It is subject to about the same influence of absorbing media and low emissivity as is the radiation pyrometer.

Corrections for transmission factor and emissivity cannot be readily calculated and must be determined separately for each type of photoelectric pyrometer. The reason is that the instrument is neither a total-radiation pyrometer nor a monochromatic optical pyrometer, and neither the Stefan–Boltzmann law nor the Wien radiation law directly applies. In general, corrections for low transmission factor or low emissivity will be less than that for the radiation pyrometer.

Optical Pyrometers

The optical pyrometer is widely employed for accurate measurement of the temperature of furnaces, molten metals, and other heated materials. It is primarily used in the range of 1000 to 5000°F. Most optical pyrometers are manually operated and thereby are somewhat limited in their application.

The optical pyrometer operates with nearly monochromatic light. The wavelength is usually a band about 0.01 micron wide at 0.65 micron in the red portion of the visible spectrum. The principles employed in most optical pyrometers are shown in Fig. 5–13. Radiation from the target surface is focused by the lens onto a screen. The screen is viewed through a red-filter glass so that only wavelengths of about 0.65 micron are seen. For comparison a calibrated tungsten lamp is used. Its radiation is focused on a screen and also viewed through a red-filter glass. The two screens are then compared by eye. As the eye is very sensitive to differences in brightness, comparison match of the two screens can be made to within 1 per cent. Because of the red filter, color blindness of an operator does not affect the operation.

For varying the brightness of the comparison screen image there are basically two methods. In the first method, the current through the standard lamp is adjusted by changing the value of circuit resistance, as shown in Fig. 5–13. The intensity of the standard source is thereby varied, and the brightness of the standard screen is matched against the brightness of the other screen. Calibration for temperature is achieved by calibrating either the variable resistance or a milliammeter measuring the current to the lamp. In the second method, the lamp is maintained at a constant intensity while the intensity of the unknown radiation is varied to match the standard source. The optical wedge of absorbing material in the upper portion of Fig. 5–13 is used for this purpose. As it is moved up and down, its variable thickness causes variations in brightness of the screen, and the position of the wedge is calibrated for temperature.

There are more than nine different combinations of these principles used in industrial optical pyrometers, and it will not be possible to describe them all. Each kind has its particular

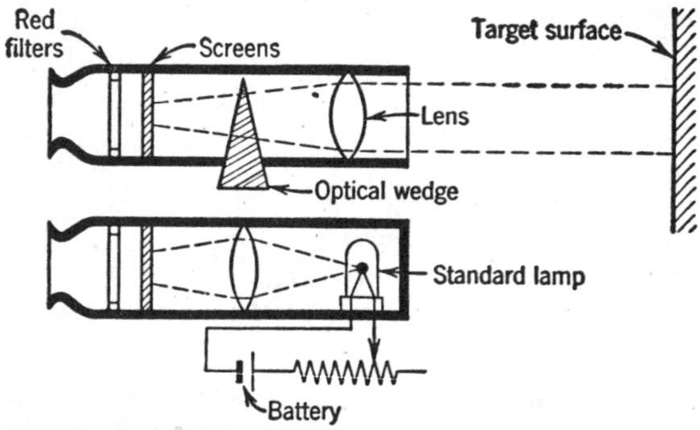

FIG. 5-13. Principles Used in Optical Pyrometer

features and advantages. To illustrate one of these, the Leeds & Northrup Company's optical pyrometer is shown schematically in Fig. 5-14. It uses the variable-intensity-source method

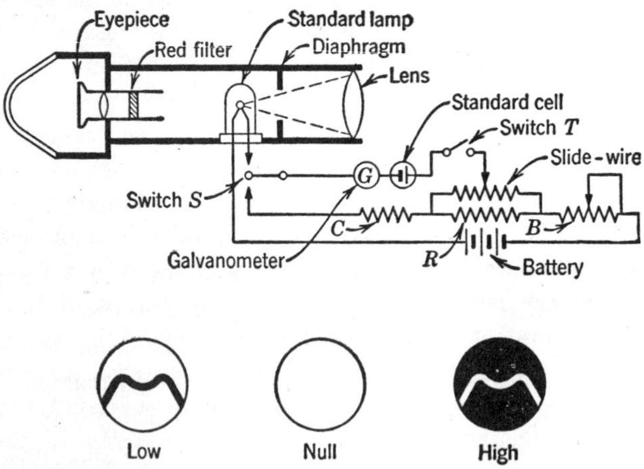

FIG. 5-14. Schematic Arrangement—Leeds & Northrup Company Optical Pyrometer

with a standard tungsten lamp and a potentiometer-type circuit for adjustment and measurement of current. The lens focuses the radiation from the target surface on the diaphragm and the lamp filament. The lamp filament appears against a background of the target surface, as illustrated by the small pictures in the diagram. The whole is viewed through the eyepiece and the red filter. In using the pyrometer, the pyrometer tube is sighted on the target surface. The current through the standard lamp is adjusted by closing switch *S* and adjusting the slide-wire contactor at *B*. The pyrometer tube can then be lowered. The switch *T* is next closed, and the slide-wire contactor is moved until the galvanometer indicates zero deflection. At this point the voltage drop of the lamp current through resistor *R* and the slide-wire shunt is equal to the emf of the standard cell. The slide-wire is generally calibrated directly in terms of temperature.

The operation of determining a temperature can be made in a matter of seconds. The accuracy of measurement is excellent, and temperatures in the usual operating range can be determined to within about $\pm 5.0°F$.

An optical pyrometer (Pyrometer Instrument Company) illustrating the method of varying the intensity from the target surface is shown in Fig. 5–15. The lens focuses the radiation from the target surface through the photoscreenic wedge and onto the upper prism. The standard lamp is located in the line of view through the eyepiece and the red filter and appears as shown in the small diagrams in the figure. The current through the lamp filament is set to a standard value indicated on the milliammeter. In operation the pyrometer tube is pointed toward the target surface, and the photoscreenic wedge is rotated until a brightness match is obtained. The position of the photoscreenic wedge is calibrated in terms of temperature. Temperature readings may be made very quickly. The accuracy of reading is good, considering the simplicity of the instrument.

Among other types of optical pyrometers, the Wanner polarizing type employs the standard lamp at constant intensity, and variation in intensity of radiation from the target surface is accomplished by polarization; the Foote and Fisher type uses the standard lamp at constant intensity and varies the intensity of radiation from the target surface by a rotating glass wedge of varying thickness; the Féry type also uses a standard lamp at constant intensity and changes the intensity of radiation from the target surface by the linear motion of a wedge of black glass.

There are in general only two factors influencing the use of the optical pyrometer: the effect of absorbing media and the spectral emissivity of the target surface. Ambient-temperature changes have only a negligible effect on the accuracy of most optical pyrometers. Distance between the optical pyrometer

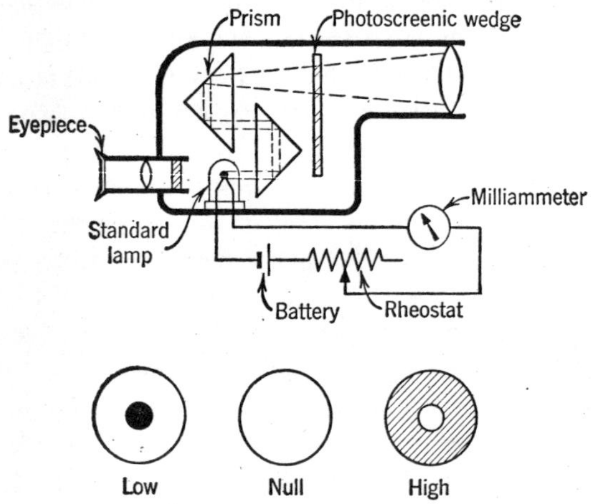

FIG. 5-15. Schematic Arrangement—Pyrometer Instrument Company Optical Pyrometer

and the target surface has little effect, as long as sufficient target area is obtained to fill most of the field of view. The standard lamp undergoes a change of calibration of about 1°C in several hundred hours of operation, but at high operating temperatures it is necessary to replace the lamp more frequently.

The effect of absorbing media is to lower the reading of the optical pyrometer, because the apparent brightness of the target surface is lowered by the absorption of radiation by gases, smoke, dust, and so on. Water vapor, carbon dioxide, and some other gases absorb in the infrared portion of the spectrum and, therefore, have little effect on optical-pyrometer readings, as long as their density or concentration is relatively small. It is better, however, to avoid such conditions if it is possible. In some applications it is desirable to use an absorbing glass between the target surface and the pyrometer either to measure temperatures above the range of the instrument or to protect the mechanism. If the transmission factor of the absorbing glass at 0.65 micron wavelengths is known, the correction may be determined by using Wien's law of radiation. This law is a simplification of Planck's radiation law and is

$$W_\lambda = \frac{C_1}{\lambda^5} \frac{1}{e^{C_2/\lambda T}} \tag{5–10}$$

For the optical pyrometer at 0.65 micron wavelengths the exponential is large compared to one, and the one in the denominator of Planck's radiation law is therefore neglected. The apparent temperature T_a is measured by the optical pyrometer,

$$W'_\lambda = \frac{C_1}{\lambda^5} \frac{1}{e^{C_2/\lambda T_a}} \tag{5–11}$$

The transmission factor R is

$$\tau = \frac{W'_\lambda}{W_\lambda} \tag{5–12}$$

Combining these equations gives

$$T = \frac{1}{\dfrac{1}{T_a} + \dfrac{\lambda}{C_2} \ln \tau} \tag{5–13}$$

and the correction for any absorbing media is found. These corrections can be made more easily by using Table A–17 (appendix).

The influence of spectral emissivity of the target surface on the use of the optical pyrometer is the same as for the radiation pyrometer so that the discussion need not be repeated. In general, variations of spectral emissivity of surfaces at 0.65 micron is not quite so large as variation of total emissivity under the same conditions. The correction for the optical pyrometer is therefore not quite so great. We may investigate the effect of spectral emissity e_λ less than one by means of Wien's law, because

$$e_\lambda = \frac{W'_\lambda}{W_\lambda} \tag{5–14}$$

where the prime indicates the spectral intensity of radiation from a nonblack body. This equation is the same as that for transmission factor, and the results of equation 5–13 may be used:

$$T = \frac{1}{\dfrac{1}{T_a} + \dfrac{\lambda}{C_2} \ln e_\lambda} \tag{5–15}$$

Problems

5–1. Using the Stefan–Boltzmann law, calculate the total radiant energy in British thermal units per minute from a black body of 1 sq in. area at 4000°F.

5–2. A well-oxidized brass thermal well for a thermocouple is used in still air. If it is assumed that the speed of response of the well is proportional to the radiant energy received, by what factor will its response be improved if it is covered with carbon black?

5–3. A radiation pyrometer has a 20:1 distance factor. The radiation receiver is 68 in. from the target surface. What diameter of target surface must be used?

5–4. The lens of a radiation pyrometer is clouded so that the transmission factor is 0.8. The instrument indicates 2700°F. What is the true temperature?

5–5. A radiation pyrometer sighted on the wall of a furnace reads 2300°F. When it is sighted into the furnace on a hot metal of low emissivity, it reads 2400°F. Is this possible? If so, what could be the cause? If not, why?

5–6. An optical pyrometer sighted on melting platinum reads 2960°F. The spectral emissivity is 0.38. What is the static error of the pyrometer?

5–7. A method of measuring spectral emissivity is to sight an optical pyrometer into a crack or cavity in a furnace wall, take a reading, then sight onto the wall near the crack, and take another reading. If the first reading is T_c°F and the second reading T_w°F, what is the emissivity?

5–8. The temperature of a furnace wall is T_1°F, as determined by an optical pyrometer sighted through an absorbing screen. When a second screen is used, the instrument reads T_2°F. Derive a relation for the true temperature of the wall and the transmission factor of the absorbing screen.

5–9. A metal of spectral emissivity 0.8 is at 1500°F inside a furnace and can be considered a black body. The metal and furnace are, of course, opaque. If the furnace walls are at (*a*) 1500°F (*b*) 2000°F, what temperature will be indicated by an optical pyrometer sighted on the metal?

5–10. Prove that for small values of λT Wien's law and Planck's law of radiation approach each other in value. What is the difference at 0.65 micron at 1000, 2000, and 3000°F?

5-11. Heated billets l ft long pass along a roller conveyer at a velocity V fps. The billets are separated by l ft. A radiation pyrometer sighted on the billets has a first-order time constant of T sec. Find an expression for the maximum temperature indicated For an allowabie error of 5 per cent, find an expression for the velocity V. What serious error is involved in this measurement?

5-12. Some radiation pyrometers respond to a sudden target temperature change along a first-order curve up to 96 to 98 per cent of the final value and then gradually drift into balance at a very slow rate. Write a brief explanation for this behavior.

6· Methods for Composition Analysis

One of the greatest advances in instrumentation in the past few years is in the application of direct and quantitative methods of composition measurement in the continuous-processing industries. This has resulted from the requirements of the chemical and petroleum industries for continuous analysis of process streams. However, even metals and other solids are analyzed and their composition determined by automatic means. These newer methods are more direct and have greater accuracy and reliability than older methods, which depended on slow or tedious chemical analysis or on unreliable temperature and pressure data. In addition, the newer methods are much faster and in nearly all cases are applicable to continuous processes.

Composition analyses through atomic or molecular or crystalline characteristics constitute positive methods, because atoms or molecules arranged in a given configuration uniquely identify the substance. Thus it is possible to analyze substances where the quantity, number, and identity of components are unknown.

Composition analyses through such physical characteristics as thermal conductivity, specific gravity, and refractive index are not positive methods, since some substances or combinations of substances produce the same magnitude in the measured physical characteristic. Consequently, these means are limited to analysis of combinations in which the composition is limited in extent or pattern.

Composition analysis is also made by special measurement techniques such as analyzing hydrogen-ion concentration (pH), amount of moisture in gases (humidity), or amount of moisture in solid materials. These methods are, of course, restricted to the particular physical characteristic being measured, and they depend on the uniqueness of this particular characteristic.

Many of the measurement techniques[1] used in composition analysis are specialized and are not available as yet in the form of industrial recording and controlling equipment. Consequently, it is only possible to review a few of the most common methods.

Spectroscopic Analysis

The rapidly growing importance of spectroscopic methods in industrial practice requires that all engineers be familiar with the principles and uses of these highly specialized tools of the physicist. Actually, the application of spectroscopy to plant operation is the closest approach yet made to the problem of direct measurement and control of product quality.

Atomic and molecular spectra serve to identify the fundamental properties of different atoms of a substance. Atoms are composed of a nucleus, having a net positive electric charge, and a relatively large mass, surrounded by a number of negatively charged electrons of small mass. The atomic number of the element is determined by the total number of electrons surrounding the nucleus and therefore by the nuclear charge. The atomic number varies from one upward, 92 being the highest commonly identified.

The electrons surround the nucleus in so-called shells with a discrete number in each shell. The number of electrons in each shell is limited, and the energy state of each electron corresponds to a particular configuration of electrons. If the atom is in a ground state, it has the lowest possible energy, and a given electron configuration is determined. If the atom is excited to a higher-energy state, as by light radiation or any other means, the electrons absorb energy by shifting to a new configuration. However, the shift in level of the electrons is limited to discrete energy "jumps," and therefore only discrete amounts of energy can be absorbed. We may consider, for example, that

[1] See *Anal. Chem.*, **21**, no. 1 (Jan. 1949), First Annual Review of Analytical Chemistry Fundamental Analysis.

an electron jumps from a low-energy orbit to a higher-energy orbit with the absorption of a particular increment of energy. The converse is also true; that is, by a radiation of energy from the atom, the electrons shift from a higher-energy configuration to one of lower energy. Furthermore, since two atoms of different elements do not have the same configuration of electrons, it may be shown that the change in configuration associated with giving or receiving energy is not identical.

Through spectroscopy the atoms of a substance may be excited, may change configuration, and will absorb or radiate energy at unique frequencies. Therefore, a means is available

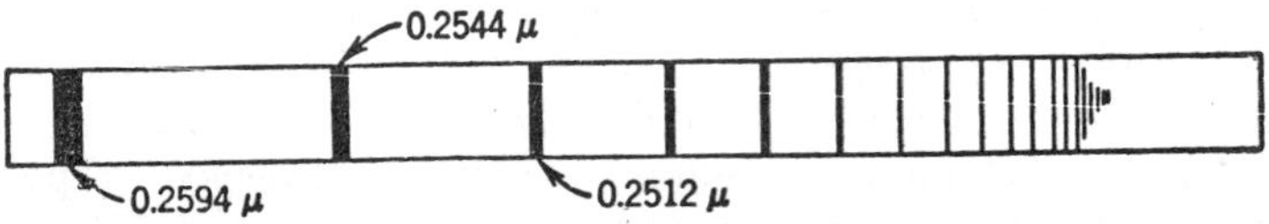

FIG. 6-1. Absorption Spectrum of Sodium (starting with the fifth line of the principal series)

for identifying the existence of a particular atom of a substance. This has been compared to a "fingerprint" in its uniqueness, although actually it is more individual.

As an example, suppose that a light beam containing all frequencies is passed through a thin film of sodium and then through a refracting prism. The prism breaks the light beam into a spectrum which can be recorded on a photographic plate. A portion of such a spectrum is shown in Fig. 6-1. Notice that several lines indicate that certain frequencies are now missing from the light, having been absorbed by the sodium. Each of these frequencies corresponds to particular changes of energy level within the atom. Thus, any compound containing sodium will produce these particular lines in an absorption spectrum.

Line spectra are caused by emission or absorption by atoms of discrete wavelengths of radiation, thus causing lines. If, however, we obtain an absorption spectrum from a molecule or a molecular compound, a band spectrum is observed. That is, many different wavelengths are absorbed but in varying degree. Thus a plot of radiation intensity versus wavelength would reveal a curve with alternate peaks and valleys having many different slopes.

Absorption spectroscopy is illustrated in the previous example; that is, when infrared, ultraviolet, or X-ray radiation is passed through the sample of unknown material, certain frequencies of the radiation are absorbed. This is determined by separating the radiation into a spectrum and measuring the intensity of radiation at each frequency. Absorption lines are dark lines on a light background on a photographic negative.

Emission spectroscopy is accomplished by placing the sample of unknown substance in a flame or in an arc. This causes emission of radiation which is characteristic of each substance. By collecting the emitted radiation and separating it by refraction or diffraction, the intensity of radiation at each frequency can be measured. An emission spectrum for an element is very similar to that of Fig. 6–1, except, of course, that each line indicates an emitted frequency, not an absorbed frequency. Emission lines are bright lines on a dark background on a photographic negative.

Fluorescence spectroscopy, actually a form of emission spectroscopy, employs the photoluminescence or fluorescence that results from excitation of a substance by absorption of light. This excitation causes an emission of light which, by measurement of its spectral distribution, can be related to the composition of the substance. This method of analysis is commonly employed in determination of vitamin content and in analysis of some pharmaceutical compounds. It is of course necessary that the substance being analyzed be capable of fluorescence.

Mass spectroscopy is yet another means of identifying substances. In this method a sample of the unknown substance, usually a vapor or gas, is bombarded by an electron beam in an evacuated chamber. The bombardment produces ions by "knocking off" an electron from each of the various kinds of atoms present in the substance. These ions are then accelerated into a circular path, the radius of which varies with the mass of the ions. Thereby the ions are sorted into beams. By simply measuring the strength of each beam, all the various kinds of atoms can be counted and identified, thereby enabling the composition of the substance to be found. Since each different substance is composed of different quantities and kinds of atoms, the identification is positive.

Absorption and emission spectrometers, as they are termed in this book, are called variously spectrophotometers, photometers, and spectrographs. Spectrometers are classed by several methods. First, they may be classed by the wavelength of radiation employed: for example, infrared, visible band, ultraviolet, and X ray. Second, they may be distinguished by the means of producing radiation: that is, absorption or emission. Third, they are classed by the method used to separate the beam, as in prism (refracting) spectrometers and grating (diffracting) spectrometers.

Absorption Spectroscopy

Infrared-absorption spectrometers operate from radiation in the range from 1 to 25 microns wavelength (see Fig. 6–2), the band usually called heat radiation. Infrared spectrometers are useful in analysis of almost all substances.[2] In petroleum refining and synthetic-rubber production they can be used for analysis of hydrocarbons and gases. In pharmaceutical work, such as penicillin production, they are used for determining structure. In chemical plants they are used for analyzing both organic and inorganic compounds.

An infrared spectrometer (The Perkin-Elmer Corporation) is shown schematically in Fig. 6–3. Similar devices are made by National Technical Laboratories (Beckman); Baird Associates; The Gaertner Instrument Company; and others. With reference to the figure, the infrared radiation is provided by an electrically heated resistor (the source). The beam passes to a mirror, then to a parabolic mirror, and, after passing through the sample cell, falls on slit 1. The sample cell contains the unknown substance. Slit 1 simply serves to form a beam of very narrow width. The beam then passes to the collimating mirror, where it is rendered parallel, and then through the prism, where it is refracted. The wavelength mirror reflects the beam back through the prism, where it is once more refracted. The beam is next reflected by the collimating mirror to a plane mirror and then to slit 2. The beam is finally reflected to the thermocouple, where the beam intensity is measured.

[2] See V. Z. Williams, "Infrared Instrumentation and Techniques," *Rev. Sci. Instruments,* **19**, no. 3 (Mar. 1948).

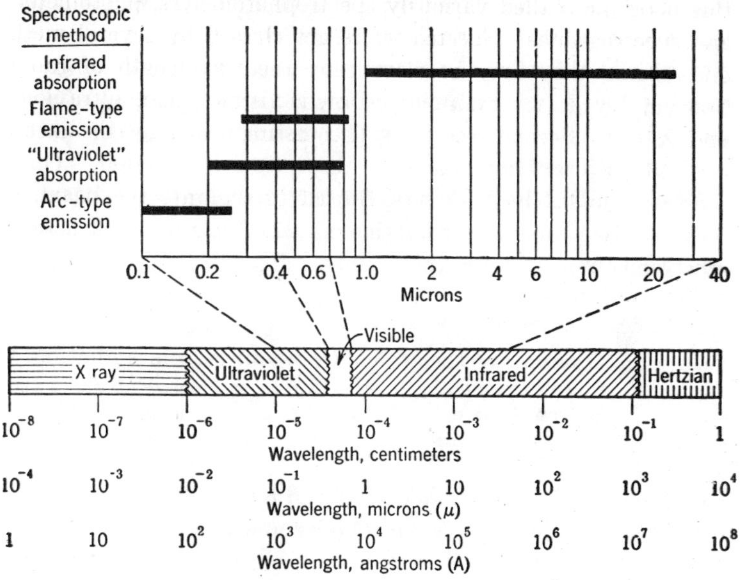

FIG. 6-2. Electromagnetic Spectrum and Useful Band of Industrial Spectrometers (see also Fig. 5-1)

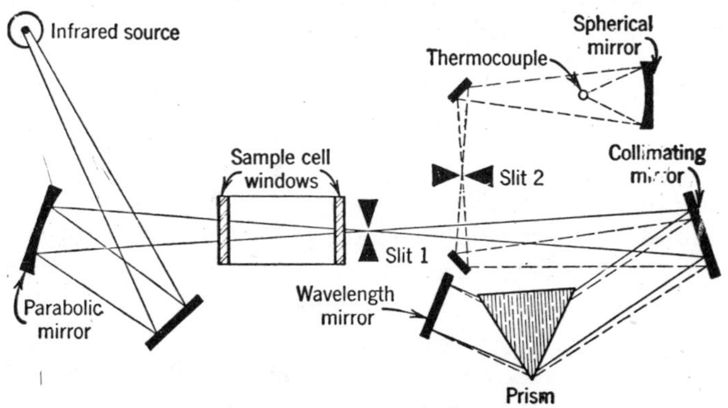

FIG. 6-3. Optical System—The Perkin-Elmer Corporation Infrared Spectrometer

The wavelength of the infrared radiation which is passed to the thermocouple is determined by the angle of setting of the wavelength mirror, since this mirror reflects the refracted beam and governs which part of the beam falls on slit 2. Thus, by slowly rotating the wavelength mirror, wavelength spectrum can be scanned. The thermocouple will measure the intensity of the beam at each wavelength.

As to constructional details, the infrared source provides approximately black-body radiation at 1000°C, and the source is controlled by a constant power input. The prisms are available in four materials:

	Useful for
Lithium fluoride	1.0 to 5.5 microns
Calcium fluoride	2.0 to 8.5 microns
Sodium chloride	3.0 to 15.0 microns
Potassium bromide	15.0 to 25.0 microns

The selection of the prism is determined by the wavelength range most adaptable to the substance under analysis. A temperature compensator (not shown in Fig. 6–3) employs a mirror positioned by a bimetallic strip to correct for the temperature coefficient of refractive index of the prism. The thermocouple is bismuth–bismuth tin, and it is enclosed in a highly evacuated housing having a potassium bromide window. An ordinary automatic-balance potentiometer is used to record spectrographic data.

The Baird Associates infrared spectrophotometer employs a compensating cell, so that the transmission of the sample substance is directly compared to the transmission of a "pure" substance contained in the compensating cell. Differential analysis of this kind is particularly adapted to analysis of slight impurities or slight variations from desired concentrations. By leaving the compensating cell empty, ordinary spectroscopic measurements are made.

A typical spectrogram of pure carbon disulfide is shown in Fig. 6–4. Note that carbon disulfide absorbs particularly at about 4.4, 4.6, 6.4, 7.0, and 11.6 microns wavelength. This diagram is unique in pattern and would not be produced by any other substance; even different concentrations of carbon disulfide in other liquids would not produce the same diagram.

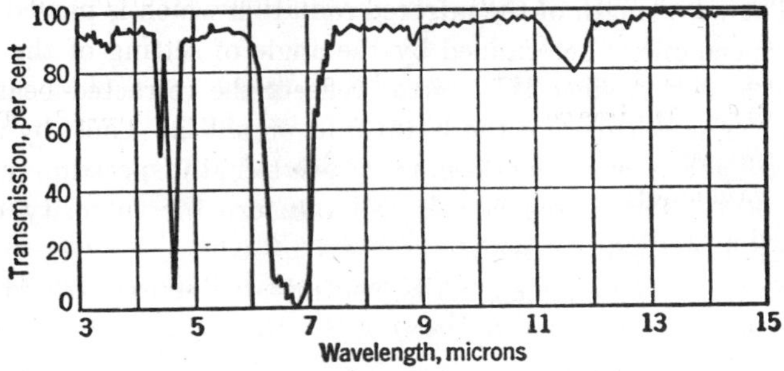

FIG. 6–4. Infrared Spectrogram of Pure Carbon Disulfide (*Courtesy Baird Associates*)

The concentration of various components of the sample substance can be determined from the amount of absorption at any wavelength. The basis of this determination is Beer's law:

$$C = \frac{1}{\alpha x} \log_{10} \frac{I_0}{I_x} \tag{6-1}$$

where C = concentration of substance
α = absorption factor of substance
x = thickness of sample (along optical path)
I_0 = intensity beam before sample
I_x = intensity beam after sample

The constants in the foregoing equation are α, x, and I_0. These are found by a trial calibration of the spectrometer, using a known sample of a substance with a known concentration. Thus, the concentration may be determined at any one wavelength in the spectrogram. In Fig. 6–4, for example, suppose the carbon disulfide contained a dissolved liquid which was transparent in the infrared. Then, at 0 per cent concentration of carbon disulfide, a continuous line at 100 per cent would be obtained in the spectrogram. At 100 per cent concentration of carbon disulfide the spectrogram of Fig. 6–4 is obtained. Therefore, at intermediate concentrations, the transmission reading at any wavelength would lie between these two values. If the liquid (or liquids) in solution is not transparent, then it is only necessary to select one wavelength at which the liquid does not absorb, and the concentration can be calculated. In some cases this cannot be done or is not feasible, and calculations involving all components must be made.

Not only can analysis be made by studying a complete spectrogram, but also, as is evident from Beer's law, it is possible to perform continuous analysis by operating the spectrometer at one wavelength. This wavelength can be selected so that the particular analysis performed is the most sensitive. The sample cell of the spectrometer is then arranged for continuous flow of the unknown fluid. The Baird Associates gas analyzer is an

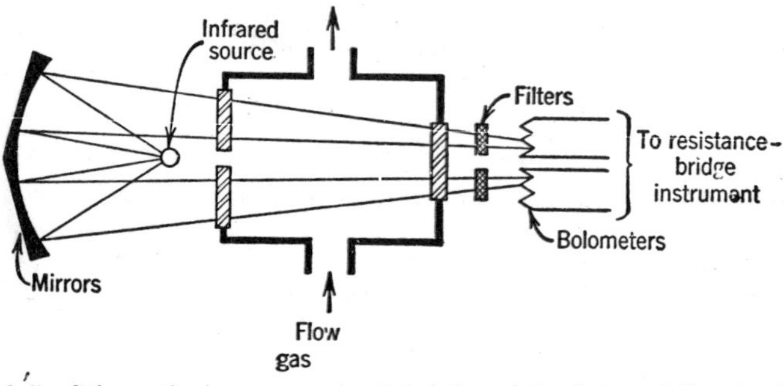

FIG. 6–5. Schematic Arrangement—Baird Associates Infrared Gas Analyzer

example of this method (see Fig. 6–5). The gas absorbs at a particular wavelength, which is passed by the upper filter but not by the lower filter. The amount of radiation obtained by each bolometer then depends on the concentration of the absorbing gas. This simple method can be used for a wide variety of gases, except such as oxygen, hydrogen, and nitrogen that do not absorb in the infrared region. Very similar devices of simple construction can also be employed for liquids.

Ultraviolet-absorption spectrometers operate in the range from 0.2 to 0.8 micron, which actually includes the visible band. Instruments of this type are widely used for nearly all types of analysis. Determinations for vitamin components, for benzene and toluene in petroleum distillates, for butadiene in saturated hydrocarbons, for ethylene and carbonyl compounds, and for polyunsaturates in fatty acids are miscellaneous examples.

The National Technical Laboratories (Beckman) photoelectric quartz spectrometer, although it can be made to operate from 0.2 micron in the ultraviolet to 2.0 microns in the infrared, is usually termed an ultraviolet-absorption spectrometer. The Applied Physics Corporation also makes a spectrometer of this type operating in the range 0.2 to 0.8 micron. The optical path of the Beckman spectrometer is shown in Fig. 6–6. An image of the light source is focused by the condensing mirror and by the flat mirror on the entrance slit. The portion of the spectrum passing through the exit slit then passes through the sample cell and falls on the phototube. The intensity of the beam is measured by a suitable amplifier connected to an automatic-balance

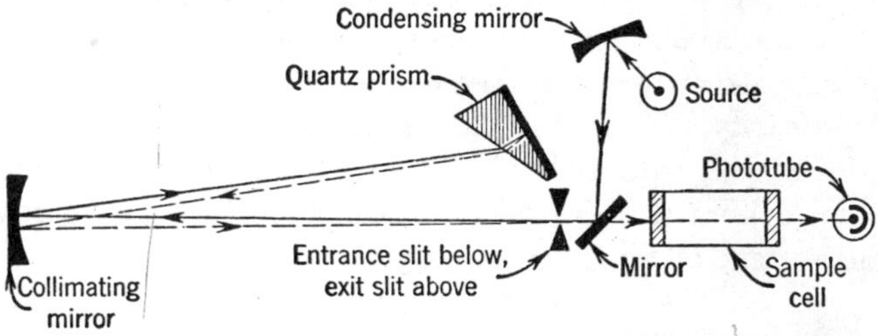

FIG. 6–6. Optical System—National Technical Laboratories (Beckman) Ultraviolet Spectrometer

potentiometer instrument. The particular wavelength passed through the sample cell is determined by the angle of setting of the prism. The angle of the prism determines which part of the refracted beam falls on the exit slit.

The source is a tungsten lamp for use in the (visible) region from 0.3 to 1.0 microns and is a hydrogen-discharge lamp for use in the ultraviolet region from 0.22 to 0.35 micron. Other sources are also available. A quartz prism is employed. A blue(ultraviolet)-sensitive phototube is used for the 0.2- to 0.65-micron region, and a red(infrared)-sensitive phototube is selected for the 0.6- to 1.0-micron region. The sample cell can be arranged for continuous flow of gas and liquid samples.

Ultraviolet-absorption spectroscopy is otherwise similar to infrared spectroscopy. The instrument may be used either for complete analysis of samples by running a complete spectrogram or for continuous analysis of process streams by operating at a fixed wavelength critical to the analysis being made.

X-ray-absorption spectrometers can also be employed for composition analysis of solids, liquids, and gases. Typical examples are the measurement of sulfur content of oil, lead content of gasoline, metal content of glass, halogen content of plastics, ash content of coal, and filler content of wood and rubber.

The General Electric X-ray spectrometer is an example of one type of instrument. The source of radiation is an X-ray tube providing a wavelength range from 0.4 to 1.5 A. Two beams are taken from the source; one beam passes through a sample cell to a fluorescent screen. The other beam passes through a variable-thickness aluminum attenuator and then through a second sample cell to the same fluorescent screen. An amplifier is used to measure the difference in intensity of the two beams, and the aluminum attenuator is adjusted until the two beams have the same intensity. Thus, the difference in absorption of the two samples can be expressed in terms of the thickness of aluminum. By trial calibration, the thickness of aluminum can be related to the content of critical component.

Emission Spectroscopy

The emission spectrometer can be employed for analysis of metallic compounds. In fact, emission spectroscopy led to the discovery of cesium and rubidium. Arc-type emission spectrometers are used for magnesium, aluminum, and steel analyses. The flame-type emission spectrometer can be used for analysis of inorganic compounds containing sodium and potassium, and other elements such as boron, calcium, cesium, cobalt, copper, iron, lithium, magnesium, manganese, and nickel.

The emission spectrometer of the Applied Research Laboratories and the Harry W. Dietert Company employs an arc spark for producing radiation emission and a grating for diffraction. The latter is used for diffraction of the emitted radiation into a spectrum. The intensity of the radiation at various wavelengths in the ultraviolet region of about 800 to 2500 A is measured by a multiplier phototube to operate a direct-reading instrument. (Baird Associates also makes a grating-type emission spectrometer.)

The flame spectrometer of the Perkin-Elmer Corporation is a small compact instrument, specifically adapted to measurement of concentrations of sodium or potassium between about 15 and 1000 ppm. Its usefulness lies in soil, water, and clinical analysis. This instrument operates by atomizing an aqueous solution and introducing it into a burner. A lens system then concentrates the emitted light through filters on a phototube. An amplifier operates an indicating instrument to indicate concentrations. Calibration is made by running several known solutions.

A flame photometer attachment may be used with the National Technical Laboratories (Beckman) spectrometer previously discussed. Spectral lines of various elements from 2800 to 8500 A in the visible region may be selected for analysis. The equipment consists of an atomizing sprayer for introducing the solution into the illuminating gas and oxygen flame. The image of the flame is then focused on the entrance slit of the spectrometer shown by Fig. 6–6, and the spectral-line intensity is measured by the phototube, amplifier, and recording instrument.

Since emission spectroscopy is mainly employed in analysis of elements, the measurements are made of intensity of spectral lines. In many applications, it is sufficient to measure the intensity of a particular line, and by empirical calibration the intensity of the line is proportional to the concentration of the element in the solution. For complete analysis of substances, it becomes necessary to record the complete spectrum and, by analysis of the wavelength and intensity of many lines, determine the concentration of unknowns.

Fluorescence (emission) spectrometers are useful in analysis of some pharmaceuticals and vitamin compounds where it is necessary to avoid photochemical deterioration. For example, light of sufficiently strong intensity will break down riboflavin. Briefly, the substance is irradiated with light of desired wavelength, and the absorbed energy causes the substance to emit (fluoresce) radiation at wavelengths peculiar to the substance. By measuring the intensity of this radiation the concentration of components in the substance can be identified.

The National Technical Laboratories (Beckman) ultraviolet spectrometer previously described can be used for such analyses by adding an accessory set of equipment. The North American Philips Company, fluorescence-analysis unit operates from an X-ray source and is useful in quantitative analysis of elemental components of metals and alloys. It is adaptable for measurements of elements from titanium (no. 22) through tin (no. 50). It operates by exposing the substance to X rays, and fluorescence of the element as X rays of various wavelengths can be measured. The fluoresced rays then pass through a collimating system to a crystal, where individual rays at various wavelengths are diffracted at specific angles by the atomic planes of the crystal. The cone of diffracted rays is then measured by a Geiger counter, and a spectrum is recorded. Calibration is made by suitable standards.

Mass Spectroscopy

The mass spectrometer is most useful in analyzing aliphatic and cyclic hydrocarbon compounds, such as those produced in aviation-gasoline and synthetic-rubber manufacture. In petroleum refining, light gases, olefins, paraffins, wet gas, propylene, and so on can be analyzed. In chemical manufacturing, liquid and gaseous pharmaceuticals, organic compounds having adequate vapor pressure, and inorganic gases and vapors may be analyzed.

The Consolidated Engineering Corporation mass spectrometer which is typical is shown schematically in Fig. 6–7. (Westinghouse Corporation and General Electric Company also produce mass spectrometers.) The sample in a gas or vapor state at about 40 microns Hg absolute pressure is introduced to the ionization chamber. The pressure in the ionization chamber and tube assembly is maintained even lower by a vacuum pump. The gas is bombarded by an electron beam from the filament in the chamber. The gas (or vapor) is transformed into ions, neutral fragments, and electrons. The electrons are collected at the plate in the ionization chamber. The neutral fragments are pumped out of the system. The ions, being positively charged, are attracted toward the two slits by the field caused by the electric potential of the slit pieces. The ions are accelerated through both slits and are formed into a parallel beam. Since the circular tube lies in a magnetic field of fixed direction, the ion beam is made to deflect in a circular path, the radius of which depends on the mass and the velocity of the ion. Thus, several beams are formed, each consisting of ions of a specific mass. The beam to strike the collector can be selected by varying the accelerating voltage at the two slits. The charges given to the collector plate by the ions are amplified and recorded by a d-c potential instrument. By automatically varying the accelerating voltage, a spectrum of ion beam intensity versus ion mass is recorded.

The analysis of the spectrum results in the determination of the composition of the gas or vapor. The spectrum is compared

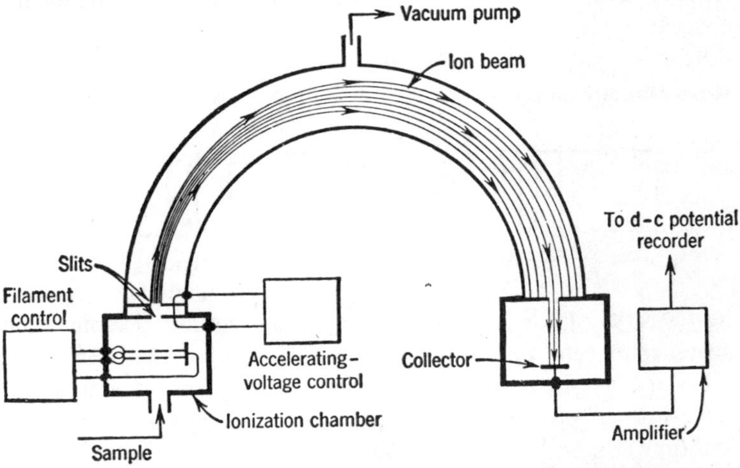

FIG. 6-7. Schematic Arrangement—Consolidated Engineering Corporation Mass Spectrometer

to the spectrum of a pure substance, and mole fractions of the components are obtained by direct proportion. When there is no mass peak in the spectrum for a component in question, then simultaneous algebraic equations must be used. For this analysis electric computers can be employed.

The mass spectrometer can separate masses differing by one part in 300 and, in some cases, those differing by one part in 1200. Thus, one use of the mass spectrometer is in analyzing for isotopes. An advantage of the mass spectrometer is that it can be employed for analyses using extremely small samples, and thus it is useful in microchemistry.

Analysis of Solids by X-Ray Diffraction

In many crystalline and amorphous solids the atoms of each element of the substance lie in regular spacings and form a so-called space lattice. If X-ray radiation of one frequency (monochromatic) is directed into a small sample of the solid at a specified angle, the radiation emerges from the solid in several beams, because some of the radiation is reflected in a given direction by certain rows (planes) of atoms, and some is reflected in another direction by other rows (planes). Therefore, a photograph of the diffracted beams will indicate certain arrangement and spacing of the atoms of the substance. Figure 6–8 illustrates the concentric circle pattern of these diffracted beams. The X-ray-diffraction pattern is unique for any solid because the arrangement and spacing of the atoms in the molecule and the molecules

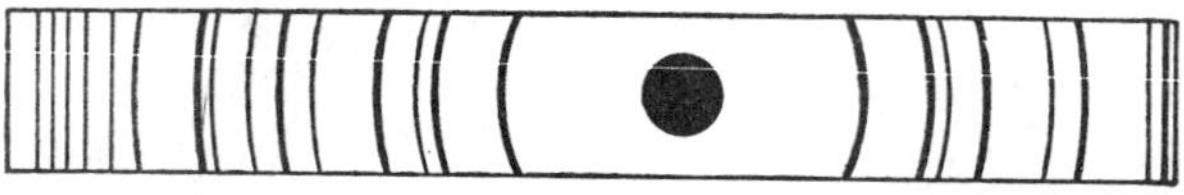

FIG. 6–8. An X-Ray-Diffraction Pattern

in the crystal are different for different solids. Liquids and gases cannot be analyzed by X-ray-diffraction methods, since they usually do not have a lattice arrangement of atoms and molecules at ordinary temperatures and pressures. X-ray diffraction is useful in analysis such as control of dyes and fillers in rubber, selection of manganese ores in battery manufacture, and selection of particle size of clay used in ceramics.

The North American Philips Company X-ray-diffraction instrument employs a Geiger counter to count X-ray radiation diffracted at the specimen. The diffraction cone is scanned by rotating the counter tube across the cone. The counter tube is connected through an amplifier, and the beam intensity versus dimensional spacing can be recorded by a potentiometer-type instrument. Analysis of the diffraction pattern (recorded in a spectrum-like manner) is made by reference to patterns obtained from pure substances. The concentration of components is then calculated by simple proportion.

Color Measurement by Spectrometers

Spectrometers employed in the visible band are mainly useful for color control of plastics, paints, dyes, ceramics, and so on. For translucent materials such as colored liquids and glasses, the usual ultraviolet-absorption spectrometer can be arranged to measure the intensity of light absorbed or transmitted at various wavelengths in the visible region from 0.4 micron (violet) to 0.7 micron (red). The spectrogram would therefore indicate the transmittance in per cent at each wavelength when the substance in question is irradiated by white light.

For opaque materials such as plastics, powdered pigments, and paper, it becomes necessary to measure the reflectance of the surface. The General Electric Company spectrometer is specifically adapted to measure the color of reflected light, but it can also be used for measuring transmitted light. This instrument employs a projection lamp as a source. The beam of white light is then passed through a collimating lens to a double prism, where it is refracted through a slit to a second prism and then into an integrating sphere. The reflectance is measured by a phototube and recorded. The integrating sphere is arranged so that the sample may be compared to a standard (or to another sample for differential comparison) and so that the spectral or gloss component may be included or excluded. The gloss of surfaces is caused by the light being reflected in a beam as by a mirror.

A diffuse reflectance attachment is available for use with the National Technical Laboratories (Beckman) ultraviolet-absorption spectrometer, which of course is operated in the visible band. This attachment is arranged to direct the monochromatic beam of light from the exit slit (see Fig. 6–6) down on the sample which replaces the sample cell. The rays leave the surface at about 45° angle and are focused on a diffusing screen. A phototube is used to measure the intensity of the beam on the screen. An instrument of this type measures the diffuse reflectance over a narrow range of angle of reflection and does not produce the same reading as the instrument above, in which an integrating sphere is used to measure the reflectance over a wide angle. Consequently, in color specifications it is often necessary to identify the instrument on which desired specifications are based.

With such instruments the color of reflecting surfaces and transmitting materials can be explicitly specified by measuring the intensity at each wavelength. For critical color specification as in paints and dyes, the human eye is not reliable. For example, Fig. 6–9 shows the spectral distribution of a color that would probably be identified by the eye as apple green.

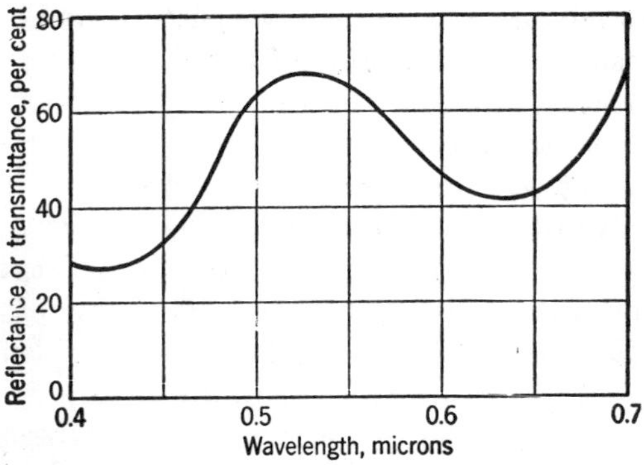

FIG. 6–9. Spectral-Energy Distribution of a Green Color

Gas Analysis by Thermal Conductivity

Methods for determining quantitatively the composition of gases in a mixture are important in all kinds of industrial operations. Thermal-conductivity measurement is very commonly employed in carbon dioxide recorders for combustion control. It is also used for determining the amount of hydrogen in nitrogen, air, and oxygen, and for determination of ammonia, sulfur dioxide, hydrocarbons, carbon monoxide, hydrogen sulfide, and argon. In addition, it is used infrequently for determining the amount of water vapor in air (humidity measurement).

Thermal-conductivity determination of efficiency of combustion of fuels is a very common application. In such cases it is not always necessary to know the exact quantity of each component, such as carbon dioxide, carbon monoxide, nitrogen, and hydrogen. Often the instrument can be calibrated directly in terms of fuel–air mixture for proper combustion, in terms of reducing–oxidizing effect, or even in terms of oxygen excess or deficiency. Carbon dioxide meters (thermal-conductivity type) often are used in this manner, simply indicating an arbitrary per cent of carbon dioxide which produces efficient combustion.

Thermal-conductivity measurement is a very simple means of analyzing certain gas mixtures. For example, a mixture of hydrogen and dry air produces a relative thermal conductivity from 1.0 for 100 per cent air to about 6.98 for 100 per cent hydrogen. Therefore, thermal conductivity of the mixture of two gases can be related to the concentration of each component. It is apparent that, in a mixture of more than two gases, thermal conductivity is a measure of concentration of two components only if all other components are constant in amount. Thus, the use of thermal conductivity for gas analysis does not necessarily distinguish selectively among various components.

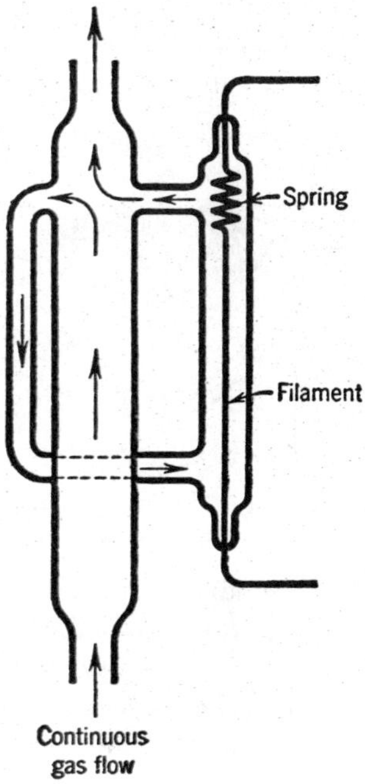

Fig. 6–10. Thermal-Conductivity Cell—Leeds & Northrup Company

It is assumed that water vapor in a gas mixture is counted as one component. Thereby, a mixture of oxygen, water vapor, and hydrogen would have three components. In thermal-conductivity measurement of multicomponent mixtures where one component is water vapor, the amount of water vapor must be constant. This is usually accomplished by either drying the gas by desiccation before measurement or by saturating the gas with a wet wick before measurement. Either method requires some maintenance to insure constant conditions. Also it is usually assumed that the thermal conductivity of a gas mixture is a linear function of the thermal conductivity and concentration of each component. However, this is not a valid assumption for such mixtures as air–water, air–ammonia, and air–butane, which do not combine linearly.

A thermal-conductivity cell (Leeds & Northrup Company) is shown in Fig. 6–10. The unit is constructed of glass. The right-hand tube contains a platinum filament held under constant tension by a spring. The filament and spring are glass-coated. The gas flow occurs at a constant rate by natural convection through the filament cell. When a current flows through the filament, the heating and consequent temperature rise of the filament depend on the rate at which heat is conducted away from the filament. This depends mainly on the thermal conductivity of the gas. Thermal-conductivity cells by the Gow-Mac Instrument Company and the Davis Emergency Equipment Company employ similar principles but are embodied in a metal cell block.

A simple bridge circuit is used with the thermal-conductivity cell, as shown in Fig. 6–11. Either a-c or d-c potential supply is used, and the bridge may be employed in a deflectional-potentiometer system. An automatic-balance Wheatstone-bridge system is also used. The reference cell may be sealed with a gas such as dry air or saturated air or any other gas desired for reference. It is important that both cells be subjected to the same ambient-temperature conditions, so that the only difference measured in the bridge is that due to the differing thermal conductivity of the reference gas and the measured gas.

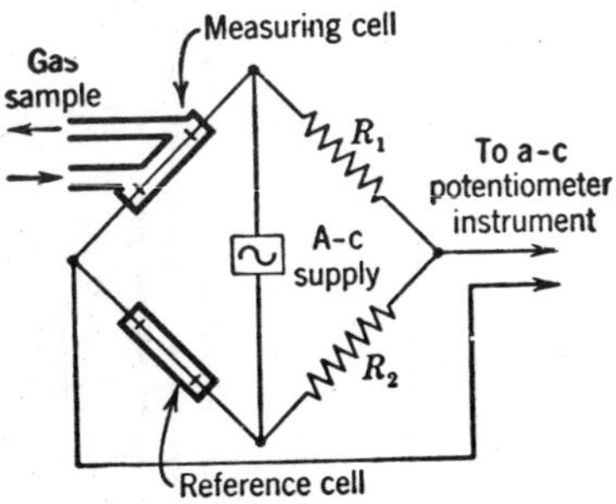

Fig. 6–11. Thermal-Conductivity Resistance Bridge

There are several methods of using thermal-conductivity cells other than the simple one of Fig. 6–11. For differential measurement of one component in a multicomponent mixture, it is possible to pass the gas through the measuring cell, then through a reactor or absorber to remove one component, and then through the reference cell. This method increases the selectivity and accuracy of measurement. Another common method is to employ two measuring cells in opposite legs of the bridge and two reference cells in the other legs of the bridge. This method also increases the accuracy of measurement where the changes in thermal conductivity are small.

Analysis of Moisture in Gases (Humidity)

Gases always contain a greater or lesser quantity of liquid vapor. Air, for example, contains water vapor. The part of the total gas pressure exerted by the vapor is, of course, the vapor pressure.

The definitions of terms for humidity are often contradictory in current engineering use. The following are patterned after the ASA definitions:

1. Absolute humidity H—weight of vapor per unit of weight of dry gas.

2. Specific humidity H_s—weight of vapor per unit of weight of mixture.

3. Density of water vapor ρ_H—weight of vapor per unit of volume of space. This is identical with the weight of water vapor per unit of volume of dry gas. Sometimes this may also be termed absolute humidity.

4. Relative humidity H_R—ratio of an actual partial pressure of vapor in the gas to the saturation partial pressure. If the vapor is considered an ideal gas, this is identical with the ratio of the density of vapor ρ_H to the density of vapor at saturation ρ_s.

The calculation of humidity is accomplished through the ideal-gas law and by the necessary assumption that the mixture of the gas and vapor is an ideal gas. The partial pressure of one of the constituents of a mixture is the pressure that would exist if that constituent alone occupied the volume of the mixture at the same temperature. Thus,

$$W_v = \frac{p_v v}{R_v T} \tag{6–2}$$

and

$$W_g = \frac{p_g v}{R_g T} \tag{6–3}$$

where W = weight of constituent
p = absolute partial pressure of constituent
v = volume
T = absolute temperature
R = gas constant for a given gas

The absolute humidity is, therefore,

$$H = \frac{W_v}{W_g} = \frac{R_g}{R_v} \cdot \frac{p_v}{p_g} \tag{6–4}$$

The density of water vapor is

$$\rho_H = \frac{p_v}{R_v T} \tag{6–5}$$

The relative humidity is

$$H_R = \frac{p_v}{p_{\text{sat}}} \cong \frac{\rho_H}{\rho_{\text{sat}}} \tag{6–6}$$

The value of saturation pressure may be found in the steam tables. Further relationships between absolute, specific, and relative humidities can be found from a psychrometric chart or table.

The *dew point* is the saturation temperature of the mixture at the corresponding vapor pressure. If the gas is cooled at constant pressure to the dew point, condensation of vapor will begin. The relation between these quantities is ordinarily found from a psychrometric chart. Also formulas relating these quantities can be used.

The dry-bulb–wet-bulb temperatures are sufficient to determine humidity of a gas mixture. The dry-bulb temperature is the temperature of the gas mixture as measured by an ordinary temperature-measuring element such as a thermometer. The wet-bulb temperature is measured by a thermometer in which the bulb is covered by a wick fully wet with the same liquid existing in the vapor phase in the mixture. For air, the wet-bulb wick would be wetted by water.

The difference between dry-bulb temperature and wet-bulb temperature is called the wet-bulb depression. By means of a psychrometric chart, the absolute humidity, relative humidity, and dew point can be determined from the dry-bulb and wet-bulb temperatures. When the dry-bulb and wet-bulb temperatures are the same, the gas is saturated, this temperature is the dew point, and the relative humidity is 100 per cent.

Psychrometer Method for Moisture in Gases

The dry-bulb–wet-bulb thermometer uses the pressure-spring thermometer with two separately filled thermal systems, Fig. 6–12. The dry bulb without a thermal well is supported in the open. The wet bulb is covered by a wick and is supported in a moving stream of the gas. The wick is maintained wet by a reservoir or supply of liquid (water in the case of air). The dry bulb and the wet bulb must be mounted adjacent to each other, so that substantially the same gas stream is sampled.

A resistance thermometer with two resistance bulbs and a two-point recording instrument is often used in place of a pressure-

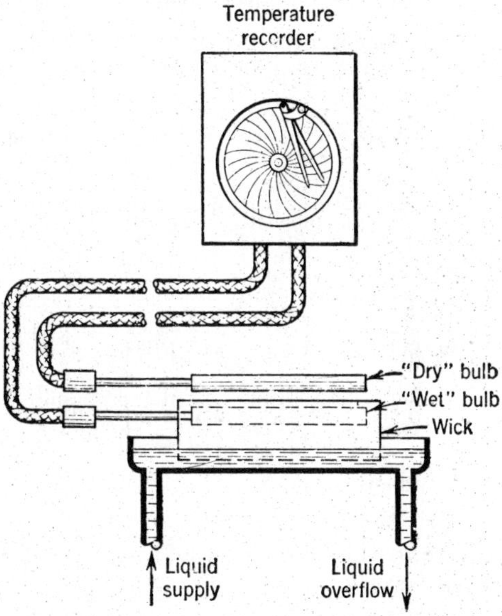

FIG. 6-12. Wet-Bulb and Dry-Bulb Thermometer (psychrometer)

spring thermometer. In other respects the use of the psychrometer of either type is the same.

Application of the psychrometer is affected by two factors. First, the dry bulb must actually measure the true temperature of the gas. The dry bulb must be installed so that radiation effects to and from the bulb are minimized. When the bulb installation is made in a duct, a radiation shield around the dry bulb may be necessary, if the duct walls are higher or lower in temperature than the gas stream.

Second, the wet bulb must be installed in a gas stream having at least 900 fpm speed. In ducts, the natural speed is often sufficient. In the open, the natural speed is hardly ever this high. In this case, a small blower is used to maintain the required velocity of gas past the wet bulb. In some cases the wet bulb may require a radiation shield.

The method of supplying liquid is important to the accurate measurement of wet-bulb temperature. The bulb should not be running wet (drenched) by a liquid spray, nor should it be too dry. In many cases, a little experimentation in the rate of supply of liquid may be required to establish the proper rate. Too little or too much liquid usually produces a positive static error. The temperature of the liquid should be the same as the dry-bulb temperature. A porous ceramic sleeve is sometimes used over the wet bulb in place of a cloth wick, especially where the cloth wick may deteriorate.

The psychrometer method is normally used for air at dry-bulb temperatures between 32 and 212°F. Although the thermometer itself may be accurate to within ½ per cent, the measurement of true wet-bulb temperature is probably not more accurate than 1 per cent unless the complete installation is thoroughly checked.

For control of moisture in gases using a psychrometric method the following facts are evident. For control of relative humidity it is necessary to measure and regulate two of the three quantities: dry-bulb temperature, wet-bulb temperature, and wet-bulb depression. There are few psychrometric instruments arranged to indicate relative humidity directly.

Hygrometer Method for Moisture in Gases

This method employs the linear contraction and expansion of a hygroscopic material such as hair, wood, animal membrane, or paper. The instrument is calibrated to read directly in terms of relative humidity.

The most commonly used elements are the hair element and the wood element. The hair element is made from a bundle of human hairs set in a fixture with one end fixed. The other end is arranged to operate directly the pointer of the instrument. The element is maintained under light tension by a spring. A wood element is simply made in the form of a prismatical bar several inches long. Applewood is often used. One end of the bar is held fixed, and the other end operates an instrument pointer directly. Another way of making the wood element is to build up wood fibers on a nickel-wire screen, the element wound in the form of a spiral as in a bourdon pressure element. The rotation of the free end of the spiral operates the instrument pointer.

The accuracy of the mechanical hygrometer is not very high. Hair and wood elements are somewhat temperature-sensitive; that is, the indication of relative humidity is affected by changes in dry-bulb temperature, even if the relative humidity is constant. Also there is some calibration drift caused by aging effects. The calibration of a mechanical hygrometer is usually checked by the psychrometer method. Mechanical hygrometers are most satisfactory for measuring relative humidity of air when the dry-bulb temperature does not vary widely and when the relative humidity is between 30 and 90 per cent. The mechanical hygrometer is slow in response; time constants of several minutes are encountered.

An electric hygrometer is the Amnico-Dunmore instrument of the American Instrument Company. The hygroscopic element is composed of a polystyrene cylinder wound with a double winding of precious-metal wire. The wires are precision-spaced and end at a pin connection in the base of the unit. The surface of the cylinder is coated with a film of partially hydrolyzed polyvinyl acetate containing a small percentage of lithium chloride. Lithium chloride is hygroscopic and gains or loses moisture until it is in equilibrium with the partial pressure of the liquid vapor in the gas surrounding the element. The electric conductivity between the wires of the double winding depends on the moisture content of the film.

By measuring the electrical resistance of the element with a form of the Wheatstone-bridge instrument, the relative humidity can be indicated. Reference to a psychrometric chart will show that relative humidity can be found if partial pressure (or dew point) and dry-bulb temperature are known. Therefore, this instrument used as a hygrometer is temperature-sensitive and requires correction for changes in dry-bulb temperature.

The advantage of the electric hygrometer is its fast response. The moisture equilibrium between the salt film on the element and the surrounding gas is reached very quickly. The electric hygrometer is accurate to within ±1.5 per cent, and there is little calibration shift with age.

Dew-Point Method

This method is used to measure absolute humidity, partial pressure of vapor, and dew point. The electric hygrometer described in the previous section can be used as a dew-point instrument. The Foxboro Company Dewcel operates on a similar principle for measuring dew point or absolute humidity. The Dewcel (Fig. 6–13) consists of a thin metal tube covered with a glass cloth saturated with lithium chloride. A double winding of silver wire is made over the glass cloth. A power supply of a

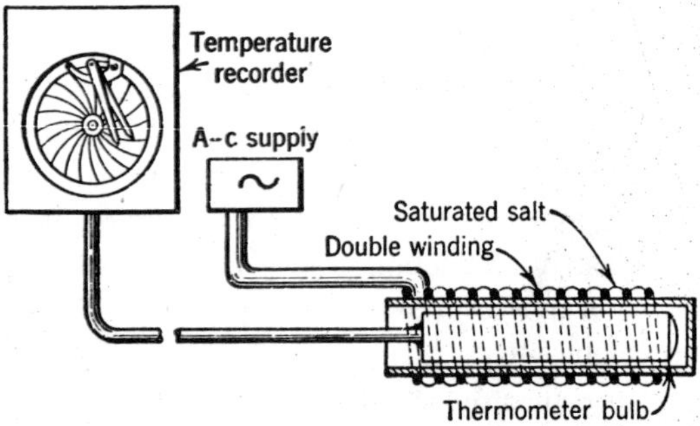

FIG. 6–13. Dew-Point Recorder—The Foxboro Company

few volts is connected to the winding. The electric conductivity between wires is directly proportional to the moisture in the salt taken on from the surrounding gas. When the moisture is low, there is a small current flow, and the temperature rise of the cell due to Joule heating is low. When the moisture is high, the current flow is high and the temperature of the cell is high. Thus, a thermometer bulb placed inside the metal tube will measure cell temperature and, therefore, absolute humidity or dew point. This instrument can be used for measuring dew point in the range −20 to 160°F. The accuracy is approximately ±2 per cent. The main advantages of this type of dew-point instrument are its simplicity and high speed of response.

A second method for measuring dew point is to determine the temperature at which moisture condenses by impinging a jet of the gas on a mirror and ascertaining the temperature of the mirror at which the surface becomes cloudy. This method is shown in principle in Fig. 6–14. Light is reflected by the mirror to the phototube which controls the heater and refrigerator. If there is no fog on the mirror, the phototube is energized, and the amplifier turns off the heater and turns on the refrigerator. This cools the mirror until a fog forms on its surface. Less light falls on the phototube, and the amplifiers turn the refrigerator off and the heater on. The cycle then continues. Thus, the temperature of the mirror cycles just above and below the temperature

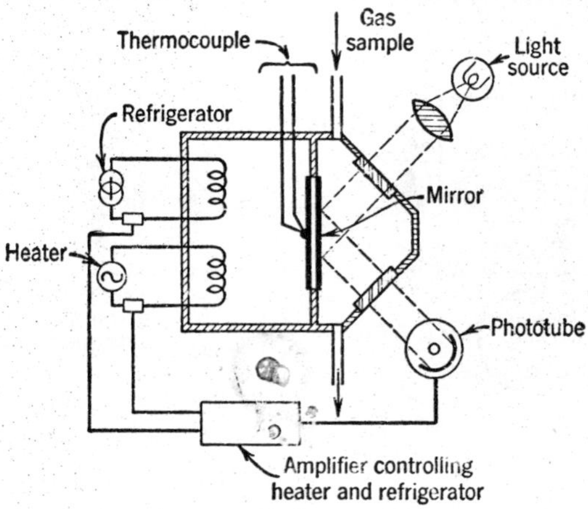

FIG. 6–14. Schematic Arrangement of Dew-Point Recorder (similar to Surface Combustion Corporation and General Electric Company units)

necessary to condense moisture from the gas. A potentiometer-type instrument is used to measure the temperature of the mirror and indicate the dew point. Gas flows continuously through the mirror chamber.

The Surface Combustion Corporation dew-point recorder operates on this principle. It includes a voltage regulator for the light source, complete cycle control for the temperature of the mirror, and a means of excluding the temperature cycle from the recorder, so that only changes in dew point are indicated. Since dirt in the gas may deposit on the mirror, this instrument includes an automatic cleaning device. It is used in the dew-point-temperature range from ambient down to $-40°F$.

The General Electric Company dew-point recorder also operates on this principle. It includes a compensating phototube to exclude variation in light source, cycle control for the mirror temperature, and periodic automatic flashing of the mirror to remove condensables from the mirror chamber. This instrument is used in the dew-point-temperature range from ambient down to −90°F.

Measurement of Moisture in Paper, Textiles, and Lumber

A major problem in drying paper and textiles is obtaining a uniform amount of moisture in the finished product. In paper

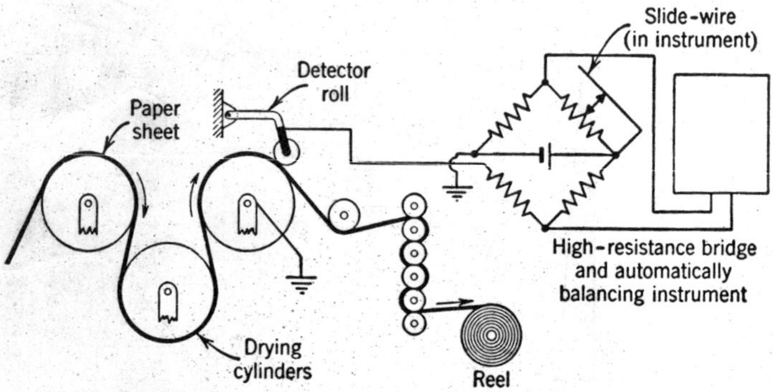

FIG. 6–15. Moisture Measurement in Paper—The Brown Instrument Company Moist-O-Graph

the amount of moisture determines the weight and quality of the paper. In textiles proper moisture produces uniform weave and better finish. Two methods for measuring moisture will be illustrated.

The Moist-O-Graph of The Brown Instrument Company operates by measuring the electrical resistance through the paper. A detecting roll (Fig. 6–15) is placed against the paper on the drying cylinder. By means of the bridge circuit the resistance across the paper is measured and indicated at the recorder, which is similar to an automatic-resistance-thermometer instrument. The resistance of the paper varies from about 25,000 ohms to 50 megohms for various papers with various moisture content. Since temperature also affects the resistance, the temperature at the point of measurement must be reasonably constant. Calibration in terms of arbitrary moisture content is established by sampling and constructing a calibration curve. Generally the calibration curve is different for each grade of paper. The paper speed may be up to 1000 fpm. The Moist-O-Graph system is also used in an entirely similar manner on textile-drying machines for measuring moisture content of yarns and cloth.

The Verigraph of The Foxboro Company is likewise used for measuring moisture content of paper and textiles. The detecting element is a rectangular box mounted with the open side in contact with the moving material. The air inside the box comes

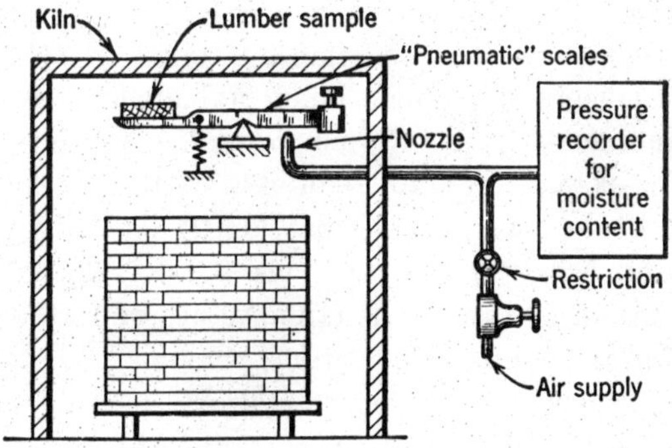

FIG. 6–16. Moisture Measurement in Lumber—The Foxboro Company Kilnboy

to a moisture equilibrium with the material moving underneath it, so that the humidity in the box is indicative of the moisture content of the moving material. A hygroscopic element is mounted in the box and consists of an electric capacitor with a hygroscopic dielectric. The variation in capacitance caused by changes in moisture content of the material is measured by a high-frequency electric system and indicated by a recording instrument. Calibration is accomplished by sampling.

One of the simplest means of determining moisture content in materials is to weigh a known volume of material during the drying operation. Naturally this method operates best for materials being dried in batch operations in ovens or kilns. One example of this method is the Kilnboy of The Foxboro Company. This device is used for measuring moisture content of kiln-dried lumber. Figure 6–16 shows a simple beam scale installed in the kiln. The scale holds a sample of the lumber to be dried. The deflection of the beam as the lumber loses moisture is detected by the simple pneumatic-nozzle system, and the resulting air pressure is indicated at the receiver. The receiver is calibrated in terms of weight (density) of the sample or is calibrated to show moisture content.

*p*H Ion Concentration Measurement

The measurement of *p*H or acid–base characteristics of liquids is necessary in many condensate systems as a check on corrosion, in pharmaceutical production for maximum yield, in food manufacture to reduce spoilage and improve taste, in ore reduction by the flotation process, and in innumerable chemical processes.

*p*H measurement is generally employed for aqueous solutions. In water, for example, a stable equilibrium at a given temperature exists between water molecules (HOH), positive hydrogen ions (H), and negative hydroxyl ions (OH). The presence of these ions in water can be determined experimentally and at 77°F is found to be 1.0×10^{-7} grams per liter. The number of hydrogen ions times the number of hydroxyl ions equals a number known as the dissociation constant, which for water is 1.0×10^{-14} at 72°F.

The *p*H scale is set up so as to avoid the handling of such small numbers as follows:

$$pH = \log_{10} \frac{1}{h} \tag{6–7}$$

where *p*H is the scale value, and h is the hydrogen-ion concentration. Thus neutral water has a *p*H of 7.0. Reference to Fig. 6–17 indicates the logarithmic quality of the *p*H scale. A change of *p*H of one unit represents a change of concentration by a ratio of 10. For example, a solution with a *p*H of 8 is 10 times more alkaline than a solution with a *p*H of 7. The practical limits of the scale are 0 and 14. A *p*H of 0 corresponds to an acid solution of unit strength, and a *p*H of 14 corresponds to a basic solution of unit strength.

The most common industrial method of measuring *p*H is by glass-cell and calomel-cell electrodes used with a potentiometer instrument. In brief, this method requires that an electrode be immersed in the solution. An electric potential is produced at the electrode which forms an electrolytic half-cell. This is the

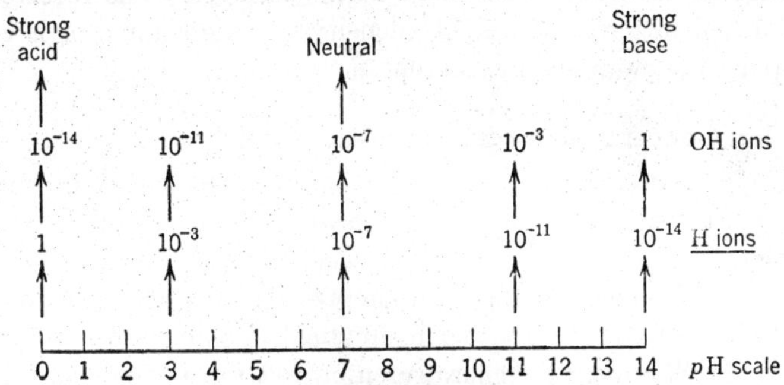

FIG. 6–17. The *p*H Scale

measuring cell. A second electrode is required to provide a standard potential and to complete the cell. This is the reference cell. The algebraic sum of the potentials of the two half-cells is proportional to the concentration of hydrogen ions in the solution.

The glass electrode is the measuring electrode in common use; it is indicated in Fig. 6–18. This electrode operates on the principle that a potential is observed between two solutions of different hydrogen-ion concentration when they are separated by a thin glass wall. This potential is a function of the two concentrations. A buffer solution is contained in the permanently sealed glass electrode, which is surrounded by the solution whose *p*H is being measured. The buffer solution in the electrode has a constant hydrogen-ion concentration. The potential at the electrode therefore depends on the hydrogen-ion concentration of the measured solution.

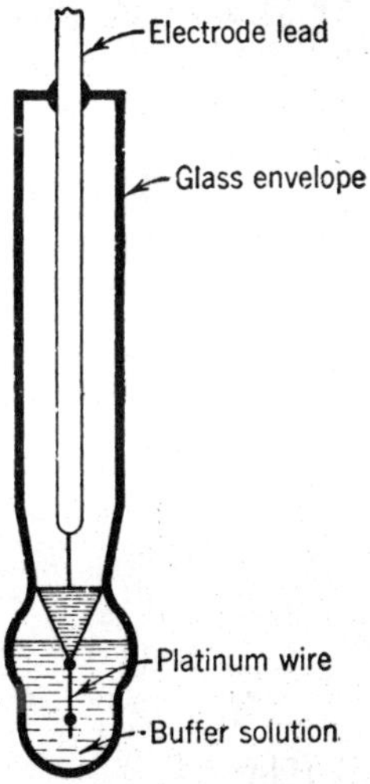

FIG. 6–18. Measuring (glass) Electrode for *p*H Measurement

The calomel electrode (Fig. 6–19) is in common use as a reference electrode. The calomel (mercury and mercurous chloride) is contained in the inner tube and covers a platinum wire. The potassium chloride, a saturated solution, is in liquid contact with the measured solution, which surrounds the reference electrode. The liquid junction is provided by a small hole in the electrode, over which a ground-glass cap is placed, the junction to the outside provided by the liquid film in the glass joint. In some cases a capillary is arranged to provide the liquid junction. Thus, the electrolyte (potassium chloride) slowly escapes from the electrode into the measured liquid. The potential at the reference electrode is constant.

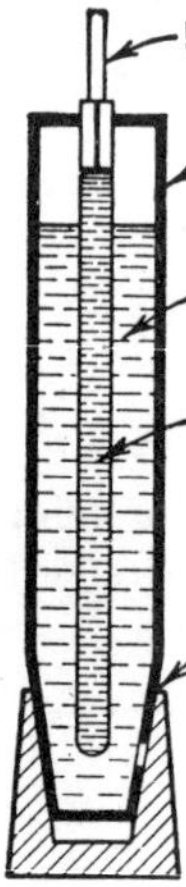

FIG. 6–19. Calomel Reference Electrode for *p*H Measurement

The operation of the electrodes depends on the electrical resistivity of the glass. Consequently, a temperature error must be compensated, since the electrical resistivity is proportional to temperature. This is accomplished by means of a resistance-thermometer element connected into the electrode circuit. Therefore, the industrial *p*H electrode assembly consists of three units: the glass and calomel electrodes and a resistance-thermometer element, all contained in a suitable housing, which is immersed in a tank or in a stream of the liquid. They can be installed directly in liquid pipe lines. The glass and calomel electrodes can be used at liquid temperatures between 34 and 212°F and up to 30 psi gage pressure.

All standard glass electrodes are subject to negative static errors when sodium ions are present in alkaline solutions. The error is serious above approximately 9 *p*H and depends on the temperature and the sodium-ion concentration. In such applications it is necessary to select a glass electrode constructed from a special-type glass.

The cell potential is measured by means of a potentiometer-type instrument operated from an amplifier connected to both electrodes. The amplifier is required because the cell potential is very small and because practically no current through the cell can be allowed.

Other measuring cells besides the glass electrode are infrequently used. The hydrogen electrode is a laboratory standard but is susceptible to contamination and requires a continuous supply of pure hydrogen. The quinhydrone electrode is limited in usefulness because it requires escape of quinone and hydroquinone into the measured solution. The antimony electrode has some uses but is subject to contamination and serious errors in some solutions.

Special Methods of Analysis

This section includes several methods for composition analysis. Although some of them are in very common industrial use, they are generally limited to specific kinds of analysis of particular substances.

Heat of combustion methods are used to determine the presence of combustible gases of any nature by burning the gas at a filament and measuring the temperature rise of the filament. Carbon monoxide and hydrogen gases near furnaces, internal-combustion engines, and gas plants are found in this manner.

The cell of the gas analyzer contains a catalyst which acts to promote combustion (or oxidation) of the combustible components of the gas. A platinum filament employed in a bridge circuit (as in Fig. 6–11) is used to measure the heat liberated by combustion. In other types a thermopile is used to measure the heat of combustion. In either case the instrument, a resistance-bridge or potentiometer indicator or recorder, can be calibrated in terms of per cent of combustibles present. The gas sample is continuously fed to the analysis cell either by an aspirator pump or by a turboblower. The gas sample is usually passed through a dryer to remove moisture.

Chemical methods may be employed for detecting many gases. The method will, of course, vary with each gas. Two examples are the measurement of concentration of hydrogen sulfide and the measurement of carbon dioxide content of combustion gases.

The Rubicon Company analyzer for hydrogen sulfide is shown in principle in Fig. 6–20. The paper tape is treated with lead acetate and other chemicals and is arranged to pass through the gas chamber. A light and photocell are arranged to measure the reflectance of the paper tape before and after the tape passes through the gas chamber. If hydrogen sulfide is present in the gas chamber, lead sulfide (brown in color) is deposited on the tape, thus lowering its reflectance. The amount of hydrogen sulfide present is proportional to the loss in reflectance of the tape. The two photocells are connected in a balanced bridge circuit, and an indicator or recorder can be operated. Accessory

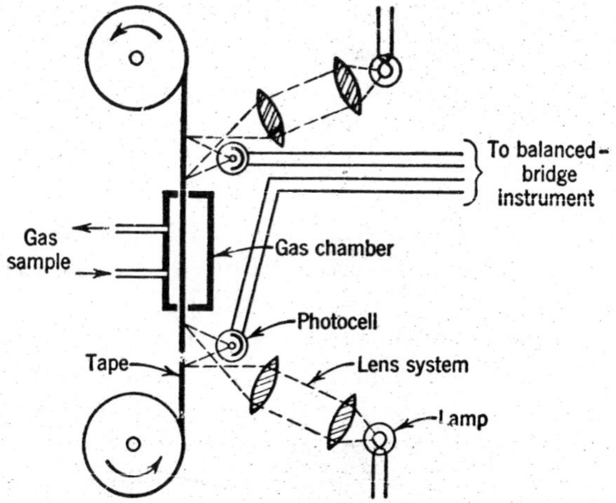

FIG. 6–20. Hydrogen Sulfide Detection System—Rubicon Company

equipment is used to supply the gas to the chamber at a constant flow rate, pressure, temperature, and moisture content.

The *automatic Orsat* is used as a carbon dioxide meter. Briefly, this device is operated on a continuous cycle to extract a given volume of the gas, which has been washed, cleaned, and temperature-controlled. The gas is then passed through a solution of caustic potash where the carbon monoxide is absorbed. The remaining gas then passes to a displacement meter, usually a bell-type oil-sealed gage, and the volume of the remaining gas is the difference between the given volume and the carbon dioxide content.

The *magnetic-susceptibility method* of gas analysis is employed mainly in measuring oxygen content of gases. Oxygen exhibits a relatively strong paramagnetic effect; that is, it is polarized in a magnetic field. Of most common gases, only nitrous oxide and oxygen have appreciable paramagnetism. The Hays Corporation oxygen meter is shown schematically in Fig. 6–21. Both the measuring cell and the comparison cell contain identical resistance wires. Thus, thermal conductivity and convection effects are balanced. The measuring cell, however, is situated in the field of the permanent magnet. The gas sample passes across the bottom of the two cells. The gas is drawn into the sample tube by either an aspirator pump or a centrifugal blower. If the gas contains oxygen, the latter is attracted to the

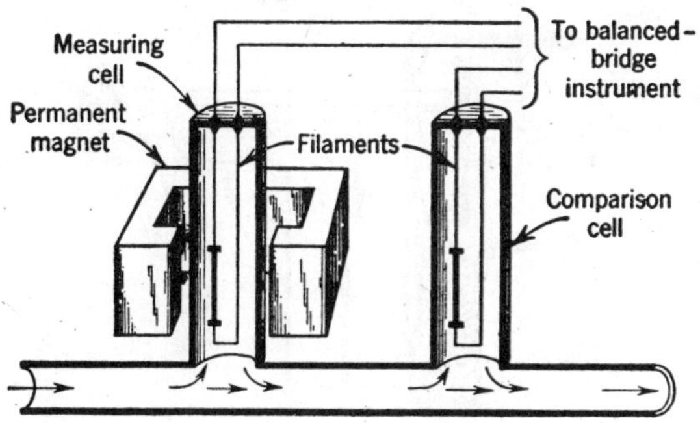

FIG. 6–21. Oxygen Analysis Cell Based on Paramagnetic Effect—Hays Corporation

magnetic field in the measuring tube and comes close to the heated filament. As the oxygen is heated, its polarity becomes less strong, and the molecules are pushed down into the sample tube by more incoming oxygen. The greater the amount of oxygen in the sample, the stronger the circulation into the measuring cell becomes. The measuring filament is therefore cooled, and its resistance is proportional to the amount of oxygen. The two cells are arranged in a balanced-bridge circuit for operation of an indicator or recorder.

The *polarizing-cell method* (Mine Safety Appliances Company) for measuring oxygen content of gas mixtures employs a primary cell with one metal electrode and one hollow carbon electrode in an electrolyte. The hydrogen given off at the carbon electrode causes polarization of the cell which decreases the cell emf. The gas sample is passed to the hollow carbon electrode, and oxygen in the sample diffuses through the carbon to combine with the hydrogen. This results in less polarization of the cell and increases the emf of the cell. Thus, the amount of oxygen present in the gas sample is proportional to the emf of the cell. The cell is surrounded by a thermostatic bath to eliminate temperature variations. The emf of the cell is measured by a self-balancing potentiometer instrument.

The *polarograph* is useful in analyzing substances that are subject to electrolytic oxidation and reduction. This includes analyses for nearly all metals in nonferrous alloys, steels, magnesium alloys, zinc alloys, and ores. Briefly, the polaro-

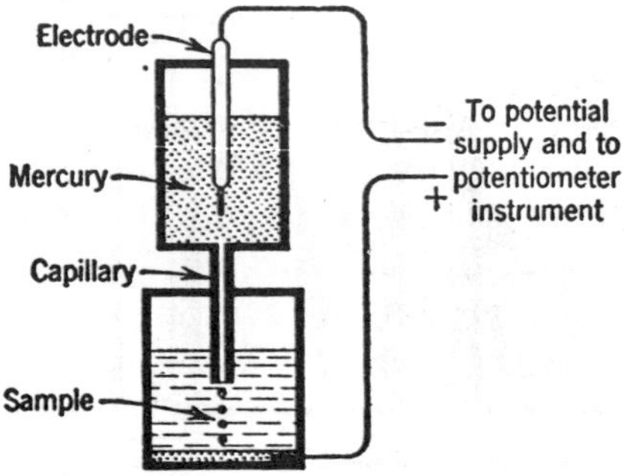

FIG. 6–22. Schematic Arrangement—Polarograph

graph method (Fig. 6–22) consists of dropping mercury into the solution containing the substance analyzed and recording the current passing through the electrode. The voltage across the electrode begins at 0 and is gradually raised to about 5 volts. The current rises in steps because of the attraction of ions of various components to the electrode. The polarogram is a plot of current versus voltage. The voltage of each step shows the presence of certain elements, and the height of the steps is proportional to the concentration of the element.

The *dilatometer* is an instrument for recording the temperature–time curve and temperature–expansion curve for a sample of metal being heated and cooled. By the shape of the temperature–dilation curve at critical points it is possible to determine the composition of steel within a reasonable degree of accuracy. The equipment consists of a fixed platform, on which is placed a sample of steel rod about 3 in. long. Expansion and contraction of the sample is transmitted mechanically to a recording chart by the differential motion of a quartz rod placed against the sample and a quartz rod placed against the platform.

The *refractometer* is an instrument for measuring the index of refraction of substances. The refractive index is a physical characteristic of substances in which light rays passing from air into the substance are deflected: that is, refracted from its continued path. By this means different substances can be identified, because every substance has its own particular index of refraction. This method is commonly used for composition analysis of gases and liquids, where the number and identity of components are known but the concentrations are not. A linear relationship between refractive index and concentration of two components in a mixture is generally obtained. If more than two components are present, then it is generally necessary that all components but two be fixed in proportion.

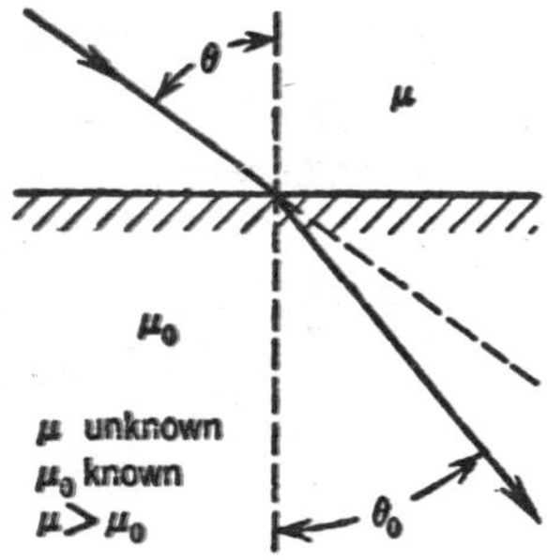

FIG. 6-23. Refraction of a Light Beam

Consider, for example, a mixture of water and sucrose (cane sugar). The index of refraction of water at 68°F is 1.333. The index of refraction of an 85 per cent solution of sucrose in water is 1.503. At other concentrations the index is nearly linear in proportion. Thus, by measuring the index of refraction the concentration of sugar solutions is obtained. Other applications of this method are found in the oil, fat, and food industries. It is also used in the alcoholic industry.

In the refractometer the index of refraction is measured at the critical angle, as shown in Fig. 6-23. The index of refraction is

$$\mu = \frac{\sin \theta}{\sin \theta_0} \tag{6-8}$$

and the critical angle is given by the particular angle of incidence θ for which reflection takes place at the interface. Thus, for any angle of incidence θ, the refracted rays are confined to the critical angle θ_0. By measuring the critical angle θ_0 the index of refraction can be found. For example, if a telescope is directed along the critical angle toward the interface, a sharp division between the light and dark field of view is seen.

The refractometer is usually a manually operated instrument, although it can be made automatic by employing a light-sensitive photocell in place of the telescope. The temperature coefficient of index of refraction is appreciable, and, therefore, the optical system of the refractometer should be enclosed in a thermostatically controlled housing. The refractometer generally employs monochromatic light in order to avoid dispersion caused by the variation of refractive index with wavelength.

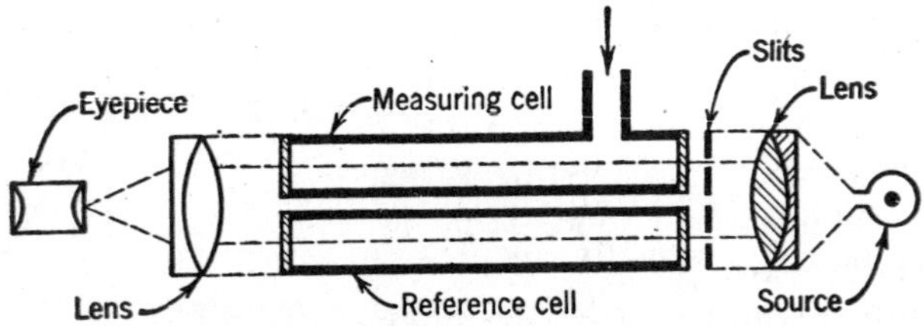

FIG. 6–24. Rayleigh Interferometer

The *interferometer* operates on the principle of superposition of light waves. The optical system of one instrument is shown in Fig. 6–24. The source may be a mercury-arc lamp, and its light is directed to the collimating lens. The light then falls on the two slits, and at this point a filter eliminates all light but the mercury green line. The two parallel beams then pass through the cells and through the lens and are focused at the eyepiece. If the fluids in the measuring cell and the reference cell are the same, then the optical path of each beam is the same. Thus, the vibrations of the light waves focused at the eyepiece are in phase, and the light beam is observed through the eyepiece. If the composition of the fluid in the measuring cell is different from that of the fluid in the reference cell, the optical-path length is changed if the index of refraction changes. The optical path length is μL where μ is the index of refraction of the fluid and L is the physical length of the path. Because the optical path is longer, there are more wavelengths of light in the cell containing a fluid of higher index of refraction. Thus, if the optical paths are different by an equivalent of one-half wavelength of the light used, the light waves will be out of phase at the eyepiece, and no light is observed. For a shift of each full wavelength, alternate light and dark fringes are observed at the eyepiece. The relationship for the difference in index of refraction is

$$(\mu_m - \mu_r) = \frac{n\lambda}{L} \tag{6-9}$$

where μ_m = index of refraction of measured fluid
μ_r = index of refraction of reference fluid
n = number of fringes shifted
λ = wavelength of light used
L = path (cell) length

This method of measurement is extremely sensitive, and changes in refractive index of 1 ppm in liquids and one part per 10 million in gases are not at all difficult to determine. The measuring cell for liquids is usually about 1 cm long, and the measuring cell for gases is about 100 cm long. The reference cell is generally of the same length and is filled with a fluid of desired composition for reference to the measured fluid.

Problems

6–1. Air absorbs strongly at and below 1900 A. In what type spectrometer would this be a serious effect?

6–2. Fluorite absorbs strongly at and below 1250 A. At these wavelengths mirrors would be substituted for lenses in a spectrometer. What could be used in the place of a prism?

6–3. Explain briefly why a prism refracts light.

6–4. On a photographic positive would flame spectrum lines appear as dark lines or as bright lines?

6–5. The amount of light absorbed in passing through a translucent substance is proportional to the amount falling on a surface at a given depth from the face and also proportional to the depth of the surface from the face. From this statement derive Beer's law.

6–6. A thermal-conductivity meter is to be used for measuring the concentration of methane in a mixture of methane, hydrogen, and nitrogen. The nitrogen concentration is constant at 15 per cent by volume. Find an expression relating thermal conductivity of the mixture to concentration in per cent of methane, if a linear combination is assumed. Thermal conductivity: methane, 1.27; hydrogen, 6.98; nitrogen, 0.995, relative to air at 32°F.

6-7. In measuring relative humidity by the psychrometer method the dry-bulb temperature is 100°F and the wet-bulb depression is 10°F. Find the relative humidity and absolute humidity. What is the dew point?

6-8. An oven-drying operation is sensitive to the amount of moisture contained in the incoming air to the dryer. What instrument would you suggest be used to indicate this effect?

6-9. What methods of analysis can be used for determining composition of:

(*a*) Steels.
(*b*) Hydrocarbon mixtures.
(*c*) Mixtures of ammonia in air.

6-10. List the methods of composition analysis that provide continuous indication of composition.

6-11. Give an example showing the use of *p*H measurement to indicate composition of a liquid solution. For the example chosen find the relation between composition and *p*H value of the solution.

6-12. Specific gravity can be used to indicate composition. For example, the specific gravity of toluene is 0.882 and of carbon disulfide 1.293. For a column of the liquid mixture 5 ft high, find the static pressure in inches of water at the bottom of the column versus the per cent by volume of toluene.

6-13. Dalton's law of partial pressures is important in humidity measurements. State this law. How is pressure of a gas mixture determined by this law?

6-14. Describe the heat-transfer process at the wet bulb of a psychrometer, and thereby deduce what the temperature equilibrium at the wet bulb means.

6-15. An arc spectrum obtained in an experiment shows strong lines at 2852.129, 2881.578, 3247.54, 3961.527, and 4030.755 A. What was the substance analyzed, and what elements did it contain? Are these emission lines or absorption lines? (*The Handbook of Chemistry and Physics* gives data on spectra.)

6-16. On a day when the barometric pressure is 28.60 in. Hg, the dew-point temperature is 50°F, and the dry-bulb temperature 70°F, compute the relative humidity.

7· Mechanical Measurements

The measurement of quantities involving the fundamental units of length, mass, and time is often required in industrial processing and manufacturing. Combinations of these three fundamental quantities occur as displacement, velocity, acceleration, work, power, and energy.

The *fundamental unit* of *length* is the foot. The standard unit of length is the yard which is defined by reference to the international meter. A prototype of the international meter in the form of a platinum–iridium bar is deposited at the National Bureau of Standards. The relation between the yard and the meter is

$$1 \text{ yard} = \tfrac{3600}{3937} \text{ meter}$$

The meter is nearly equal to one millionth of the distance at sea level on the surface of the earth from the equator to a pole.

The *fundamental* and *standard unit* of *mass* is the pound avoirdupois. It is defined in relation to the prototype international kilogram at the National Bureau of Standards as

$$1 \text{ lb} = 453.5924277 \text{ grams}$$

The kilogram is as closely as possible equal to the mass of a cube of pure water at 4°C and 760 mm Hg pressure, the sides of the cube measuring $\frac{1}{10}$ meter.

Mass, weight, and force must be carefully distinguished. Mass is defined by a quantity of matter as is illustrated by its definition. *Weight* is the pull on a quantity of matter expressed in standard gravitational force units. The latter sense is used in weighing.

The *gravitational unit* of *force* is defined as the force required to give a unit standard mass an acceleration equal to that produced by the local gravitational pull of the earth. However, local variations in the gravitational constant occur, and the local value of a force as determined on a scale must be corrected by

$$f = \frac{gw}{g_0}$$

where f = value of force, pounds
g = local acceleration due to gravity
g_0 = standard acceleration due to gravity = 32.1739 ft/sec^2
W = weight measurement of force, pounds

In most engineering work the variation of the gravitational constant g can be neglected. The greatest variation in the gravitational constant g in the United States and Canada is 0.352 per cent. Variation with altitude also occurs.

The *dynamical unit* of *force* is defined as the force required to produce an acceleration of 32.1739 ft per sec per sec on a unit standard mass. This force is measured in pounds. Another *dynamical unit* of *force* is defined as the force required to produce an acceleration of unity on a unit standard mass. This force is measured in poundals. Either dynamical unit of force is invariable and is not affected by the value of local gravitational constant.

The *fundamental unit* of *time* is the second and is, by definition, 1/86,400 part of the mean solar day. The standard of time at the National Bureau of Standards is the Riefler clock, a totally enclosed clock with a compensated pendulum.[1] The Riefler clock is corrected by astronomical observation. A time signal, accurate to within 0.02 sec per day, is sent out through the Arlington Radio Station at the U. S. Naval Observatory.

Displacement Gages

Measurements of distances and displacements can be made by employing rules and tapes. Distances or displacements in fractions of an inch can be determined with dial gages, micrometer

[1] See also H. Lyons, "The Atomic Clock," *Instruments,* **22,** no. 2 (Feb. 1949).

calipers, and vernier calipers. These methods are essentially manual. Automatic methods of measurement of displacements are, however, in common industrial use.

The *pneumatic displacement gage* is shown in principle in Fig. 7-1. Although its most popular use is in connection with go and no-go measurements in the production of manufactured parts (inspection), it is also convenient in making continuous measurements of small displacements such as might be required in measuring thickness of moving sheet or strip materials. The

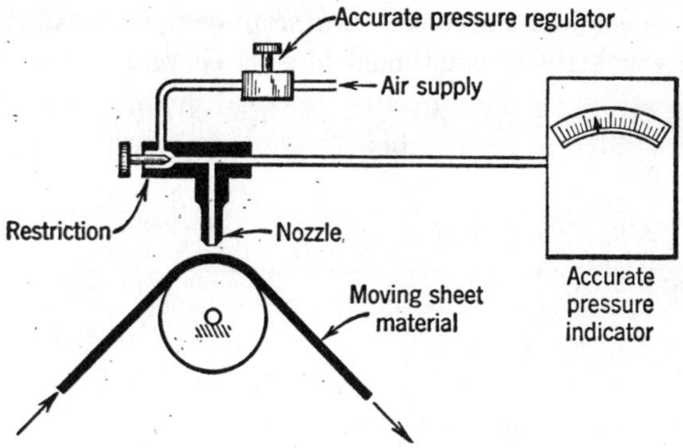

Fig. 7-1. Pneumatic Displacement Gage in Principle

operation is that of an open nozzle, calibrated for nozzle back pressure versus distance of obstruction in front of the nozzle. The nozzle tip need not touch the material passing near it. The change in nozzle back pressure caused by covering and uncovering the nozzle is about 2 to 10 or 20 psi. The nozzle back pressure may be measured with a mercury manometer or with an accurate bourdon-tube gage. The indicator is usually calibrated in 0.001 in. or similar units. Another method of calibrating the pneumatic gage is to measure the air flow through the nozzle for various distances of the obstruction in front of the nozzle. With this method a small air-flow indicator (a rotameter) is inserted in the nozzle line. The height of the float in the rotameter is read on an adjacent scale calibrated in 0.001 in. or smaller units.

With the pneumatic system, magnification up to 100,000 can be achieved. That is, displacements in the region of 0.00001 in. may be measured, and displacements as small as 0.000001 in.

can be detected. The maximum measurable displacement is generally about 0.010 in., although linkage arrangements can be made to allow greater displacements to be measured. The accuracy of the pneumatic system, once it is calibrated, depends on the accuracy of operation of the supply pressure regulator, and regulators for this service are available. The advantages of the pneumatic-gage method are that it is simple and inexpensive, does not require physical contact in performing the measurement, and does not impose a force on the material or body on which it is used.

Several electrical methods [2] are employed in measuring displacements. All such methods depend on converting the displacement into a change of some electrical quantity such as voltage or resistance so as to employ an electric instrument for indication.

The *electric displacement gage* can be used for measuring thickness of moving sheet or strip materials, as shown in Fig. 7–2. The roller of the gaging rod rests directly on the moving sheet with a force of about 5 to 10 oz. As the gaging rod moves vertically, the armature is moved closer to one or other of the coils. This unbalances the circuit, and the resulting a-c unbalance is rectified to a direct current which can be indicated at the microammeter. The ammeter is then calibrated in terms of thickness of the moving sheet. A record of sheet thickness can be obtained by using a recording instrument.

The electric gage makes possible magnification up to about 11,000; that is, measurement of displacement as small as 0.00001 in. can be made. Displacements as large as 0.10 in. can be measured. The gage arrangement shown in Fig. 7–2 can be used in continuously measuring the thickness of steel or other metal strip passing over rolls. For use on thin, nonmagnetic, conductive materials, an arrangement is made whereby the armature is omitted and the strip passes directly between the coils. For measuring thickness of paper sheet, the two coils are mounted in a detecting head in such a manner that the face of both coils is held in close proximity to the moving sheet at a point where the sheet passes over a steel roll. Variation of

[2] See *Precision Measurement in the Metal Working Industry,* Vol. 2, Syracuse University Press, 1944.

thickness of the paper sheets moves the coils *axially* with respect to the steel roll, and the microammeter is calibrated in terms of sheet thickness. The advantages of the electric displacement gage are that it is simple and inexpensive and its accuracy is not appreciably affected by grease films or other heavy-fluid films on the surface of the sheet.

The *electric strain gage* [3] can be used to measure displacement if sufficient force is available to provide the displacement. The

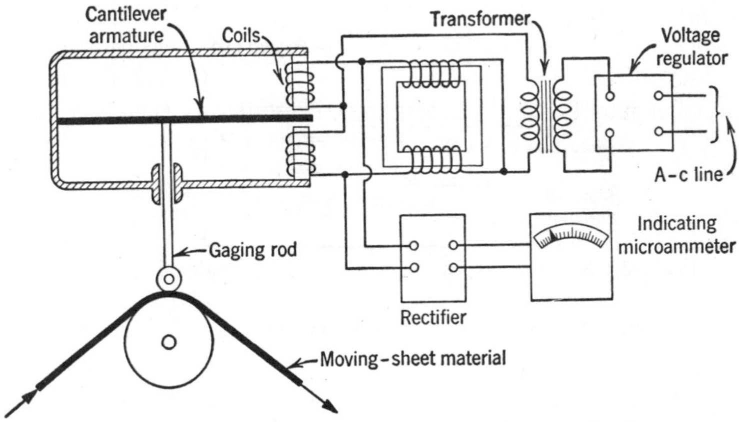

FIG. 7–2. Electric Displacement Gage—Pratt & Whitney

strain gage is very commonly used to determine strain in a member under load. One form of the strain gage, called the unbonded type, is shown in Fig. 7–3 representing the Transducer of Statham Laboratories. The fixed frame and the movable armature are mechanically connected by four sets of wire elements under initial tension. When a force acts on the armature, a displacement results, and two of the wire elements are elongated while the other two become shortened. The displacement of the armature is proportional to the force applied and to the modulus of the wire material. The change in electrical resistance of the elements is proportional to the displacement, and the unit can therefore be calibrated in displacement units versus resistance. The unit can be arranged to operate with an indi-

[3] See M. M. Frocht, *Photoelasticity,* Vol. 1, John Wiley & Sons, 1941, and J. P. Den Hartog, *Strength of Materials,* McGraw-Hill Book Co., 1949.

cating Wheatstone-bridge instrument very similar to that used for the resistance thermometer. Ambient-temperature changes have little effect, because the unit can be connected into the bridge circuit in such a manner that the thermal change of resistance of each element opposes the other. The calibration of the unit is affected by the slight change in elastic modulus with temperature. The nominal resistance of the unit may be from about 50 to 1000 ohms. Displacements up to about 0.002 in. may be measured. The force required at this displacement is from ½ oz. to 4.0 lb.

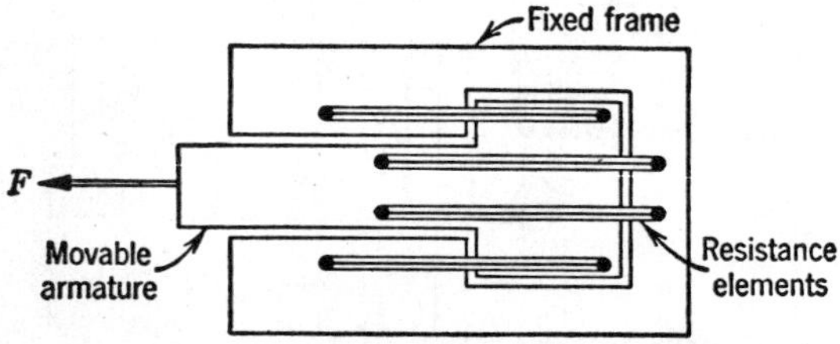

FIG. 7–3. Electric Displacement Gage or Unbonded Strain Gage

The bonded form of the electric strain-gage element consists simply of a resistance wire set in a cloth or paper strip and bonded by a special glue or cement to the member under load. The wire material may be Constantan, Manganin, Nichrome, or even steel, depending on the magnitude of the strain. The effective length of the gage may be from about ¼ to 1½ in., and the nominal resistance may be from 50 to 5000 ohms. Because the temperature coefficient of resistivity of all wires except Manganin is appreciable, it is necessary either to correct the strain readings if the ambient temperature changes, or, as is more common, to connect two elements into a Wheatstone bridge so as to nullify temperature effects.

The application of the strain gage can be discussed only briefly, and for greater detail the student is referred to the two references previously mentioned and to the *Transactions* of the Society for Experimental Stress Analysis. If in a member under the load, the direction of principal stress is known but the magnitude is unknown, the single element in Fig. 7–4 is used. If the direction and magnitude of principal stress are not known then the delta, or star arrangement, called a rosette, Fig. 7–4, is used. By calculation both the magnitude and direction of the stress may be found. Stresses on the inside of a sheet or plate may be found by strain measurements made with a rosette gage bonded to the outside of the sheet. This technique is very useful in pressure vessels and the like. Strains as small as 0.000001 in. can be measured. The maximum strain that can be measured depends on the type of wire and the length of the element. The calibration of a strain-gage element is usually given in terms of

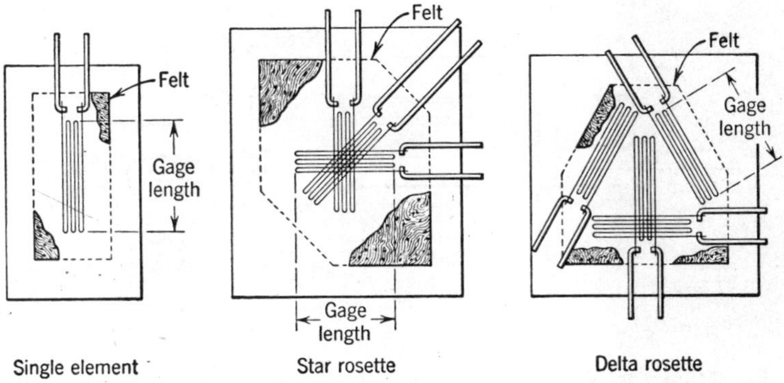

FIG. 7–4. Bonded-Type Strain-Gage Elements (*Reproduced by permission of the Baldwin Locomotive Works*)

a gage factor which equals the ratio of the per cent change of resistance to the per cent change in strain. This factor is about 2 to 4 for most strain-gage elements. The strain can be indicated by any Wheatstone-bridge instrument; for example, a resistance-thermometer instrument is used as an automatic recorder. The strain-gage element is also useful in measuring amplitude of vibrations in a member acted on by fluctuating loads, as in propellers and turbine blades. For frequencies up to 30,000 cps the strain-gage element is usually capacitance-coupled to an amplifier, and an oscilloscope is used for indication.

The *variable differential transformer* (Schaevitz Engineering), another electric device for measuring displacement, is shown schematically in Fig. 7–5. It consists simply of a magnetic iron core positioned freely in a double-wound coil. When an alternating voltage is applied to the primary coil and the armature is centered in the coils, no voltage appears at the output. When the core is displaced, the reactance of the secondary coil is unbalanced, and the alternating voltage at the output is proportional to the displacement of the core. The differential transformer is small in size and can be used for measuring displacements from about 0.001 to 2.0 in. It may be used for measuring amplitude of vibrations with frequencies up to about 20,000 cps. The input voltage to the coils must be maintained constant by a voltage regulator. When a continuous recording of displacement

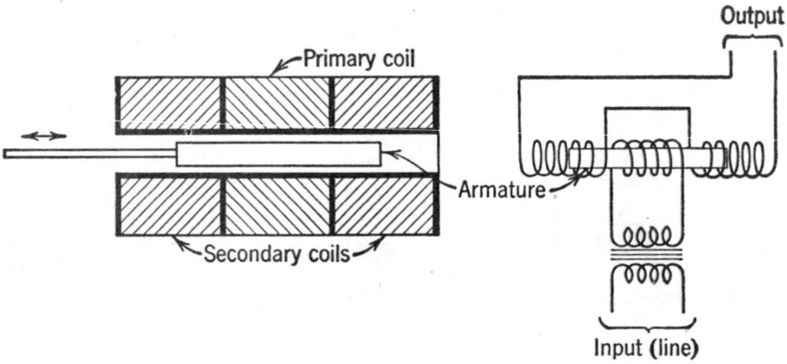

FIG. 7–5. The Variable Differential Transformer—Schaevitz Engineering

is required, the a-c output can be measured by an automatic-balance instrument in an inductance-bridge circuit. The displacement-measuring gage of Wiancko Engineering is somewhat similar and consists of two variable inductance coils mounted on a core with two air gaps. The length of one gap is increased and the other decreased by displacement of an armature. The inductance coils may be connected into two arms of a bridge. An a-c supply is used for the bridge.

The *synchro system* of measuring displacement is also useful in applications where very low power is available for operation. For example, the Servograph of Servo-Tek Products Company employs a transmitting and receiving synchro system.

The synchro system employs a synchrogenerator and a synchromotor connected as shown in Fig. 7–6. The synchrogenerator comprises a two-pole rotor with a single winding connected through slip rings to a-c line. The stator contains three coils with distributed windings 120° apart. The rotor current induces alternating voltages in each of the three stator coils, and these three voltages uniquely identify the angular position of the rotor in a 360° circle. The synchromotor operates similarly, and the angular position of its output shaft is determined by the three voltages at the stator coils. Thus, when the rotors of the motor and generator are not aligned, there is a resultant voltage at the motor stator coils, and a torque is caused on the motor rotor. When the motor and generator rotors are aligned, there is no resultant voltage at the motor stator coils, and the system is in balance. The generator and motor must be con-

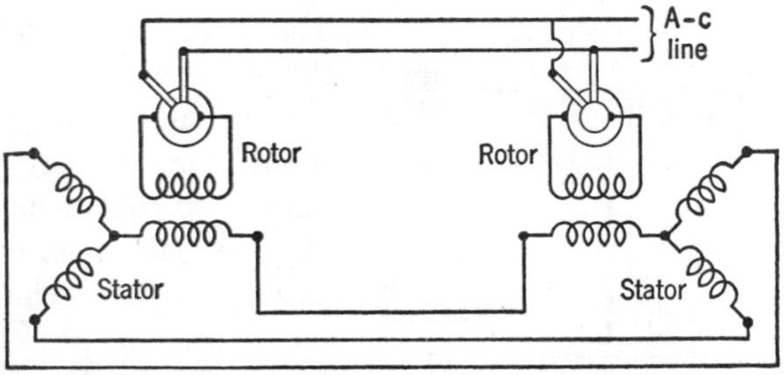

FIG. 7-6. The Synchro (selsyn) Transmission System

nected to the same a-c line, or at least to lines having the same phase and frequency.

It is evident that the synchro system is not a power amplifier, since a like torque exists at both the motor and generator. For this reason an amplifier is usually required to drive the generator by means of a reversible motor from the unbalanced stator voltages. In this case a synchrocontrol transformer may be used to determine the extent of the unbalance voltages in the stator coils of the synchrocontrol transformer and the synchrogenerator. In most synchro systems not employing an amplifier, a torque up to $\frac{1}{5}$ oz-in. may be required to operate the synchrogenerator. Measurement of angular displacement up to 360° can be made, and by suitable gearing greater displacements are measured.

Measurement of Weight, Force, and Torque

Weighing devices or scales are used for determining the magnitude of a weight or force and also in measuring torque. Scales operate on the principle of comparing the state of equilibrium of two forces. The comparison is made between the force due to gravity acting on a standard mass and the force due to gravity acting on the unknown mass. The kinds of scales are the balance, the indicating scale, and the automatic balance.

The *balance* is not ordinarily indicating because it must be manually balanced. As shown in Fig. 7–7, the balance, in simple form, consists of pans hung at equal arms from a beam. Whether the scale is used for weighing small or large masses, the unknown weight is determined by balancing against a standard mass, the weight of which is equal to the unknown weight. This assumes, of course, that the balance arms are equal. If the balance arms are not equal, this fact may be determined by balancing masses on the scale, and, when the masses are interchanged on the pans, the arm that descends is the longer of the two. The accuracy of a balance depends on using carefully calibrated masses. Any industrial laboratory should have a set of standard kilogram masses as secondary standards. In addition, it is necessary that the balance be in good condition, particularly that the knife-edges be sharp and clean.

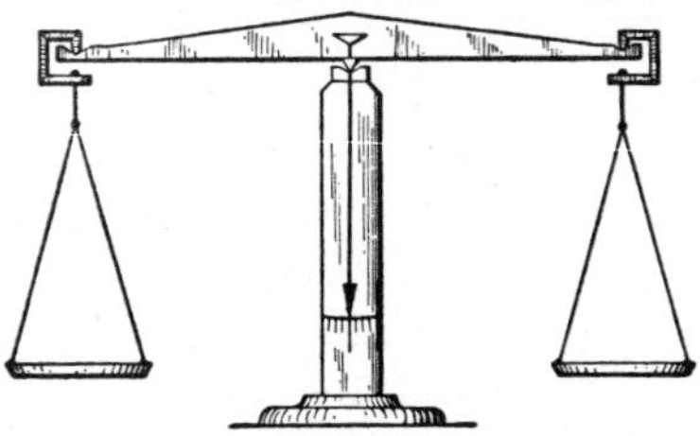

Fig. 7–7. The Balance

Most scales used in commercial and industrial work employ an arrangement similar to the balance and steelyard, as indicated by Fig. 7–8. The balance consists of the lower parallelogram-bar arrangement and is employed because the operation of the scale is not affected by the position of the unknown mass on the platform.

Beam balances and platform balances in small sizes can be calibrated with standard masses. One-pound and fifty-pound standard masses can be used for this purpose. Larger scales must usually be calibrated with a known mass of water. It is important that the knife-edges, pivots, and bearings of such scales be maintained in good condition. Lever ratios as high as 10,000 to 1 are employed in larger scales, and high accuracy of weighing depends on accurately located and friction-free bearings. Scales of this type are sometimes sensitive to ambient-temperature errors.

An *indicating scale* (spring type) is shown in Fig. 7–9. The unknown weight applied to the free end of the spring causes a

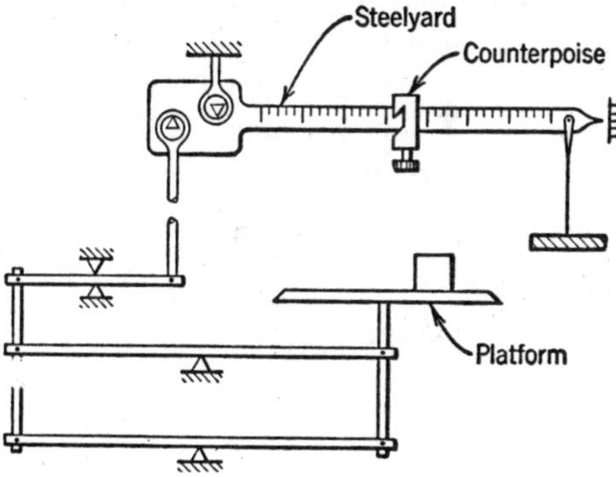

FIG. 7–8. The Commercial Balance in Principle

displacement, which is indicated by the pointer. A tape-and-drum movement is used to operate the pointer. Often a pinion-and-rack movement is employed with the rack spring-loaded to

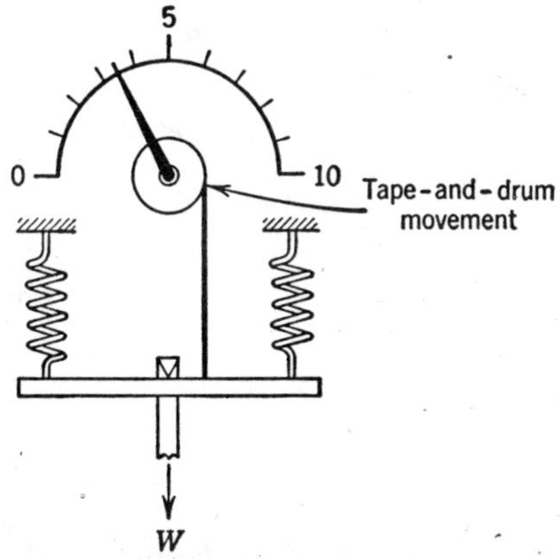

FIG. 7–9. The Spring Scale in Principle

eliminate backlash. The spring scale is calibrated, using a known mass. The accuracy of the spring scale depends on the precision and reproducibility of the spring. Most often the spring is temperature-compensated.

An *indicating scale* (variable-arm type) is shown in Fig. 7–10. The unknown weight is balanced by a known mass acting at a variable moment arm. The wheel-and-tape movement is used in order to be able to introduce a cam so as to correct the scale calibration to a linear shape. This is required, because the moment arm varies as the sine of the angle of displacement. The calibration of the variable-arm scale is not affected by local variation in gravitational constant, because the effect on both the known and the unknown mass is the same.

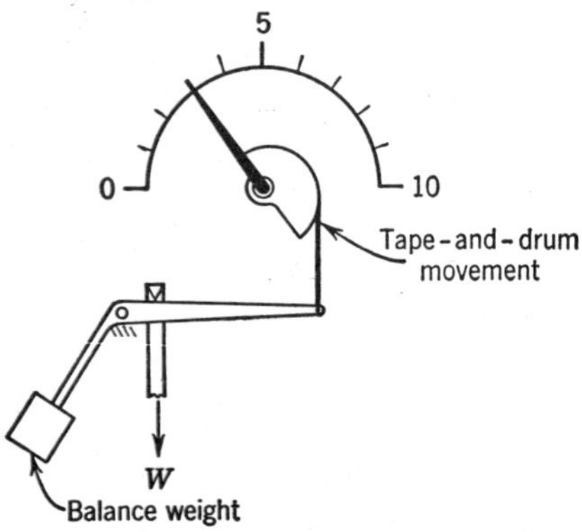

Fig. 7–10. The Variable-Arm Scale in Principle

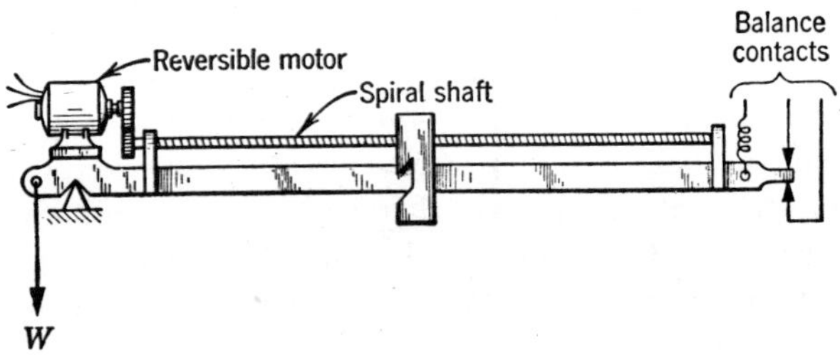

Fig. 7–11. The Automatic Balance in Principle

The automatic balance has been constructed with many different schemes of operation; one simple electrical method is shown in Fig. 7–11. An electric-contact arrangement at the end of the steelyard is so connected as to drive the reversible motor, thus moving the counterpoise along the beam by rotating the spiral shaft. When the steelyard rests at the free space between the contacts, the motor stops, and the scale may be read. Automatic balances are in common use in testing machines and in continuous-weighing devices.

Force is most commonly measured by scales and weighing devices. Several limitations of the method of measuring force are apparent. First, the force must be reasonably constant in value or else changing gradually and continuously in one direction. The ordinary balance is not well suited to the measurement of fluctuating forces and particularly forces that reverse in direction. The mass of the moving parts results in a very low natural frequency and consequent lag in accurate indication. Second, the line of action of the force must be perpendicular to the platform of the scales; otherwise, only the cosine component is measured. Third and rarely, the measurement may require correction for local variation in gravitational constant. Within these limitations, forces may be measured with reasonable accuracy. The thrust and torque output of various types of motors and engines is measured in this manner.

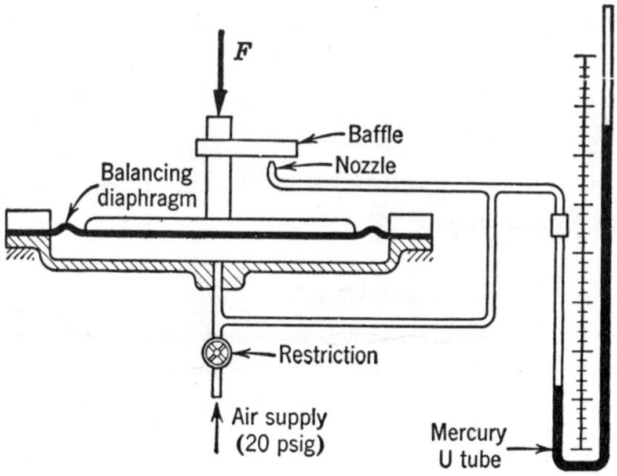

FIG. 7–12. Pneumatic-Force Meter

The force meter or thrust meter is employed for the measurement of forces varying with time in magnitude and sense. The *pneumatic-force meter* is shown in Fig. 7–12. Here the variable downward force is balanced by the upward force of the air pressure against the effective area of the diaphragm. The balance is automatic. Suppose that the force F increases. The diaphragm rod then moves downward, and the baffle covers the nozzle. The nozzle back pressure increases, and the resulting greater upward force at the diaphragm returns the diaphragm to a balanced position. The force magnitude is indicated by the height of the mercury column. The magnitude of the unknown force is determined from the effective area and the pressure at the diaphragm. Greater accuracy through null-balance operation is effected by inserting an amplifying pilot in the nozzle line. These meters may be used for measuring reversing forces by employing two such devices, one above the other. Such

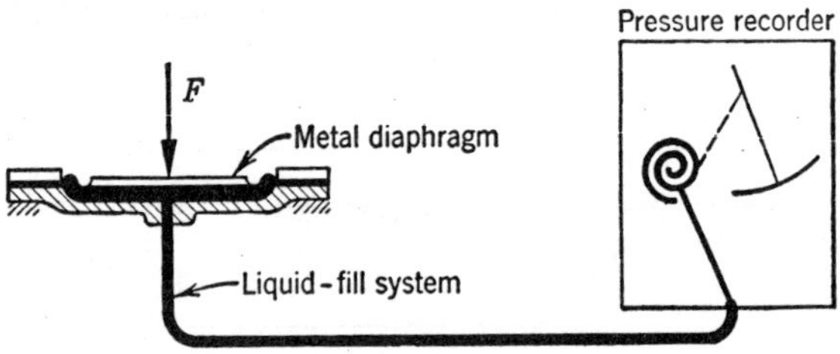

Fig. 7-13. Hydrostatic-Force Meter

meters measure forces from a few pounds up to several tons in magnitude.

The *hydrostatic-force meter* is shown in Fig. 7–13. In this system the force acts against the diaphragm, creating a fluid pressure in the diaphragm chamber. The fluid pressure equals the force magnitude divided by the effective area of the diaphragm. This pressure is then indicated by the bourdon pressure-spring gage, and the calibration can be made in units of force. The diaphragm has a sufficiently low modulus that substantially all the force is transmitted to the liquid in the chamber. The displacement of the diaphragm is very small, because the pressure spring in the indicating gage requires very little volume of fluid for full-scale motion. Such meters may be used for measurement of forces as high as 30,000 lb.

Force measurement by electrical means is made possible by converting the force into a displacement at an elastic element and then measuring the displacement. Thus the electric strain gage, either the unbonded or the bonded type, may be employed for measuring small forces, because the resistance wire itself may comprise the elastic element. The gage previously shown in Fig. 7–3 is for measuring forces from about ½ oz to 4.0 lb. Direct use of the strain-gage element as the elastic element is limited to small forces.

The *stress ring* is one form of an elastic element serving to convert the force into a displacement. The stress ring in Fig. 7–14 consists of a ring of forged steel. Forces applied across the ring cause a deflection in that direction, and the deflection can be measured by almost any of the displacement gages discussed in the previous section. Forces from 1 to 100,000 lb are measured in this manner. The change in elastic modulus with temperature is about 1 per cent for an ambient-temperature change of 50°F, so that correction may be required for wide changes in temperature. In measuring forces of smaller magnitude special alloys having negligible temperature coefficient of modulus can be employed in the stress ring. Obviously the ring is not the only structural member that can be used for an elastic element. The cantilever beam can be employed in a similar manner, and the circular shaft can be used for torque measurement. In measuring the power transmitted by shafts, it is possible to employ bonded strain gages to measure the angular twist of the shaft, thereby allowing the statical moment to be calculated.

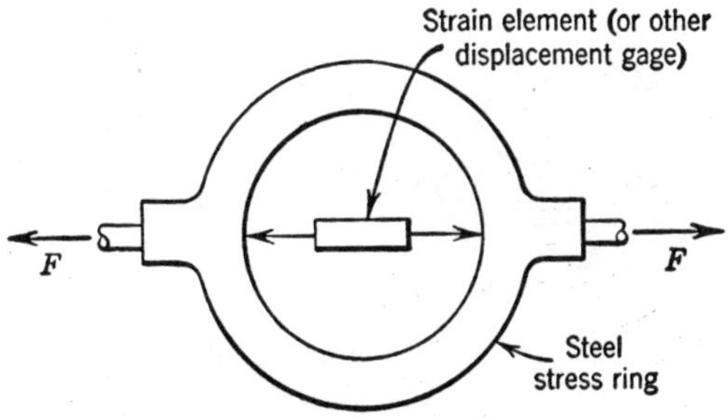

FIG. 7–14. The Stress Ring for Force Measurement

The Amsler Polar Planimeter

The planimeter is a mechanical device for measuring an area enclosed by finite boundaries. It is employed for the measurement of areas where exact calculation is not possible, either because the boundaries of the area cannot be expressed by simple mathematical relations or because the integrals involved cannot be easily evaluated. Numerical integration can be accomplished by using Simpson's rule or an equivalent method. However, the planimeter is very simple and easy to use in measuring areas on pressure-versus-volume diagrams (work), on flow-rate-versus-time diagrams (flow quantity), and on velocity-versus-time diagrams (displacement), and so on.

The Amsler polar planimeter is shown in Fig. 7–15. It consists of two rigid arms pivoted on each other at the bearing.

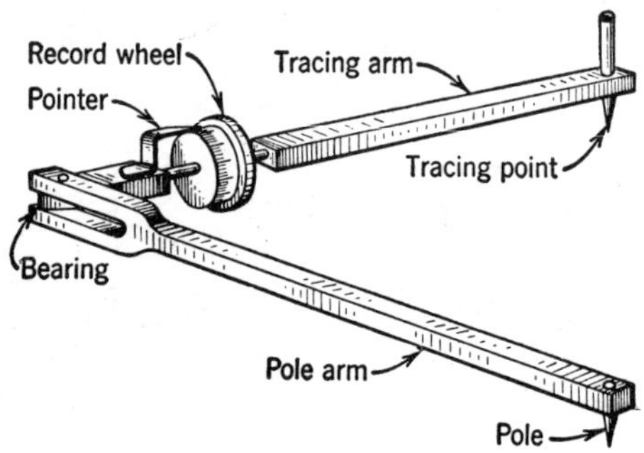

FIG. 7–15. Amsler Polar Planimeter

The record wheel is free to rotate, and the axis of the record wheel is assumed to be parallel to the axis of the tracing arm. The planimeter is used by fixing the pole at a point in the plane of the unknown area and moving the tracing point exactly once around the periphery of the area. The planimeter is supported during this operation at three points: the pole, the record wheel, and the tracing point.

Consider the area A in Fig. 7–16, and suppose that the tracing point n follows the boundary of the area A and that the inner end of the tracing arm m follows some circular arc T_1T_2. The pole is located outside the area A. Now, in an infinitesimal motion of the tracing arm from position mn to position $m'n'$, the area swept out is [4]

$$\text{Area } mm'n'n = l\,dn + \tfrac{1}{2}l^2\,d\phi \tag{7–1}$$

[4] See H. Lamb, *An Elementary Course of Infinitesimal Calculus,* Cambridge University Press, 1897.

where dn is the distance moved by the tracing arm parallel to itself, and $d\phi$ is the angle through which the tracing arm rotates. The length of the tracing arm is l. It should be noted that those areas swept out by the tracing arm in moving in a clockwise direction must be termed positive, and those areas swept out by the tracing arm in moving in a counterclockwise direction must be termed negative. This is automatically accomplished by the record wheel, since a motion of the tracing in one sense rotates

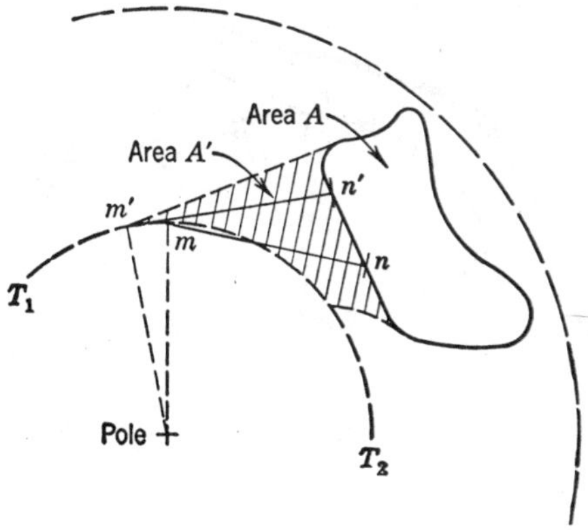

Fig. 7–16. The Polar Planimeter in Use, Pole Exterior to Area

the record wheel in one direction and a motion of the tracing arm in the opposite sense rotates the record wheel in the opposite direction. With reference to Fig. 7–16, it may easily be seen that every element of the area A' is swept out by the tracing arm once in the positive sense and once in the negative sense as the area is encircled by the tracing point. The total angle ϕ is zero, and, hence,

$$A = l\int dn + k \tag{7–2}$$

The motion of the tracing arm causes the wheel partly to roll and partly to slip over the surface on which it rests. The rotation of the wheel thus records only the motion of the tracing arm parallel to itself. Therefore,

$$d\theta = \frac{1}{\gamma} dn \tag{7–3}$$

where θ is the angular displacement of the record wheel and γ is its radius. Therefore,

$$A = l\gamma\theta + k \tag{7-4}$$

and the area A may be found by noting the total angular displacement of the record wheel.

When the pole is located inside the area A, as shown in Fig. 7-17, the total angular rotation of the tracing arm is not zero but

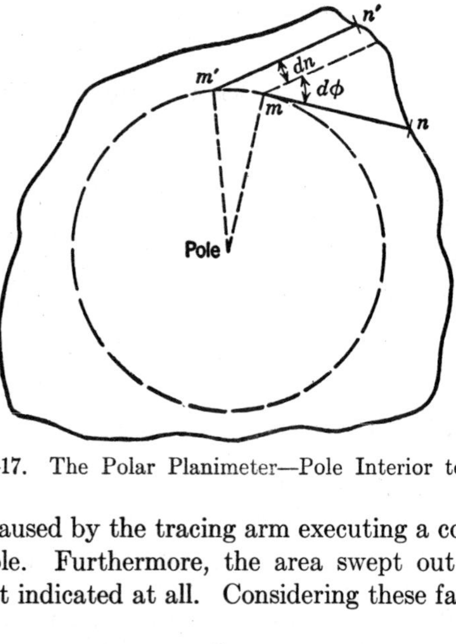

FIG. 7-17. The Polar Planimeter—Pole Interior to Area

2π. This is caused by the tracing arm executing a complete circle about the pole. Furthermore, the area swept out by the pole arm πd^2 is not indicated at all. Considering these factors,

$$A = l\int dn + \frac{l^2}{2}\int d\phi + \pi d^2 + k \tag{7-5}$$

The normal distance of motion of the tracing arm dn is recorded by the record wheel as $\gamma\, d\theta$. However, when the tracing arm executes a complete rotation, accounted for in the foregoing equation by $\int d\phi$ equals 2π, the record wheel records an additional area equivalent to $la\, d\phi$. Considering these factors,

$$A = l\gamma\theta + \pi l^2 + \pi d^2 - 2\pi la \tag{7-6}$$

We notice that the last three terms of the equation may be arranged as

$$A = l\gamma\theta + \pi Z^2 + k \tag{7-7}$$

where $Z = \sqrt{(d^2 - a^2) + (l - a)^2}$ and is the diameter of a zero circle. This circle, shown in Fig. 7–18, is the particular circle of radius Z from the pole, about which the tracing point can be moved without causing a rotation of the record wheel.

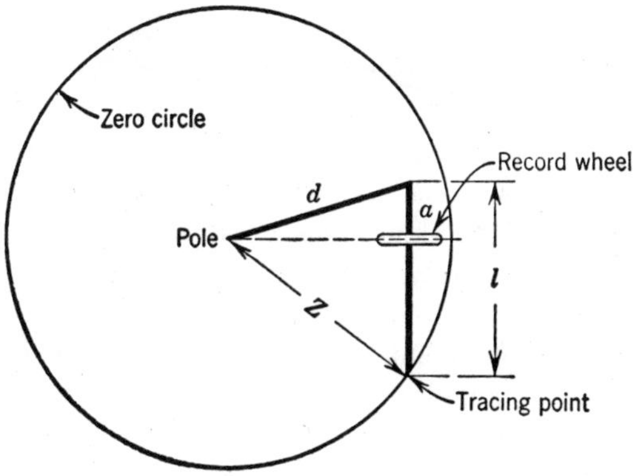

FIG. 7–18. The Zero Circle of the Polar Planimeter

Obviously, in this position there is pure slip of the record wheel. It is concluded that, when the boundary of the area being measured is wholly outside the zero circle and the pole is inside the area, the unknown area equals the area of the zero circle plus the indication of the record wheel. Also, when the boundary of the area is wholly inside the zero circle and the pole is inside the area, the unknown area equals the area of the zero circle minus the indication of the record wheel. Except in special circumstances, the polar planimeter should not be used with the pole interior to an area, the boundary of which lies on both sides of the zero circle. In this case the area is best determined by subdividing the area into partial areas.

There are several other mechanical planimeters, and most of them employ the same principles as the polar planimeter. The roller planimeter is useful in measuring areas long in length and narrow in width. Special planimeters are available for integrating the circular charts of industrial recording instruments.

The planimeter is quite accurate in the hands of a skilled operator, and measurement of area to within 1 per cent is not difficult. In interpreting the area measured by an ordinary planimeter such as the Amsler polar planimeter, it must be remembered that the coordinates of the area must be linear; otherwise, the measured area has little significance.

Velocity and Acceleration

The most accurate method of measuring speeds in the normal range is to measure the displacement in a measured time interval. The revolution counter is a device for measuring angular displacements. However, this method is not continuous. The speedometer, as employed in the automobile, measures speed but is not often used in industrial service. The air-speed indicator, as employed in the airplane, measures air speed and is simply a form of the pitot tube, which is discussed in the chapter on flow measurement.

Continuous measurement of linear velocity is nearly always made in terms of angular velocity (speed) by converting the linear motion to a rotating motion. In measuring the velocity of a moving sheet of material, for example, the sheet or strip can be made to pass between rolls, and the angular velocity of the rolls is then measured.

The d-c electric *tachometer* is universally used for measuring paper machine speeds, turbine and generator machine speeds, and conveyor-belt speeds, as indicated in Fig. 7–19. The tachometer is simply a d-c generator or magneto with a permanent-magnet field. The generated voltage is directly proportional to the speed of rotation of the armature and is usually about 6 volts at an armature speed of 1000 rpm. The generator is designed to operate between 500 and 1500 rpm for the maximum machine speed. For measuring other speeds it is necessary to use a suitable gear ratio to bring the maximum armature speed within the limits mentioned previously. The tachometer system may be employed for measuring rotative speeds between about 50 and 10,000 rpm, although theoretically any speed can be measured, as long as the machine to which the tachometer is connected has sufficient power to overcome the gear-box friction. As indicated in Fig. 7–19, the output of the tachometer is measured by an automatic-balance potentiometer. The elcctric tachometer can be used for accurate speed measurement when the brushes in the tachometer are maintained in good condition. Changes in ambient temperature usually cause an error, owing to the altered output of the tachometer.

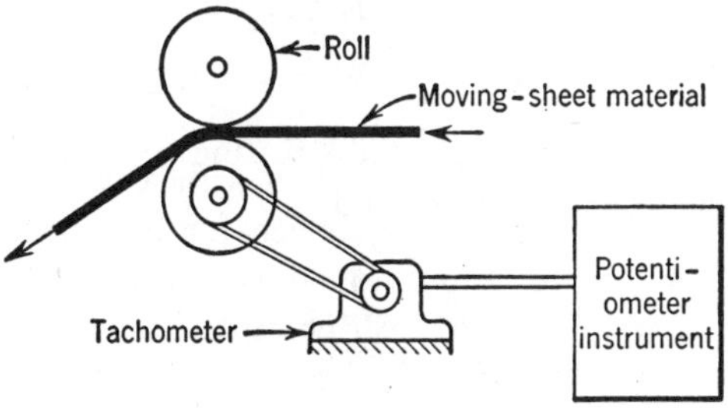

FIG. 7-19. The Tachometer for Measuring Speed of Moving Sheet Material

An *accelerometer* is a device for measuring acceleration. Such instruments are based on Newton's second law of motion relating impulse and momentum. For a mass of constant magnitude

$$F = m\ddot{x} \tag{7-8}$$

where F is a force applied to a mass m, resulting in an acceleration $\ddot{x}$ relative to axes fixed in space. It is assumed that the force acts through the center of gravity of the body.

The seismic element of Fig. 7-20 is employed in accelerometers. The mass m is supported on the frame by thin cantilever springs and is restricted to move only in the horizontal direction. Motion of the mass m is damped either by surrounding the mass with a viscous oil or by a magnetic means. To analyze the motion of the mass m, let y be the coordinate of the frame relative to fixed axes at time t and x the coordinate of the mass relative to fixed axes at time t. The equation of motion of the mass is then,

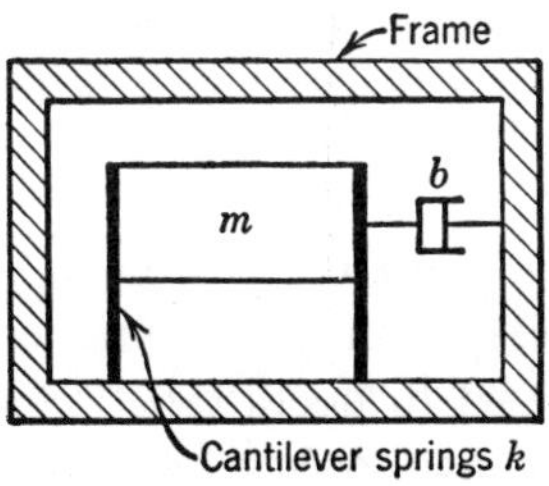

FIG. 7-20. Seismic Element of the Accelerometer

$$m\ddot{x} + b\dot{x} + Kx = Ky(t) + b\dot{y}(t) \tag{7-9}$$

where b = damping constant
K = effective gradient of springs
$y(t)$ = the motion of the accelerometer frame

Suppose that the motion of the frame is one of constant acceleration, or

$$y = \frac{\ddot{y}_0}{2} t^2 \tag{7-10}$$

where $\ddot{y}_0$ is the acceleration. Substituting in equation 7–9 and solving for the steady state, we find

$$S = \frac{1}{\omega_n^2} \ddot{y} \tag{7-11}$$

where S is the displacement of the mass relative to the frame. This relationship indicates that the displacement of the mass relative to the frame is proportional to the acceleration of the frame.

The important use of the accelerometer is in measuring acceleration as it occurs in vibrations where the acceleration is not a constant value but varies periodically in some manner. This can be investigated[5] by choosing the forcing function,

$$y = y_0 \sin \omega_f t \tag{7-12}$$

where ω_f is the forcing frequency. Substituting in equation 7–9 and solving for the steady state, we find

$$S = \frac{1}{\omega_n^2 \sqrt{(1 - \beta^2)^2 + (2\zeta\beta)^2}} \ddot{y} \tag{7-13}$$

where $\beta = (\omega_f/\omega_n)$ = frequency ratio
$\omega_n = \sqrt{K/M}$ = natural undamped frequency
$\zeta = \sqrt{b^2/4Km}$ = damping ratio

From this relationship it may be seen that the calibration of the accelerometer is different for different forcing frequencies. For this reason, and in order to be able to measure acceleration at

[5] J. P. Den Hartog, *Mechanical Vibrations,* McGraw-Hill Book Co., 1948. W. T. Thomson, *Mechanical Vibrations,* Prentice-Hall, 1948.

high forcing frequencies, it is necessary to select a high natural undamped frequency ω_n by choosing a large spring gradient K and a small mass m. Also the damping ratio should be about 0.6 of critical damping. Thomson shows that with a damping ratio of 0.6 the error is less than 0.5 per cent for a frequency ratio from 0 to 0.4.

The measurement of acceleration using the seismic element of Fig. 7–20 reduces to a measurement of displacement. Generally this is accomplished electrically. The strain element may be used for the spring, and the change in resistance of the element is proportional to the acceleration. The arrangement of this type of accelerometer is shown in Fig. 7–3 and was previously discussed as the displacement element. For low-frequency measurements the indicating or recording Wheatstone-bridge instrument can be used. For high-frequency investigations the oscilloscope or oscillograph is generally used. Accelerometers of the strain-element type can be employed up to about 800 cps and for measuring accelerations up to 500 g. Another method for measuring the relative displacement between the mass and frame is to mount a single coil on a spring-suspended core and arrange them to move inside a stationary double coil, much like the variable differential transformer previously shown in Fig. 7–5. The unbalanced voltage caused at the double coil by a displacement of the core is measured and is proportional to acceleration.

Measurement of time rate of change of acceleration $\dddot{x}$ is sometimes necessary as, for example, in determining the load-carrying qualities of vehicles. A sudden change of acceleration of a body produces a change in force acting on a body. It is this change in force that is related to the riding quality of automobiles and is important in load tiedown in aircraft, railroad cars, and trucks. The seismic element of the accelerometer (Fig. 7–20) produces a velocity of the mass relative to the frame proportional to the rate of change of acceleration. This is evident when equation 7–13 is inspected. Measurement of rate of change of acceleration, often called "jerk," can be accomplished as indicated schematically in Fig. 7–21. If a permanent magnet is employed as the seismic mass, the voltage appearing at the coil is proportional to the time rate of cutting flux lines. Therefore, the voltage at the coil is proportional to the velocity of the seismic mass. By employing an amplifier and oscillograph or other indicating means, the device can be calibrated in terms of time rate of change of acceleration.

The *vibrometer* is used for measuring amplitude of vibrations. Its operation is based on the seismic element of Fig. 7–20. If equation 7–13 is rewritten for the steady-state condition,

$$S = \frac{\beta^2}{\sqrt{(1 - \beta^2)^2 + (2\zeta\beta)^2}} y \tag{7-14}$$

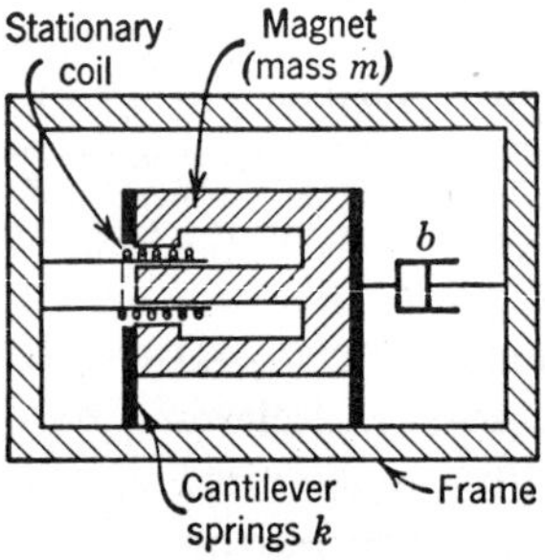

Fig. 7–21. Seismic Element for Measuring Rate of Change of Acceleration

In the vibrometer it is necessary that the undamped natural frequency ω_n of the element be made small by the selection of a small spring gradient k and a large seismic mass m. This results in the seismic mass remaining stationary when the frame vibrates at high frequencies compared to the natural frequency of the element. By employing the arrangement of Fig. 7–21 the velocity of the seismic mass is measured, and this is proportional to the velocity of the frame. This is the basis for the *velocimeter*.

Problems

7–1. A spring scale is accurately calibrated at one locality employing standard masses. The scale is then transported to another locality where the gravitational constant is different. Does the scale weigh accurately in the second locality?

7–2. The force exerted by a dynamometer arm in an engine test is measured on an accurate spring scale. Should the value of the force so determined be corrected for a variation of gravitational constant? Why?

7–3. The pneumatic-force meter in Fig. 7–12 operates with atmospheric pressure over the diaphragm. Would an error be caused by a change in barometric pressure?

7–4. In the pneumatic-displacement gage of Fig. 7–1 is the calibration changed by a change in barometric pressure? Assume that the nozzle back pressure varies between 2 and 10 psig.

7–5. If a strain-gage element is used with a gage factor of 2.0, a resistance of 120 ohms, and a length of 0.500 in., what is the change in resistance for a measured displacement of 0.005 in.?

7–6. Sketch a simple Wheatstone-bridge circuit arranged to provide temperature compensation when a strain-gage element is used for measuring displacements. Is the change in ambient temperature along the lead wires important?

7–7. Explain each of the terms in equation 7–6. If the record wheel of a polar planimeter is located at the midpoint of the tracing arm, is this equation simpler in form?

7–8. Suppose that a polar planimeter is physically constructed so that the tracing point can be brought very close to the pole. Now, if the planimeter is rotated around the pole, the record wheel is given an angular displacement and therefore indicates an area, but the area encircled by the tracing point is negligible. How is this explained?

7–9. Derive completely the equation of motion 7–9 for a seismic element. Let the forcing function for the frame be y = a constant, and solve the equation of motion for the transient state. What is the condition for critical damping? For use as an accelerometer is this damping the best? Why?

7–10. In equation 7–11 for the accelerometer is the relative displacement s given at the same time t as the acceleration $\ddot{y}$ for a steady state? In equation 7–12 is the same condition true?

7–11. In making measurements of vibration phenomena, harmonics may occur at frequencies several times as large as the fundamental frequency. Would the accelerometer likely indicate acceleration accurately for a high harmonic? Would the vibrometer likely indicate amplitude of vibration of a higher harmonic?

8· Measurement of Pressure and Vacuum

The measurement of pressure and vacuum has always been important in continuous-processing industries. In a surprising number of chemical applications, pressure measurement is more important and is easier to perform than temperature measurement. The very rapid advance of high-vacuum techniques in industry during the past few years has opened a whole new field of pressure measurement at extremely low absolute values. It is essential that the engineer be familiar with methods that allow pressure to be measured in within 1×10^{-6} psi of absolute zero. Such measurements are made continuously with standard industrial equipment in an increasing number of industries.

Pressure, Vacuum, and Head

The pressure exerted by a fluid is given in terms of force per unit area. Therefore, the unit of pressure is not a primary quantity but is a derived unit, since it is based on the unit of force. Also pressure units are usually given in terms of force, the value of which is affected by the local gravitational constant. Ordinarily, the resulting variation in pressure units is not important, but correction may be necessary if extreme accuracy is required.

Absolute pressure is the total pressure exerted by a fluid. Differential pressure is the algebraic difference in two pressures. As a special case of differential pressure, gage pressure is the algebraic difference between the total pressure exerted by a fluid and the pressure exerted by the atmosphere. Thus,

$$p_g = p_a - p_s \tag{8-1}$$

where p_g = gage pressure
p_a = absolute pressure
p_s = atmospheric pressure existing at the time and place of measurement

Vacuum is another special case of differential pressure and is defined by

$$V = p_s - p_a \tag{8-2}$$

where V is the vacuum. It is obvious that a negative vacuum and a positive gage pressure are identical.

In order to relate absolute pressure, gage pressure, and vacuum, it is always necessary to specify the atmospheric pressure existing at the time and place of measurement. The standard atmospheric pressure or barometric pressure is taken as 14.696 psia (760 mm of Hg column) at sea level and at 0°C. The actual atmospheric pressure at the time and place of measurement of gage pressure may vary considerably from the standard atmosphere, depending on barometric conditions.

Another unit of pressure is the static pressure head or the liquid-column height corresponding to a given pressure. Head and pressure are related by

$$p = \gamma h \tag{8-3}$$

where p = pressure, either absolute or gage
h = height of liquid column
γ = density of liquid

Since the density of a liquid varies considerably with temperature, the temperature should always be given if the pressure is expressed in units of liquid-column head. The variation of density of mercury and water with temperature is shown in Table A–18 (appendix). Correction to standard conditions may be made by

$$h_s = h_o \frac{\gamma_o}{\gamma_s} \tag{8-4}$$

where the subscript o refers to the conditions at the time of observation and the subscript s to the conditions at the desired standard temperature.

A unit often used for expressing high pressures is atmospheres, abbreviated atm. This is simply the pressure in pounds per square inch expressed as multiples of standard atmospheric pressure, or

$$p_{\text{atm}} = \frac{p}{p_s} \tag{8-5}$$

A convenient unit for expressing very low absolute pressures is the micron. A micron, abbreviated μ, is 0.001 mm of Hg column absolute pressure at 32°F.

The selection of one or another of the various units of pressure or head depends on the use made of the measurement. Pounds per square inch, abbreviated psig for gage pressures and psia for absolute pressures, is the most common. Head is often expressed in inches of mercury, inches of water and feet of water, abbreviated in. Hg., in. H_2O, and ft H_2O, respectively. Conversion between various pressure units is given in Table A–19 (appendix).

Liquid-Column Manometers

One of the oldest means for the measurement of pressure is the liquid-column manometer. It is the simplest, most direct, and most accurate of all pressure-measuring means. Manometers are normally limited to measurement of differential pressures less than about 30 psi; for higher pressures they become very unwieldy.

The types of manometers in common use are: the U-tube type, the enlarged-leg type, the well type, the inclined-tube type, and the ring type.

The *U-tube manometer* is shown in Fig. 8–1. The vertical tubes (the "legs") are freely connected through a metal fitting at the bottom. The assembly is partly filled with a manometer liquid such as mercury, oil, or water. If the legs of the U tube are vertical, for static balance,

$$p_2 - p_1 = \gamma h \tag{8-6}$$

where h = difference in level
γ = density of manometer liquid
p = pressure

Here it is assumed that the density of any fluid over the manometer liquid is negligible. The manometer can be used for

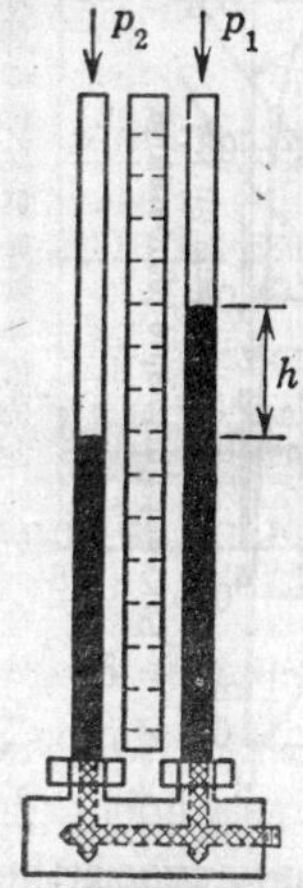

FIG. 8–1. Simple U-Tube Manometer

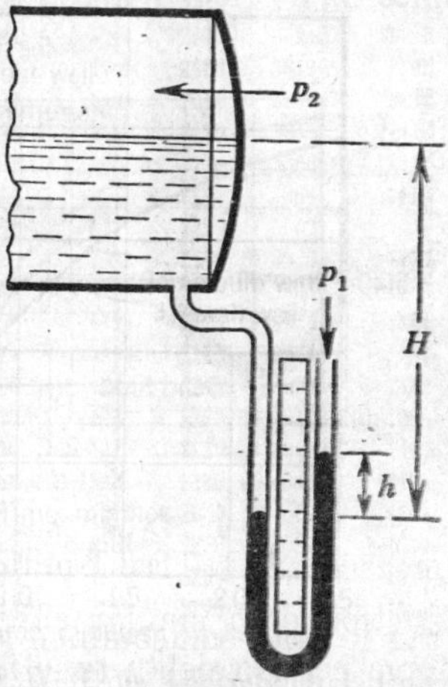

FIG. 8–2. Liquid Head over a U-Tube Manometer

the measurement of pressure differential $(p_2 - p_1)$, for the measurement of gage pressure (p_1 equals atmospheric pressure), for the measurement of absolute pressure (p_1 is zero absolute), or for the measurement of vacuum (p_2 equals atmospheric pressure).

When the fluid over the manometer liquid is of appreciable density, for example as shown in Fig. 8–2, then the additional head must be taken into account. The equation for static balance now is

$$p_2 - p_1 = h\rho - H\delta \tag{8-7}$$

where h = head of liquid over manometer
δ = density of liquid over manometer

It is evident from this relation that, when there are several immiscible liquids over the manometer liquid, the measurement of pressure using a manometer is no longer simple.

The *enlarged-leg manometer* in Fig. 8–3 is used in recording-type manometers, because it provides a means for obtaining an

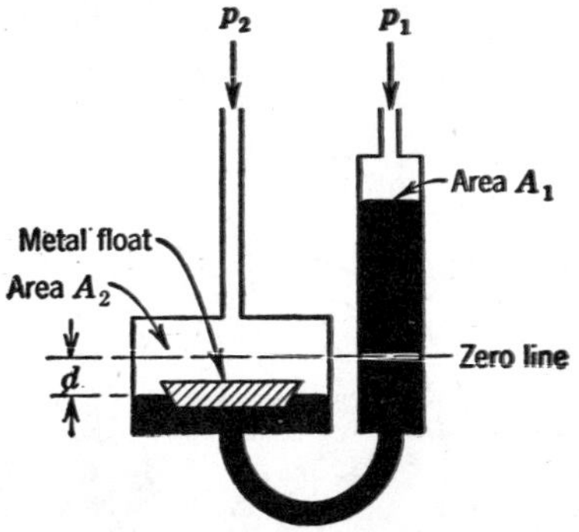

FIG. 8–3. Enlarged-Leg Float Manometer

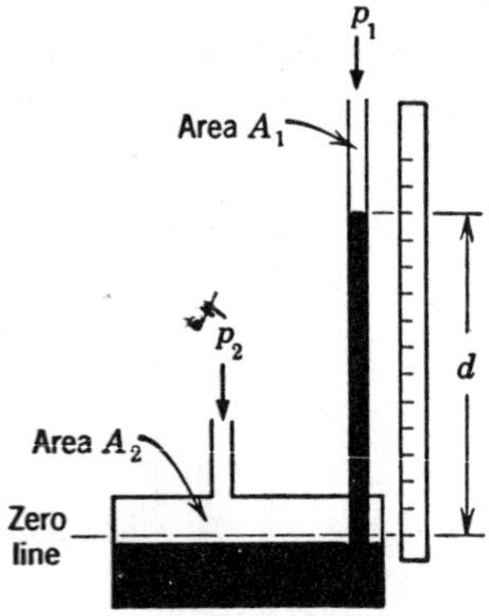

FIG. 8–4. Well Manometer

indication of the pressure as well as a means of changing the span of measurement. The equation for static balance is

$$p_2 - p_1 = \gamma\left(1 + \frac{A_2}{A_1}\right) d \tag{8–8}$$

where d = float displacement, inches from the zero line. Note that the area of both legs can be changed to provide for different span of measured pressure.

The *well manometer* shown in Fig. 8–4 is a variation of the enlarged-leg manometer, the purpose of which is to make the variation of level in the well of negligible magnitude. The head can then be read directly on a single scale instead of subtracting two distance readings as in the U-tube manometer. For static balance,

$$p_2 - p_1 = \gamma\left(1 + \frac{A_1}{A_2}\right) d \tag{8–9}$$

where d = head, inches from the zero line. If the area of the well is 500 or more times larger than the area of the vertical leg, then the error in neglecting the area term is negligible. The head relation for a properly designed well-manometer is then

$$p_2 - p_1 = \gamma h \tag{8-10}$$

when the vertical scale is placed so as to read zero when the pressure differential is zero.

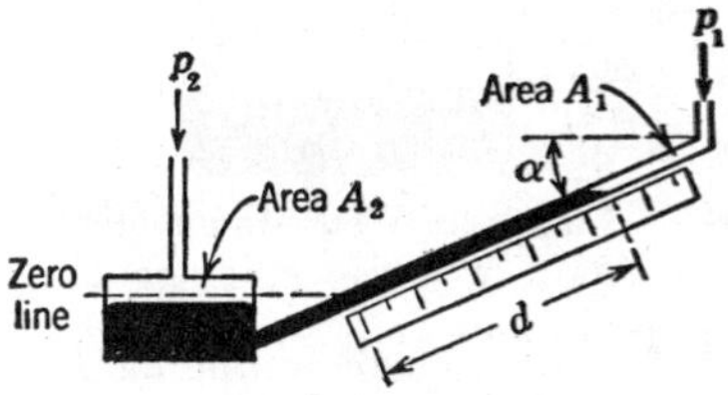

FIG. 8–5. Inclined-Tube Manometer (well type)

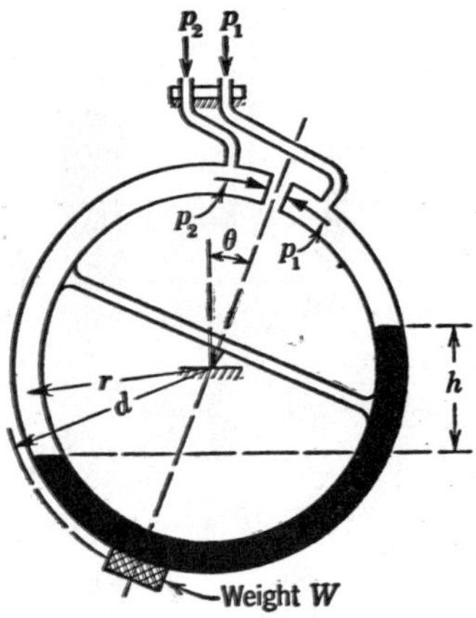

FIG. 8–6. Ring "Manometer"

The *inclined-tube manometer* is shown in Fig. 8–5. The purpose of the inclined leg is to expand the scale so that lower pressure differentials may be read more easily. For static balance,

$$p_2 - p_1 = \gamma\left(1 + \frac{A_1}{A_2}\right) d \sin\alpha \tag{8-11}$$

where d is the distance, inches along the tube through which the level moves. When the ratio of areas A_1/A_2 is negligible (well-type),

$$p_2 - p_1 = \gamma d \sin\alpha \tag{8-12}$$

The scale of the instrument can be greatly expanded by reducing the angle of the inclined leg α to a small value.

The *ring differential-pressure meter* in Fig. 8–6 is not a manometer in the true sense, but it is commonly considered so. It is made from a tube in the form of a ring which is supported near its center by a knife-edge pivot. The ring is divided into two parts by a partition or by separating the ring as is shown here. The tube is sealed by a manometer liquid. The pressures p_1 and p_2 are led to the tube through flexible connections. The rotation of the ring is caused by the difference in pressure acting against the cut portion of the tube. The rotation of the ring is limited by the counterweight W. If A is the area of the tube, then, for static balance of moments,

$$(p_2 - p_1) = \frac{d}{r} \cdot \frac{W}{A} \sin \theta \tag{8-13}$$

where it is assumed that the center of gravity of the ring itself without the weight W is at the pivot point. Note that this equation does not contain the density of the liquid. The liquid merely acts as a pressure seal. The tilting U-tube meter in Fig. 8-7 is a mercury "manometer" mounted on a pivot or knife-edge bearing similar to the ring meter. Flexible connections are made to the top of the U tube so that the rotation of the U tube is unrestricted. The pressures acting against the two ends of the U tube cause an angular deflection proportional to the applied difference in pressure.

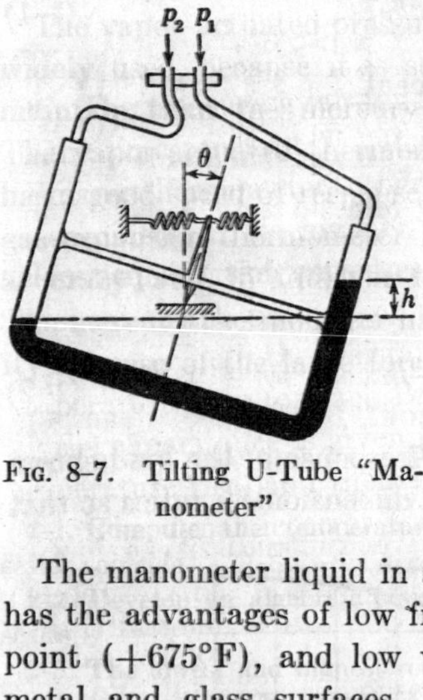

FIG. 8-7. Tilting U-Tube "Manometer"

The manometer liquid in almost universal use is mercury. It has the advantages of low freezing point (−38°F), high boiling point (+675°F), and low vapor pressure and does not "wet" metal and glass surfaces. It is insoluble in many common liquids. It has the disadvantages that it amalgamates or corrodes many metals, is expensive, and is poisonous. In some applications its high density is an advantage, whereas in others it is a disadvantage.

For low-pressure manometers water with coloring agent is used. Often an agent is added to reduce surface tension so that the capillary rise is reduced. A disadvantage of using water is its high vapor pressure. Other manometer liquids are shown in Table A–20 (appendix). Most of these have important disadvantages such as high vapor pressure or large change of density with pressure. Many are hygroscopic and are soluble in other liquids. Petroleum oils and kerosene are infrequently used.

Measuring Elements for Gage Pressure and Vacuum

For the measurement of pressures from 20,000 psi and less, the elements in common use are the pressure spring (bourdon), the bellows, and the diaphragm.

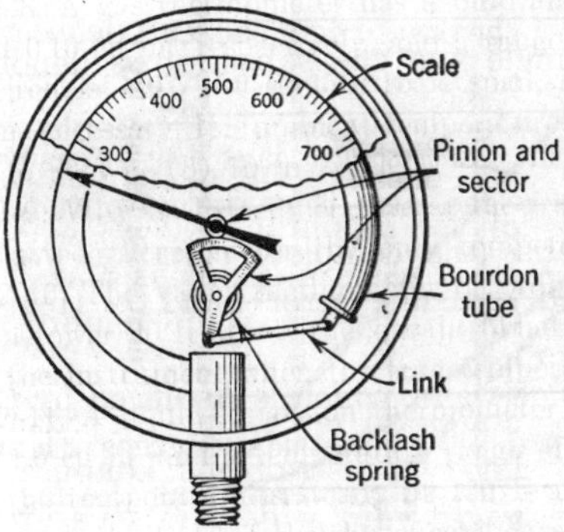

FIG. 8–8. Dial Indicating Pressure Gage

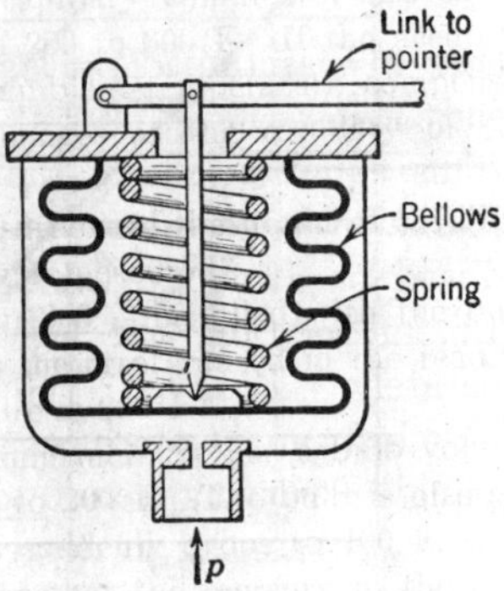

FIG. 8–9. Bellows Pressure Element

The *pressure-spring gage* employs a spiral, helix, or bourdon tube. A pressure spring of bronze is used in the pressure range up to about 600 psi, a beryllium–copper pressure spring up to about 10,000 psi, and a steel or alloy-steel pressure spring up to 10,000 psi and higher. The most common type of the pressure-spring gage is the simple dial pressure gage with a bourdon-tube element for measuring gage pressures or vacuums, as in Fig. 8–8. This type is used literally by the thousands in all industries and may be obtained in sizes from 2-in.-diameter dial up to 14-in.-diameter in many different indicating styles.

The *bellows pressure element* is shown schematically in Fig. 8–9. Generally a metallic bellows acting with the pressure on one side and a spring on the other is used. The pressure range of the system is determined mainly by the effective area of the bellows and the spring gradient. Commonly the bellows are of brass or phosphor bronze. Often the inside of the entire assembly is tinned for further protection against corrosion. Sometimes bellows of other materials such as Monel or stainless steel may be used with the arrangement shown in Fig. 8–9 somewhat modified. Bellows gages are commonly employed for measuring gage pressures or vacuums from about 5 in. of water up to 100 psi.

The *metal-diaphragm pressure element,* similar to Fig. 8–10, employs a thin flexible diaphragm of such material as brass or bronze. The force of the pressure against the effective area of the diaphragm causes a deflection of the diaphragm. In some cases the deflection of the diaphragm is opposed by the spring qualities of the diaphragm itself, and in other cases a spring is added to limit deflection of the diaphragm. The motion of the diaphragm operates an indicating- or recording-type instrument. Diaphragm gages vary widely in measuring range, depending on their construction. They are normally employed for low pressures or vacuums up to about 5 psi.

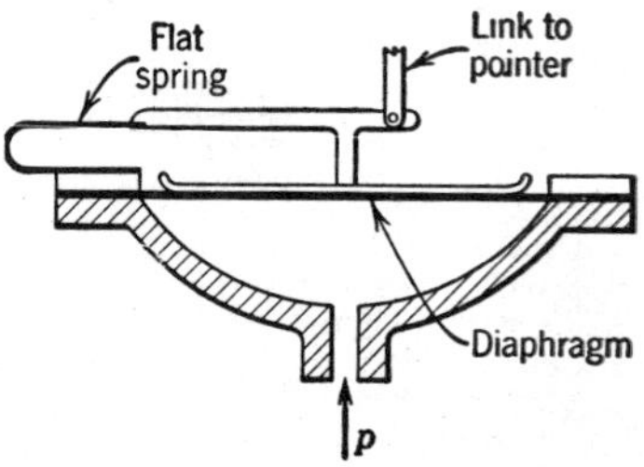

FIG. 8–10. Diaphragm Pressure Element

Instruments for indicating or recording low gage pressures or vacuums are used for measuring furnace drafts, air-duct pressures, and so on. In such applications the pressure or the vacuum does not usually exceed 20 in. of water (about 0.7 psig). The *nonmetallic-diaphragm pressure element,* similar to the element in Fig. 8–10, employs a flexible diaphragm of high-quality leather, goldbeater's skin, or a thin neoprene-like material. The pressure acts against the effective area of the diaphragm, causing the diaphragm to deflect against the force of a flat spring. The resulting displacement is multiplied by a suitable linkage, and a pointer is made to operate over a scale. The span of these gages is usually in inches of water. For higher pressures, a flexible neoprene diaphragm can be employed. Differential pressures can be measured by applying the second pressure to the other side of the diaphragm and using a sealed means of detecting motion of the diaphragm. Some types of gages for measuring differential pressure may employ two diaphragms.

The measurement of pressures in the range from 1 to 600 in. of water where the static pressure may be as high as 5000 psig or even higher is one of the most common industrial problems. In flow measurement, liquid-level measurement, and direct-pressure measurement the differential-pressure element is used. The differential-pressure element employs the principles of any one of the pressure-gage elements, such as the enlarged-leg manometer, the bell gage, the bellows element, the diaphragm element, and the ring or tilting U-tube meter.

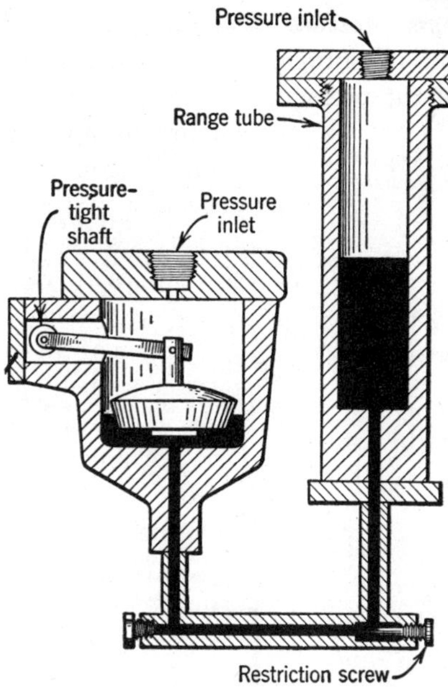

FIG. 8–11. Differential-Pressure Indicating Manometer

The *enlarged-leg mercury-manometer differential-pressure element,* shown in principle in Fig. 8–3 and in more detail in Fig. 8–11, is generally used for differential pressures from 10 to 600 in. of water at static pressures up to 5000 psig. A metal float contained in one leg of the manometer follows the surface of the mercury in responding to changes in differential pressure. The motion of the float is transmitted out of the housing by various mechanical and electrical means. For protecting against surges or sudden reversal of differential pressure, check valves (not shown) are used in the U tube. This is especially necessary to prevent blowing out the mercury. The leg opposite the float chamber is called the range tube, since by substituting tubes of various diameters the span of the meter may be altered. Damping of the manometer is accomplished by viscous damping through the restriction screw.

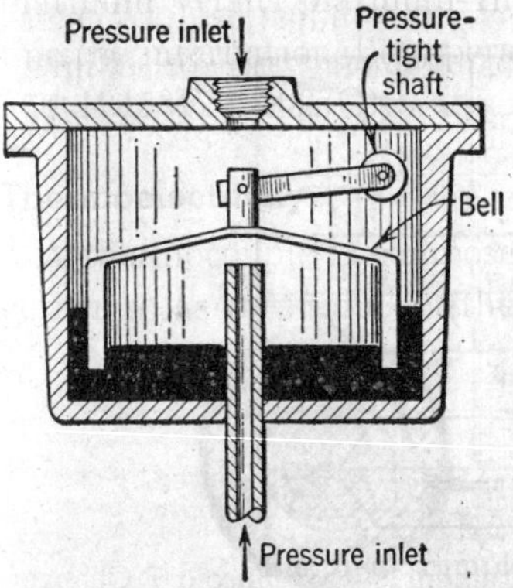

FIG. 8–12. Bell Differential-Pressure Gage

The *bell differential-pressure meter* is shown in Fig. 8–12. With this device differential pressures from about 1 to 15 in. of water may be measured. The static pressure can, in some cases, be as high as 800 psig. The bell meter is therefore employed for lower differential pressures in gases. (An exception to this is the Ledoux bell meter of the Bailey Meter Company which is discussed in the chapter on Flowmetering.) The motion of the bell is taken from the housing by various mechanical and electrical methods. The bell is arranged to close the opening at the top or bottom of the housing so that protection is afforded against overrange and reversal of pressures.

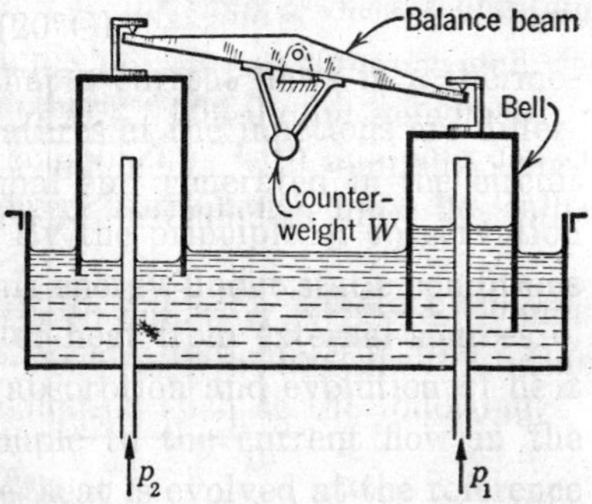

FIG. 8–13. Bell Differential-Pressure Gage

The *bell differential-pressure element* is shown in Fig. 8–13. Light metal bells are supported on a beam pivoted on knife-edges. A counterweight W balances the deflection of the system. Since the liquid, usually an oil, merely acts as a pressure seal, the static balance for small deflections is

$$(p_2 - p_1) = \frac{Wd}{lA} \sin \theta \qquad (8\text{–}14)$$

where A = area of bells
l = distance from pivot to bell support point
θ = angle of deflection of beam from horizontal
d = distance from pivot to counterweight

Note the similarity of this equation and the equation for the ring meter.

The *bellows differential-pressure meter* is indicated in Fig. 8–14. This type is used for differential pressures in the range 25 in. of water up to about 50 psi at static pressures up to 2000 psig. The differential pressure acts across the large brass or stainless-steel bellows, the resulting motion of the bellows being transmitted out of the housing by various means. The motion of the bellows is usually opposed by a spring

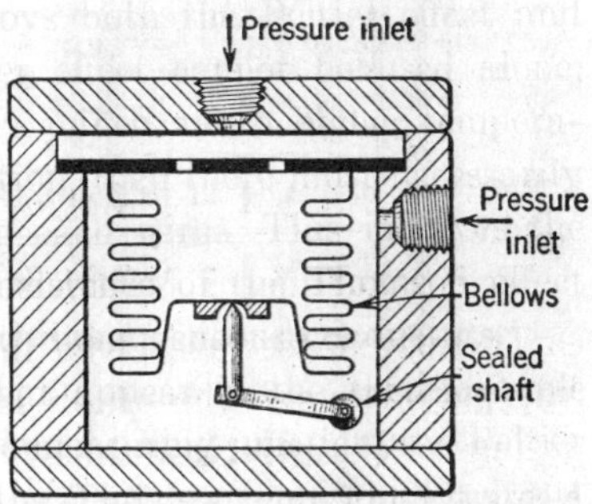

FIG. 8–14. Bellows Differential-Pressure Element

The *ring* and *tilting U-tube differential-pressure meters* are shown in principle in Figs. 8–6 and 8–7. These meters are generally used for differential pressures from 20 to about 400 in. of water at as high as 10,000 psig static pressure. The rotation of the ring is opposed by a calibrated spring, and no sealing means is necessary in operating the indicating pointer. Flexible connections to the meter tube are used. Several rings are arranged in parallel for higher differential-pressure range.

Indicating Elements for Pressure Gages

The various forms of industrial pressure gages differ mainly in the method by which the displacement of the measuring element is arranged to operate an indicating pointer. By far the most common method is the mechanical system, in which the displacement element is directly connected to an indicating pointer, as previously shown in Fig. 8–8, or to a recording pen. In most respects the indicating or recording system of the mechanical gage is the same as the recording system of the pressure-spring thermometer. However, pneumatic and electrical means are also used.

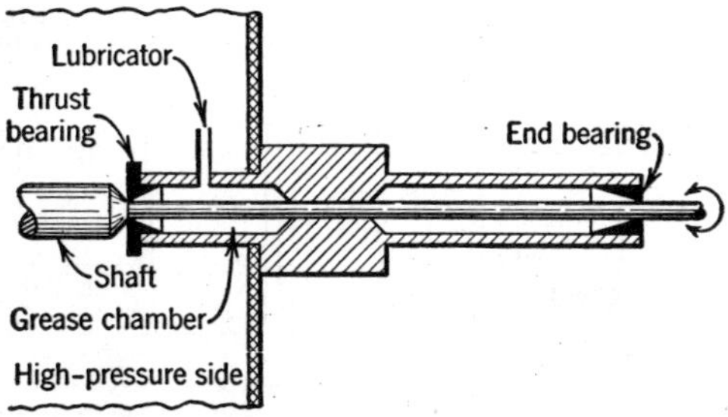

Fig. 8–15. Pressure-Tight Shaft Used for Differential-Pressure Meters

Mechanical means for transmitting motion of the float, bell, or bellows of the differential-pressure meter are the pressure-tight bearing, and the torque tube. The *pressure-tight bearing* used in nearly all types of meters is shown in Fig. 8–15. The shaft is lapped to a precision fit, and the bearing is made several inches long to provide a long path for leakage. The lubricator acts mainly as a seal, and it is remarkable that even under very high pressures no leakage and very little friction exist when the bearing has proper maintenance. The torque tube, shown in cross section in Fig. 8–16, is often employed in the bellows meter. The outer tube is made from relatively thin-metal tubing and is clamped rigidly to the meter housing. The inner shaft is rigid, and the force of the bellows causes the free end of the torque tube to twist. The inner shaft then rotates in the same direction. The torque tube also serves as a spring. Thus the motion of the bellows is transmitted through the housing without the necessity for a sealed bearing.

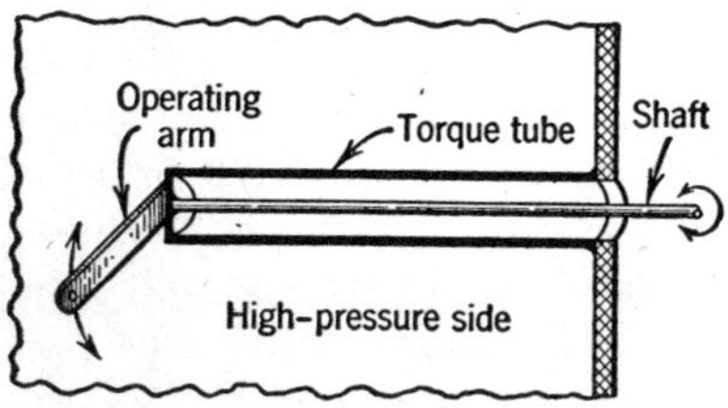

Fig. 8–16. Torque-Tube Shaft Used for Differential-Pressure Meters

Electrical means are also used for transmitting motion of the float, bell, or bellows of the differential-pressure meter. Some of these are the inductance-coil system, the resistance-rod system, and the magnetic clutch.

The *inductance-coil system* for differential-pressure meters is shown in Fig. 8–17. The float in one leg of the manometer supports an armature on the inside of a nonmagnetic tube. This tube seals the float side of the manometer. A divided inductance

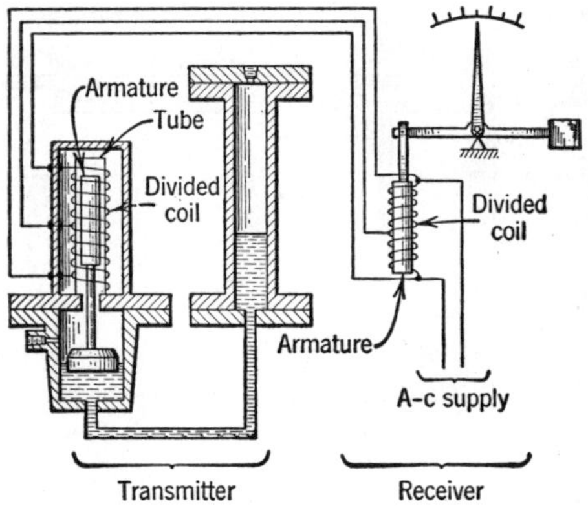

Fig. 8–17. Inductance-Bridge Pressure-Differential Meter

coil is fixed in position on the outside of the tube. The instrument or receiver contains an identical divided inductance coil, within which is suspended a counterbalanced armature. At balance the armatures are similarly positioned in each coil, and the current flow through each coil is the same. When the armature at the transmitter moves to a new position, the reactance of each side of the transmitter coil differs. This causes a different current flow in each arm of the circuit. The magnetic field at the receiver coil is redistributed, and a net force acts on the armature, tending to bring the system to balance. Thereby, for every position of the armature in the transmitter, there is a corresponding position of the receiver armature, and the motion of the float is transmitted to the receiver pointer. Variations of line voltage have no effect on calibration of the system. The system has the advantage that the meter is completely sealed. A variation of the inductance-coil system is employed, in which the balance of the bridge is detected by a galvanometer, and a reversible electric motor is used to position the receiver armature in the coil. A second variation of the inductance-coil system employs an electronic amplifier for detecting balance and a reversible electric motor for positioning the receiver armature.

The *resistance method* used by the Republic Flowmeters Company is shown in principle in Fig. 8–18. The mercury manometer is arranged with a contact-rod assembly in one leg. The contact-rod assembly includes about 100 resistance wires hung

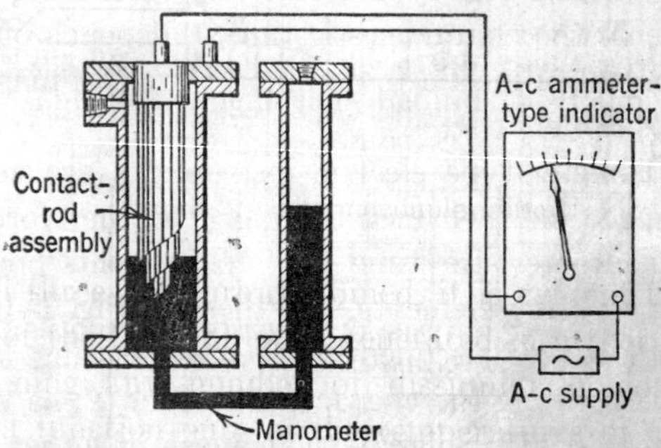

FIG. 8–18. Resistance-Rod Pressure-Differential Meter—Republic Flowmeters Company

vertically with the length of the wires selected so as to form a spiral of contact points above the surface of the mercury at zero pressure differential. As the mercury level moves upward, successive contacts are made between the wires and the surface of the mercury. Thus, the total resistance between the contact-rod head and the manometer chamber varies from maximum when none of the rods are grounded to zero when all rods are grounded. The instrument is then simply a form of ammeter suitable for recording or indicating purposes. The contact-rod assembly is pressure-tight, and no sealed bearings are required.

The *magnetic-clutch method* employed by the Hays Corporation and the Penn Industrial Instrument Company is shown in Fig. 8–19. The float in the mercury manometer positions the balls in a nonmagnetic tube below the float. The permanent magnet is self-centering on the steel ball, and the glass balls are used for additional buoyance. The movement of the float produces a corresponding rotation of the magnet arm, and a pen or pointer operates over an indicating scale. No pressure-tight bearings are therefore required.

The *pneumatic-balance differential-pressure meter* is shown schematically in Fig. 8–20. The Republic Flowmeters Company employs an arrangement of this kind, whereas the Moore Products Company uses similar principles in a bellows-type system. These meters are for differential pressures up to 400 in. of water at static pressures up to 600 psig. The mechanism is very similar to the pneumatic-balance pressure thermometer. Suppose pressure p_2 increases. This causes an upward force on the measuring diaphragm and results in a small upward deflection of the left end of the balance beam. The right end of the beam covers the nozzle through which there is a continuous flow of air. The back pressure at the nozzle then increases, and the balancing diaphragm pressure increases. The resultant upward force on the right side of the balance beam brings the balance beam to a

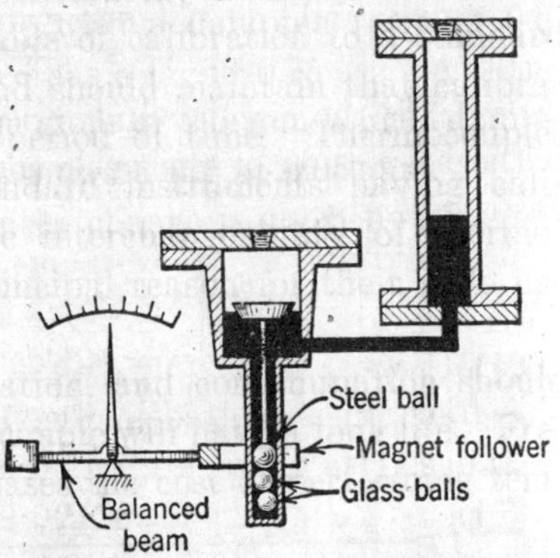

FIG. 8–19. Magnetic-Clutch Pressure-Differential Meter—Hays Corporation and Penn Industrial Instrument Company

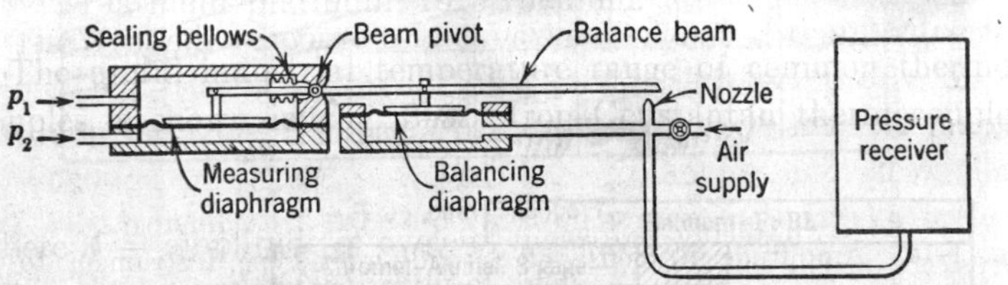

FIG. 8–20. Pneumatic-Balance Pressure-Differential Meter—Republic Flowmeters Company

force equilibrium. The balance beam is sealed at the point of entering the measuring-diaphragm chamber by a small metallic bellows. The balancing-diaphragm pressure is transmitted to the pressure receiver which is an ordinary pressure-spring instrument of about 2 to 20 psi calibration. The pressure receiver can therefore be calibrated in terms of the measured differential pressure. The pneumatic-balance pressure gage has the advantages of high speed and low dead zone. Ordinarily these pneumatic systems employ an amplifying pilot or relay between the nozzle and the balancing diaphragm.

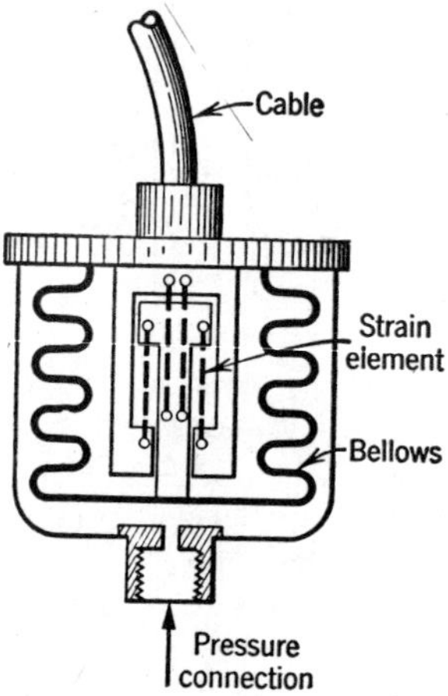

FIG. 8–21. Electric Pressure Gage

The *electric pressure gage* comprises an elastic element for converting pressure into displacement, and an electric-displacement gage for measuring the displacement. One such arrangement is shown in Fig. 8–21 where an unbonded strain gage is employed with a bellows element. It is evident that the bourdon tube or the diaphragm could be used as the elastic element and that a variable differential transformer or even a synchro system might be employed in place of the strain gage. The electric gage is made with several arrangements. For example, in the gage of Fig. 8–21 the strain-gage element serves as the elastic element, and the bellows may have a negligible modulus. The total displacement of the bellows need only be about two or three thousandths of an inch, and this small displacement is characteristic of most electric pressure gages. For measuring high pressures the bellows, or a suitable heavy diaphragm, with a high modulus is used. The strain element then serves only to measure displacement of the bellows.

The electric pressure gage can be used to measure absolute pressure if the upper side of the bellows in Fig. 8–21 is evacuated. Differential pressure can be measured by applying the second pressure to the upper side of the bellows. The strain-gage-type pressure unit is used with an automatic Wheatstone-bridge instrument for indicating and recording. If the electric gage uses other types of displacement elements, such as the variable differential transformer or the inductance-coil system, an automatic inductance-bridge instrument can be used.

Measurement of Absolute Pressure

Absolute pressure and gage pressure differ by a constant only when the barometric pressure is constant. The error in meas-

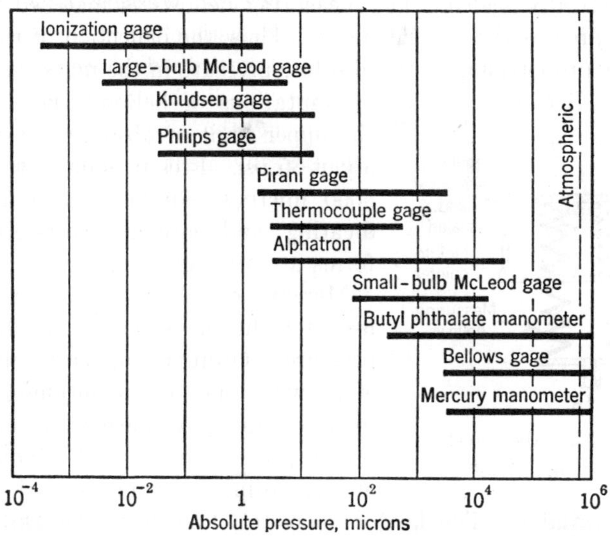

FIG. 8–22. Useful Ranges of Absolute-Pressure-Measuring Gages (*Reproduced by permission of Distillation Products Industries*)

uring absolute pressure with a gage-pressure instrument depends on the magnitude of change in barometric pressure from normal. The error is about ½ psi for every 1 in. Hg change in barometric pressure. For absolute pressures above 500 psi this error is negligible. For the measurement of absolute pressures between 50 and 500 psi manual corrections are made to the readings of a gage-pressure instrument when great accuracy is desired. For absolute pressure below 50 psi it is better to employ methods for determining absolute pressure and not rely on gage-pressure readings. Many different types of absolute pressure gages are used, as shown in Fig. 8–22. In order to illustrate the principles of these gages, the discussion is limited to the bellows gage, the McLeod gage, the Pirani gage, the thermocouple gage, and the ionization gage. All these are industrial-type instruments and are used in plant service.

The *bellows gage* is simply a double-bellows gage for differential-pressure measurement, as shown in Fig. 8–23. The upper bellows is evacuated to a pressure of about 5 microns or less and hermetically sealed. The measuring bellows is matched to the same area as the upper bellows. Thus, the deflection of the system is proportional to the absolute pressure in the measuring bellows. Because of physical limitations of bellows and springs, the upper limit of absolute pressure is about 35 psi. Pressures down to about 5000 microns can be measured. Indicating and circular-chart-recording forms are available.

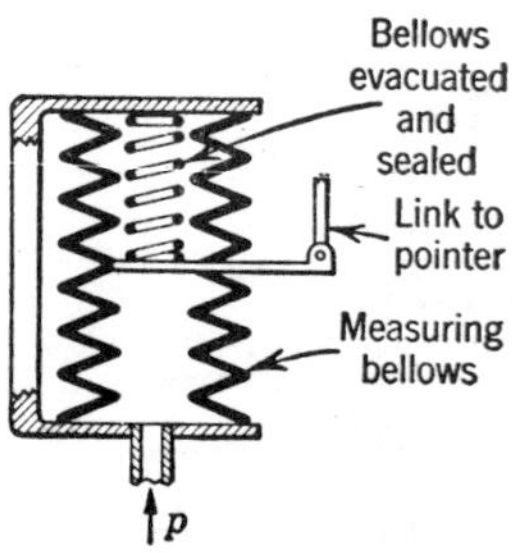

FIG. 8–23. Absolute-Pressure-Gage Element

Mercury or oil manometers may be used for the measurement of absolute pressure. Ordinary mercury manometers with one leg maintained at a high vacuum by means of vacuum pumps are sometimes used. Indicating and recording instruments of this type are available. The float manometer, known as the DuBrovin gage, is useful in the range down to 1000 microns.

The *McLeod gage* element is shown schematically in Fig. 8–24. This type of gage is generally used as a primary standard for the measurement of high vacuums. It has the advantages that it is simple to use and that its calibration is the same for nearly all noncondensable gases. The gage is filled with the gas by periodically rotating the assembly by means of a small electric motor. When the gage is returned to the vertical position, the trapped gas is compressed to a smaller volume by the action of the mercury. At this pressure the height of the mercury column in the capillary tube is directly related to the vacuum by the gas laws. The capillary tube contains a fine platinum wire, the total resistance of which depends on the height of the mercury around it. The resistance of the platinum wire is then measured by a resistance-thermometer-type instrument. With this gage, vapor traps are usually necessary to eliminate condensable vapors from the vacuum system. A few gases which deviate from the ideal-gas law at low pressures require special calibration in the McLeod gage.

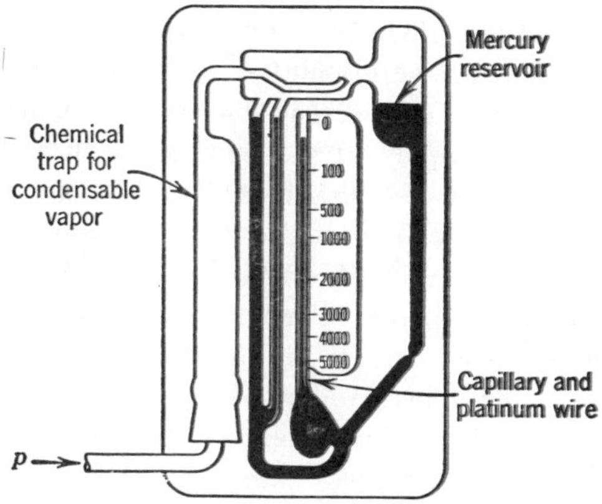

FIG. 8-24. Schematic Arrangement of McLeod Vacuum Gage—F. J. Stokes Machine Company

The *Pirani gage* and circuit are shown schematically in Fig. 8-25. It has the advantage of being simple and easy to use but

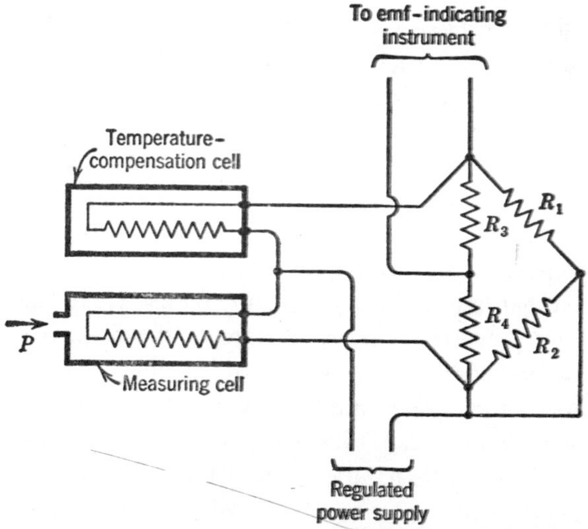

FIG. 8-25. Pirani Vacuum Gage and Circuit

has a disadvantage that its calibration depends on the type of gas in which the pressure is measured. It is useful in such gases as argon, carbon dioxide, air, water vapor, helium, acetylene, and hydrogen. This gage operates on the principle that the heat loss by conduction and convection from a heated resistance wire depends on the pressure of the gas surrounding the wire. The tungsten filament is sealed into an enclosure which can be connected to the low-pressure point. The enclosure is usually formed of hard glass. The current through the filament is maintained constant by employing the unit in a Wheatstone-bridge circuit, the other arm of which contains the temperature-compensation cell. This cell is sealed at a pressure less than 1 micron, and any changes in ambient temperature thus affect both arms of the bridge in a like manner. The unbalance voltage of the bridge is measured by an instrument similar to the automatic-balance potentiometer. The open tube of the Pirani gage can be exposed to atmospheric pressure, but this practice is not recommended, since corrosive gases affect the filament and change the calibration. In some cases, specially treated filaments are used to avoid contamination and corrosion. The Pirani-gage filament should not be operated above about 450°F if organic vapors from grease, oil, or gasket materials are present in the system. Higher filament temperatures decompose organic vapors, leaving a deposit on the filament and causing a zero shift of calibration. Pirani gages are most useful in the range 20 to 200 microns. Outside of this range calibration with each use of the gage is necessary.

The *thermocouple* (*thermal*) *gage* is shown schematically in Fig. 8–26. It is apparent that the basic principle of operation is the same as in the Pirani gage where the temperature of the heated filament depends on the heat loss by conduction and convection through the gas surrounding the filament. In the thermocouple gage, however, the temperature of the filament is measured by a fine-wire thermocouple directly attached to the filament. As the diagram indicates, a compensating cell and a measuring cell are used with the thermocouples connected in series opposition so that the compensating-cell emf is subtracted from the measuring-cell emf. The emf output of such a cell is about 20 millivolts at a pressure of 500 microns. This gage has about the same features and disadvantages as the Pirani gage.

The *ionization gage* (thermionic type) is shown in Fig. 8–27. The tube is directly connected to the point of pressure measurement. The tube is in the form of a triode with cathode, grid, and plate. The grid is maintained at a high positive potential

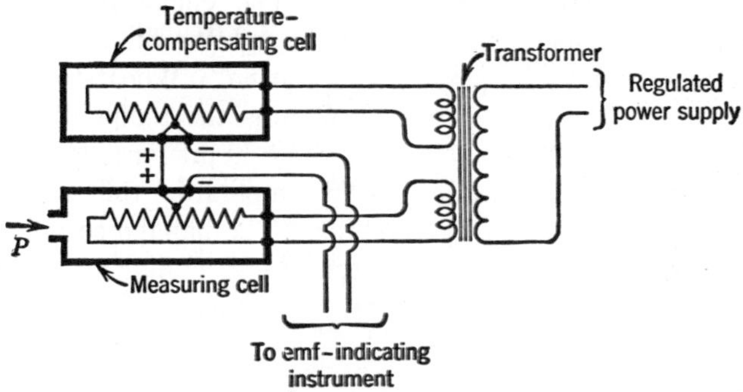

FIG. 8–26. Thermocouple Vacuum Gage and Circuit

with respect to the cathode while the plate is held at a negative potential with respect to the grid. Its principle of operation is as follows: Electrons emitted at the hot cathode by thermionic emission are attracted toward the grid, pass through the grid, and start toward the plate. Ions are formed by collision of the

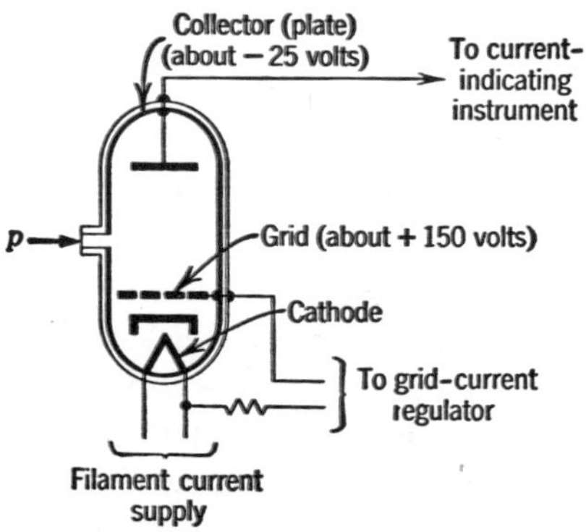

FIG. 8–27. Thermionic-Type Ionization Vacuum Gage

electrons with molecules of the gas present in the tube. These positive ions are collected on the plate, and, as a result, a positive ion current exists. The amount of ion current is proportional to the amount of gas present, provided the electron current is constant. The electrons return to the grid and are collected, and the flow of electrons is maintained constant by a grid-current regulator. The ion current is measured by an instrument similar to the automatic-balance potentiometer. With the ionization gage, low pressures can be continuously measured. At extremely low pressures (about $\frac{1}{100}$ micron) it is advisable to degas the tube: that is, to increase momentarily the filament current to a high value with the tube at low pressure, so as to heat the tube and remove all occluded and adsorbed gas from the tube walls, the elements of the tube, and the collector. This type of ionization gage cannot be used at pressure above about 2 microns because of the possibility of burning out the filament. An ionization gage is employed in conjunction with a Pirani gage to measure absolute pressure above 2 microns. A thermionic ionization gage can be calibrated for various inert gases but is not reliable with oxygen, hydrogen, carbon dioxide, and other gases that are decomposed at the hot filament.

The ionization gage (radioactive type) is very similar in principle to the thermionic type. However, the ionization is produced by collision of gas molecules and alpha-particle radiation from a small permanent radium source instead of by electron collision. The positive ions formed by collision of the alpha particles with the gas in the tube are collected and the positive ion current measured. This type has the advantage that no filament is required, so that the gage is not harmed by use at higher pressure.

Measuring Pressures in Corrosive Fluids

In nearly all pressure gages, the fluid in which the pressure is measured is conducted to the inside of the pressure-measuring element and is in direct contact with the element. This creates a problem of handling high temperature, corrosive, sludgy, or semisolid materials. The metallic materials used in the pressure-measuring element are, variously, brass, bronze, phosphor bronze, copper, beryllium–copper, steel, chrome–molybdenum steel, stainless steels, Monel, silver, and tantalum. The nonmetallic materials are leather, neoprene, rubber, and glass. Often the fluid is in direct contact with mercury. In many applications, there is no selection among these materials, because one or the other must be used for its spring properties. Some pressure-gage elements can be protected by copper-plated, nickel-plated, cadmium-plated, or tinned-metal surfaces. This is not always possible, as for example on the inside of bourdon tubes. Therefore, some other method of excluding the measured fluid must be employed.

The *single-coil siphon,* one form of which is shown in Fig. 8–28, is effective in protecting a pressure-gage element from the high

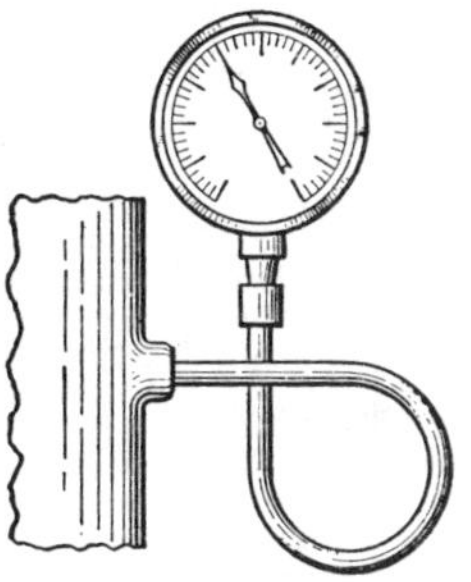

Fig. 8–28. The Steam-Gage Siphon

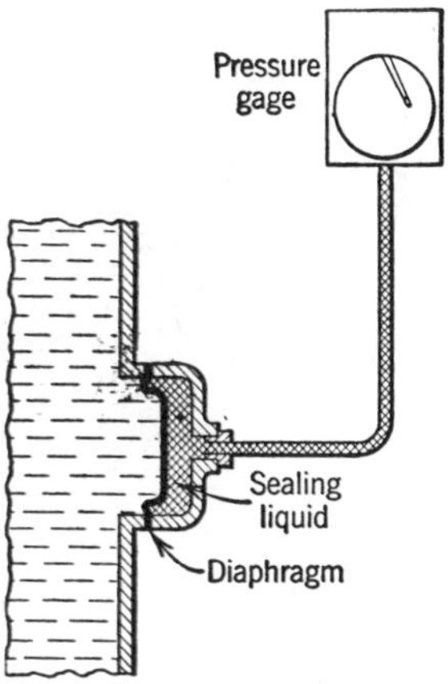

Fig. 8–29. Diaphragm Seal in Pressure Measurement

temperatures of steam. The brass coil traps condensed steam and limits the temperature rise in the gage. A siphon is necessary on all steam-pressure gages.

The *diaphragm seal* shown in Fig. 8–29 is a simple but effective device. The unit is usually made of bronze or cast iron with a neoprene or thin-metal diaphragm. The system is solidly filled with a liquid such as glycerin or oil. The area of the diaphragm is made sufficiently large that the required motion is almost negligible to produce the required volume change. The diaphragm is also quite flexible, so that the pressures on both sides of the diaphragm are equal. This translates the problem of high temperature or corrosion to the sealing unit where selection of materials is possible. It has the further advantage that the line leading to the pressure gage is always filled with clean oil of reasonable viscosity and cannot become clogged with sludge or solids. The system must be filled solidly with liquid, or a calibration error results. The bellows seal is a slightly different form of the diaphragm seal in which a metallic bellows is used in place of the diaphragm.

The *liquid seal* is illustrated in Fig. 8–30. A sealing chamber, usually of bronze or cast iron, is interposed in the gage line between the pressure gage and the point of measurement. The

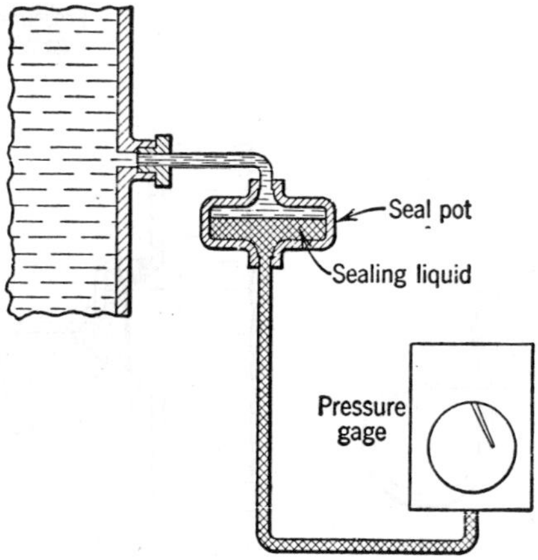

FIG. 8–30. Liquid Seal in Pressure Measurement

pressure-gage element, line, and sealing chamber are filled with a suitable liquid. When the gage and line are below the point of measurement, the sealing liquid must have a density greater than that of the fluid at the point of measurement. When the gage and line are above the point of measurement, the sealing liquid must have a density less than that of the measured fluid. The sealing chamber is made with sufficient area that the change of level of sealing liquid is small. The reason is that the pressure-gage calibration must be corrected for the additional liquid head of the sealing liquid, and this should preferably be a constant correction. Corrosion, sludge, and high-temperature problems are now transferred to the sealing chamber where they are more easily handled.

The purge system is shown in Fig. 8–31. It is useful when a small and continuous flow of air or other gas through the measuring line is sufficient to maintain the line free from sludge. Air is supplied at the necessary pressure through a constant-flow regulator, which contains an orifice and a differential-pressure-controlling means. The regulator is small, simple, and inexpensive. If the flow of air through the line is constant, then there is a constant pressure drop in the measuring line. The pressure-gage calibration can then be corrected. In most cases

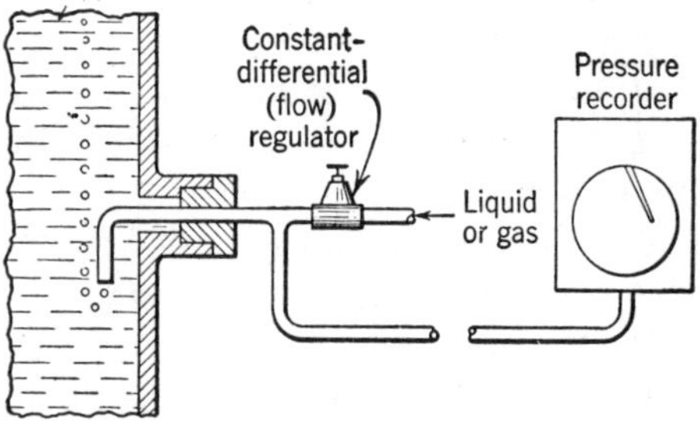

FIG. 8-31. Purge System in Pressure Measurement

the pressure drop in the measuring line is so small in comparison to the pressure being measured that no correction is required.

Static Accuracy of Pressure Gages

The accuracy of the simple dial-indicating-type pressure gage, when calibrated with dry air or with a dead-weight hydraulic tester, is usually ± 1 per cent of span above 5 per cent of span. The precision-type dial-indicating gage and the instrument-type recording gage are generally accurate to ± 0.5 per cent of span under the same calibration conditions as listed previously.

The sources of static error in pressure gages are the ambient effect, the head effect, the barometric effect, the line-loss effect, and the calibration shift. Ordinarily, the effect of ambient temperatures is negligible except infrequently in liquid-column manometers.

The ambient-temperature effect in manometers is due to cubical expansion of the manometer fluid. Ordinarily the error so caused is negligible if the ambient-temperature change is less than 30°F. The extent of the ambient error in pressure gages depends on the construction. There are two causes for ambient-temperature error: (1) the temperature coefficient of linear expansion which generally results in a zero shift of calibration, and (2) the temperature coefficient of Young's modulus which results in a span shift and zero shift of calibration. In mechanical pressure gages these effects occur at the elastic element, such as a pressure spring, bellows, or diaphragm, and at the spring if one is used. When the elastic element has negligible modulus, the effects are localized at the spring, and it is possible to select a spring having negligible temperature coefficient of expansion. The change of modulus with temperature is about 1 per cent per 75°F for many materials and is not usually compensated. In electric pressure gages using a bellows or diaphragm of negligible modulus and a strain element as shown in Fig. 8–21, there is no effect due to temperature coefficient of expansion since the measurement is made directly on the basis of force.

The head due to liquid columns in the gage line can be corrected by shifting the zero point of the pressure-gage calibration. If the liquid column in the gage line is not constant, then a corresponding error is introduced. By careful planning it is nearly always possible to avoid variable liquid columns.

The effect of barometric-pressure changes on measurement of low-gage pressures or vacuums may or may not be important, depending on the use made of the pressure measurement. In measuring draft pressure, for example, it is usually necessary that the pressure be a constant amount above or below atmospheric pressure. In other processes, the performance depends on absolute pressure. In this case, the indicated pressure must be manually corrected, or automatic barometric compensation used, or else absolute pressure measured.

The effect of leakage in the measuring lines will be to cause an error that depends on the size of the line and the size of the leak. A constant flow through the measuring line, as in purge systems, causes a constant pressure drop for which the pressure gage can be corrected. Unintentional leakage is easy to detect with liquids but more difficult to locate with colorless, odorless gases. In either case it is advisable to install a shutoff cock in the measuring line close to the point of measurement. Leakage may then be checked at periodic intervals. Leakage with any type of diaphragm or liquid seal usually results in a serious error if not soon corrected.

Calibration shift in pressure gages of the diaphragm-spring and pressure-spring types is mainly due to the gage being subjected to pressures higher than the highest calibration point. So-called overrange protection can be provided by employing the gage only in the lower part of the elastic range. This seriously limits the use of pressure gages, so that, as further protection, limit stops are usually added to the pressure-gage element. This prevents the element from deflecting more than a given amount. Even with limit stops, however, a sufficiently high pressure will permanently distort the element or cause a rupture.

Response of Mechanical Pressure Gages

The response of most mechanical pressure gages when measuring a gas pressure is typical of mass and spring systems with light damping. This is illustrated in Fig. 8–32 for a pressure-

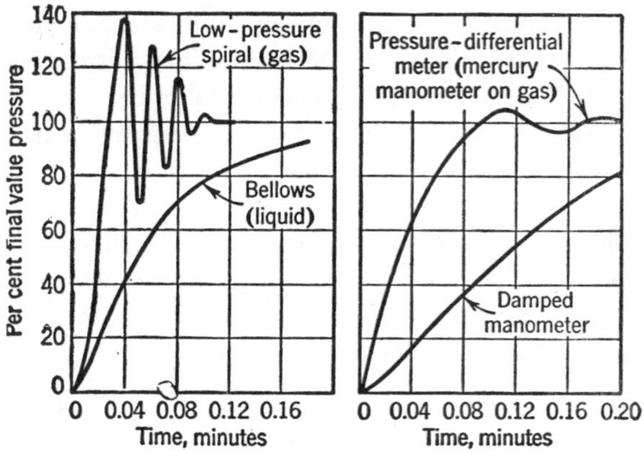

FIG. 8–32. Response of Pressure Gages

spring gage. When a liquid pressure is being measured with the pressure system liquid-filled, the response is considerably slower because of viscous damping. The measuring lag of pressure gages can usually be considered first-order type, the time constant depending entirely on the magnitude of viscous damping.

The length and size of the measuring line determines the viscous-damping effect. Since the viscosity of gases is much lower than that of most liquids, liquid-filled lines need to be larger in diameter for a given length than gas-filled lines. Measuring lines over 50 ft in length may seriously increase the measuring lag, not only because of additional damping, but also because of the inertia of the liquid when the volume change of the measuring element is large.

The response of pressure gages at high frequencies may become important in measuring rapidly fluctuating pressures such as impact pressures and supersonic air-flow pressures and shock waves. The ordinary mechanical pressure gage is not suitable because its natural frequency is so low, about 1 per second down to $\frac{1}{5}$ per second, as is seen in Fig. 8–32. The electric pressure gage, such as is shown in Fig. 8–21, can be used at frequencies up to about 3000 cps because its mass is small and the spring modulus high. For pressure measurement of this kind an electric gage cannot be equaled.

In using pressure gages it is sometimes difficult to obtain satisfactory measurement of a pressure exhibiting rapid fluctuations or pulsations, either random or periodic. In addition, the mechanism of a mechanical pressure gage usually has low mass and little if any damping, but a reasonable spring gradient. Therefore, the gage itself will have a natural frequency of oscillation. Resonance may occur when the frequency of pressure oscillation matches the natural frequency of the gage. In such applications it is necessary to dampen the pulsations before they reach the pressure gage. Most methods of damping pressure gages involve increasing the measuring lag of the pressure-gage system so that it is slow and cannot respond to the pulsations. All these methods must be employed cautiously, since in some cases the measuring lag is increased to a point where the dynamic error is excessively large.

Damping by installing a needle valve in the measuring line is not recommended, however, since the opening usually must be quite small and therefore clogs easily with dirt or sludge.

Piston snubbers are more effective for high-frequency pulsations. They operate by using the inertia of a small piston in a cylinder connected in the measuring line to reduce the pulsation. They also absorb sudden shock or surge in pressure by nearly closing the line. Much less difficulty from clogging is encountered with them since they are self-cleaning to some extent. Shock waves (water hammer) in liquid-filled lines can be absorbed by installing an air trap in the line. This is simply a chamber of about 20 cu in. volume which is partly filled with air. Shock waves are absorbed by the cushioning or spring effect of the air.

Damping of differential-pressure mercury manometers is accomplished by adjusting the damping valve in the U tube connecting the legs of the manometer. Since the mercury is normally clean, clogging is not usually a problem.

Problems

8-1. Given the following pressure measurements, convert each to absolute pressure in pounds per square inch:

(*a*) 2880 psia.
(*b*) 34.0 in. Hg gage at 80°F and normal barometer.
(*c*) 250.0 ft H_2O gage at 40°F and normal barometer.
(*d*) 200 atm absolute.
(*e*) 890 mm Hg gage at 32°F and a barometer of 790 mm Hg.
(*f*) 40,000 microns absolute.
(*g*) 20 in. Hg vacuum at 70°F and a normal barometer.
(*h*) 30,000 kg per sq cm absolute.
(*i*) 30 in. Hg gage at 60°F and a barometer of 750 mm Hg.

8-2. In the enlarged-leg manometer of Fig. 7-4, the displacement of the float is to be 1.0 in., the area of the float chamber 10 sq in., and the manometer fluid mercury. For a maximum span of 100 in. H_2O, what is the inside diameter of the range tube?

8-3. In a well-type manometer the pressure-differential measurement must be accurate to within 0.01 per cent at 15 psi differential. What must be the ratio of tube area to well area?

8-4. In the ring meter is the rotation of the ring for a given pressure differential independent of density of sealing liquid? Why?

8-5. The bellows element of Fig. 8-9 has the following data: The effective area of the bellows is 2.24 sq in., the spring rate of spring 50 lb per in., and the spring rate of the bellows 4.5 lb per in. Assuming that the spring and bellows are both at free lengths, and that the motion of the bellows rod is 0.375 in., what is the range of the instrument?

8-6. In the absolute-pressure element the upper (evacuated) bellows develops a leak, and the pressure rises to 50 microns. When a pressure of 26 in. Hg vacuum is being measured, what is the error in inches of mercury?

8-7. Using references, write a brief description of the operation of the DuBrovin (float-manometer) vacuum gage.

8-8. In the Pirani vacuum gage, the calibration depends on the kind of gas measured. On what physical effect is this based?

8–9. A diaphragm seal using an oil of specific gravity 0.8 is installed in a pressure-gage line 20 ft above the center line of the gage. What is the static correction for the gage?

8–10. A liquid seal is used with a pressure gage so that the sealing liquid surface is 3.0 ft below the pressure tap and 7.0 ft above the gage. Water condenses above the sealing liquid, which has a specific gravity of 2.0. What is the static error of the gage, if the gage calibration made with air is correct?

8–11. Show that, if damping is neglected, the period of oscillation of a U-tube manometer is $2\pi\sqrt{l/g}$ where l is the total length through the liquid from one surface to the other.

8–12. For an enlarged-leg manometer, is the period of oscillation changed by having manometer legs of different areas? Derive an expression for the natural period of oscillation, neglecting damping.

8–13. Write an exact expression for the deflection versus pressure difference for the bell pressure gage of Fig. 8–12.

9· Measurement of Head and Level

Measurement of level, volume, and weight of liquid and dry materials in a vessel is very frequently encountered, and, although it poses a seemingly simple problem, accurate measurements often become very difficult. The great variety in corrosiveness and fluidity of materials and the large size of many storage vessels along with the necessity for measuring the contents within a fraction of a gallon are obstacles to be overcome. Volume and weight measurements are the least susceptible to direct attack, and it is nearly always necessary to rely on pressure-measuring instruments.

Head, Density, and Specific Gravity

There are basically two kinds of measurements made of liquid level (or even of level of dry material). The first is *head measurement* in which the level or height is directly measured by a float that follows the surface or by a direct-contact means. The second is *pressure measurement* in which the level or height is determined from the relation,

$$h = \frac{p}{\gamma} = \frac{p}{\gamma_w G} \tag{9-1}$$

where p = pressure
γ = density of liquid
γ_w = density of water (usually at 60°F)
G = specific gravity

If the purpose of level measurement is to determine the volume of liquid contained in a vessel, then head measurement is preferable, since

$$V = \int A\, dh + C \tag{9-2}$$

where A = area of tank or vessel, and no other factors need be known. If pressure measurement must be used because of details of particular application, then,

$$V = \int \frac{A}{\gamma}\, dP + C = \int \frac{A}{\gamma_w G}\, dP + C \tag{9-3}$$

and the measurement of volume depends on density or specific gravity.

If the purpose of the level measurement is to determine the weight of liquid contained in a vessel, then pressure measurement is preferable, since

$$W = \int A\, dP + C \tag{9-4}$$

and no other factors need be known. If head measurement must be used, then,

$$W = \int A\gamma\, dh + C = \int A\gamma_w G\, dh + C \tag{9-5}$$

and the measurement of weight depends on density or specific gravity.

If pressure at a reference point in the vessel, usually the bottom outlet, is the purpose of level measurement, then, pressure measurement obviously is preferred. If head measurement must be used, then,

$$p = \gamma_h = \gamma_w G_h \tag{9-6}$$

and the measurement of pressure depends on density or specific gravity.

Therefore, the selection of one method or the other for measuring level depends on the purpose for which the measurement is made.

Density and specific gravity may also be inferred from a pressure measurement, since

$$\gamma = \frac{p}{h} \tag{9-7}$$

so that, at constant head, pressure is proportional to density. Also,

$$G = \frac{p}{\gamma_w h} \tag{9-8}$$

so that, at constant head, pressure is proportional to specific gravity.

Direct Measurement of Liquid Level

The simplest method of measuring liquid level directly is to use an ordinary sight glass or gage glass on the vessel. This method is quite effective for clean colored liquids if remote indication is not required.

Electric level devices use electrodes placed in the tank to make electric contact when the level reaches a given point: the number of electrodes equal to the number of desired points of measurement. Electric signal lights or other indicating mechanisms can then be employed at a remote point. The use of electrodes is limited, since continuous measurement of level requires a large number of electrodes. Other electric devices employ the dielectric constant of the liquid to alter the electric capacitance between two continuous electrodes. All these methods are generally limited by the characteristics of the liquids: some are nonconducting, some are explosive, and some contain solids which deposit on the electrodes.

Another electrical method of liquid-level measurement is represented by the Gagetron of the Engineering Laboratories. This device operates from a float, on which is placed a small iridium–platinum needle containing a small quantity of radium salts. The float is guided by vertical rods and follows the free surface of the liquid. Gamma rays emitted by the radium salts are detected by a Geiger–Mueller counter placed at the top and outside of the vessel containing the liquid. The intensity of radiation is proportional to the square of the distance of the float to the counter. The electric pulses at the counter are integrated and rectified to produce a direct current, which can be measured by an automatic-balance potentiometer-type instrument. Thus, the instrument can be calibrated in terms of liquid level.

A widely used method for directly measuring level is the float-and-tape method shown schematically in Fig. 9–1. It is employed in open vessels only. The float, usually of nickel-plated copper, rests on the surface of the liquid, supported by buoyant

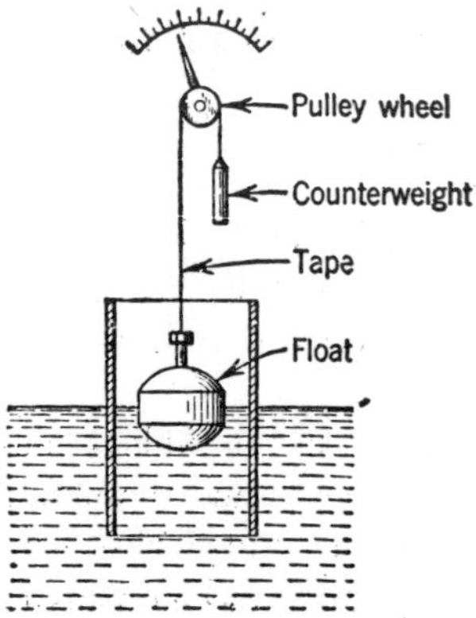

Fig. 9–1. Float-and-Tape Liquid-Level Gage

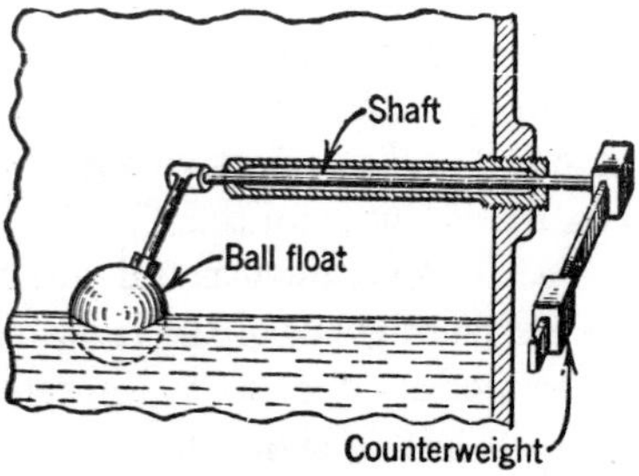

Fig. 9–2. Float-and-Shaft Liquid-Level Unit

force. The float is made with a sloping top in order to avoid building up of solid material on top of the float, thereby changing its weight. The float is connected to the drum by a thin, lightweight, flexible tape or cable. Slipping of the tape on the drum is prevented by a direct connection of the tape to the drum. By a suitable reduction in motion the pointer indicates on a scale calibrated in feet or other units. Both indicating and recording instruments of this type are available in ranges from 4 in. up to 60 ft level. These instruments must be installed close to and preferably over the vessel in which the level is measured so as not to introduce extra pulleys into the tape line.

The float-and-shaft type is shown schematically in Fig. 9–2. It is employed either in open vessels or in pressure vessels having up to 1000 psi internal pressure. The temperature in the vessel may be up to 750°F, in which case the pressure limit is lower. The float-and-shaft level device is suitable for a wide range of liquids and semiliquids. Difficulties are sometimes encountered when the liquid deposits solids on the float and when the liquid level is foaming or turbulent. The float rests on the surface of the liquid, and the motion of the float is taken through the stuffing box by the shaft. The counterweight is adjustable, so that the float can be made to ride half-submerged. This is important, since floats are usually spherical, and this is the point of maximum area. The float cage may be obtained in steel or cast iron, and the float in copper, stainless steel, nickel, or alumi-

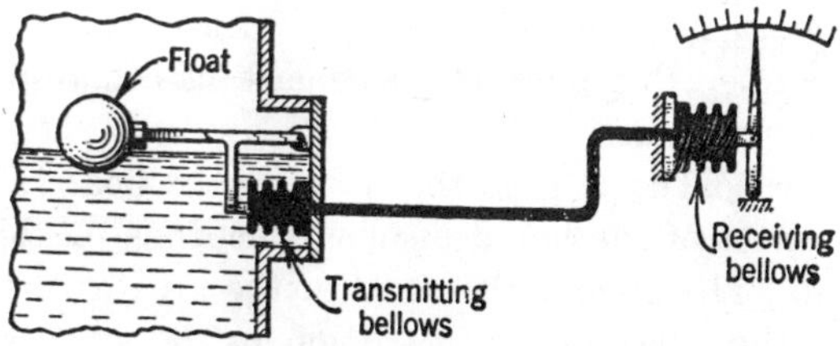

FIG. 9-3. Hydraulic Remote Transmission of Liquid Level—Liquidometer Corporation

num. The rotation of the shaft may be converted into a change of pneumatic pressure by the use of a pneumatic transmission system similar to those described in earlier chapters. Indicating and recording instruments, remotely located, may be operated from the float when pneumatic means of transmission are used. Often, the motion of the float is used to operate directly a valve for control of level or to operate directly a recording instrument. The range of these instruments is restricted, and level changes greater than 10 to 20 in. cannot generally be measured.

Several different means are in use for transmitting the arm motion of the float-and-shaft-type unit to a remote point. One of these (Liquidometer Corporation) is illustrated in Fig. 9-3. The movement of the float causes a displacement of a metallic bellows which is solidly filled with oil. The displacement of oil is carried by the tube to an identical receiving bellows at a remote point. The receiving-bellows displacement is equal to the transmitting-bellows displacement, and the instrument may be calibrated in terms of liquid level or volume. Temperature compensation for expansion of the hydraulic oil is provided. Indicating- and recording-type instruments are made.

The accuracy of these free-float-type liquid-level instruments depends almost entirely on their installation. Usually sufficient manual adjustments are provided on the instruments so that the zero and span can be properly set after installation. By individual calibration very good accuracies can be obtained if the surface of the liquid is reasonably smooth and the liquid itself clean. For volume measurement the area characteristic of the tank must be known. For weight measurement the area characteristic of the tank must be known, and the density or specific gravity must be constant. Greater measurement accuracy of volume can be achieved in tall, slender tanks than in short tanks of the same volume.

With the float-type gage the force on the float caused by the buoyant effect of the liquid displaced depends on the density of the liquid. The greater the density, the greater the force, and, therefore, the effect of static friction in the system is reduced. The shaft and other bearings require maintenance, particularly when the vessel is under pressure.

Pressure (Level) Measurements in Open Vessels

For pressure measurement of liquid level only a pressure-measuring instrument is required, provided that the density or specific gravity of the liquid is known. The most common pressure-gage systems employed for liquid-level measurement are known as: the bubbler system. the diaphragm-box system, and the air-trap system. The displacement-float system is also widely used. The following discussion is based on the measurement of water level. For liquids of other densities, the level calibration in feet or inches must be suitably modified.

The *bubbler system* which is the most popular is shown in Fig. 9–4. This system is satisfactory for nearly all liquids, even corrosive liquids and liquids containing suspended solids. Practically the only limitation to this system is that clogging of the bubbler pipe may occur on a few semisolids such as chemical slurries. A length of 1-in. pipe is lowered into the vessel to a point about 3 in. above the sediment line. Air or any suitable gas is supplied to the pipe through a valve and sight-feed bubbler so that the gas bubbles out the open end of the pipe at a rate of about 3 cu ft per hr. A liquid with a flow of about 1 gal per hr can sometimes be substituted for the air or gas supply. A pressure gage of suitable type is connected separately to the upper end of the bubbler pipe. The system operates by building up a pressure in the feed line until the gas escapes and the flow stabilizes at the rate determined by the valve or other regulator in the feed line. The pressure in the bubbler pipe necessary to cause the flow is just negligibly higher than the liquid head over the bubbler pipe. Therefore, the air, gas, or liquid

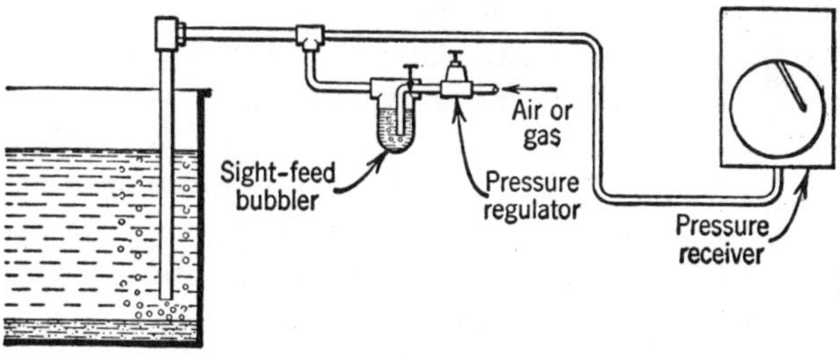

FIG. 9–4. Bubbler System for Liquid-Level Measurement

feed must be available at a pressure slightly higher than the maximum head to be measured. This pressure is measured by the pressure gage which can be calibrated in terms of head or level.

To measure the pressure, any of the instruments for pressure measurement are suitable. The range of the instrument depends on the height to be measured and the specific gravity of the liquid. Assuming a specific gravity of 1.0, the diaphragm pressure gage can be used from about 4 in. to 150 ft change in level, the differential-pressure meter (used as a static pressure gage) from about 5 in. to 400 in. level change, and a pressure-spring gage from about 150 in. to 250 ft. The pressure gage may be installed up to about 1000 ft from the vessel and can be placed either above or below the vessel. The rate of feed to the bubbler pipe should preferably be controlled with an orifice and differential-pressure regulator. Often a small rotameter is used to indicate the flow of feed. These accessories can be obtained in small sizes specifically adapted to the bubbler system. Since they are inexpensive, the improvement in accuracy through constant flow is well worth the cost.

The *diaphragm-box system* is shown in Fig. 9–5. It is very similar to the diaphragm seal used for pressure gages, with the difference that the diaphragm is very slack, thin, and flexible, and the system is filled with air. The diaphragm box is usually suspended in the vessel by a chain, or, if currents exist in the vessel, by a fixed support. The box is connected to the pressure receiver. The calculations regarding pressure, head, and density are the same as for the bubbler system. The instru-

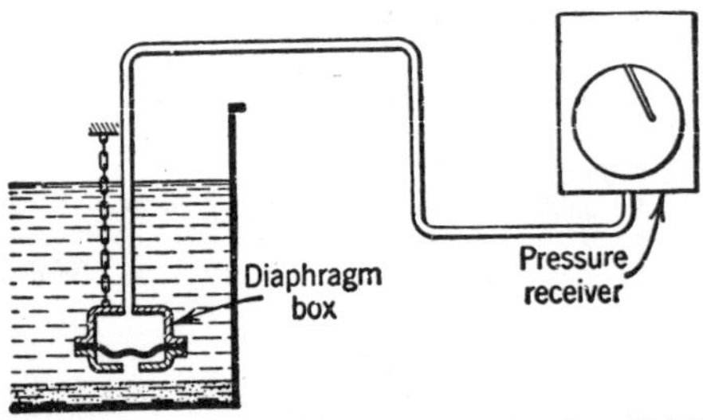

FIG. 9–5. Diaphragm-Box System for Liquid-Level Measurement

ment may be located up to about 500 ft from the vessel and either above or below it. Changes in level from about 20 in. up to 250 ft may be measured. This system is limited somewhat for use in corrosive liquids. Diaphragm boxes are made of cast iron or bronze, and diaphragms are of neoprene fabric. To avoid corrosion or attack of the diaphragm, liquid seals may be used as with any pressure gage. For liquids with suspended solids, a liquid flushing arrangement can be made to clear the open end of the diaphragm box. It is essential that the complete system be airtight since loss of air would prevent operation.

The *air-trap system* is shown in Fig. 9–6. This system is very similar to the diaphragm-box system except that no diaphragm is used; the reason is that no suitable diaphragm material can be found for some kinds of corrosive liquids. The pressure receiver operates from the pressure of the trapped air, and all other details are the same as in the diaphragm-box system. The main limitation of this type is the possibility of slow loss of air trapped in the system. For this reason, an auxiliary tap should be provided for filling the system with air as frequently as may be necessary.

The pressure measurement of liquid level is theoretically as accurate as the pressure gage itself. Weight determinations are direct when pressure measurements are used. Volume determinations depend on density or specific gravity. In this latter case, temperature variations, causing variations in density, will result in a corresponding error. Density variations or density "strata" in the column of liquid also may cause an error. In

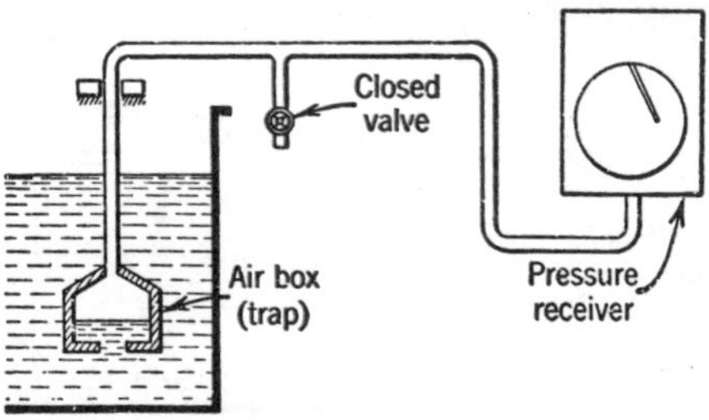

FIG. 9-6. Air-Trap System for Liquid-Level Measurement

addition, the gage reading must be corrected for various heads which may exist in the connecting line.

Level Measurement in Pressure Vessels

When a vessel contains a liquid under pressure, pressure measurement of the liquid level becomes much more difficult. For example, if a vessel contains 10 ft of water at a vessel pressure of 100 psig, a 1-ft change in head requires measuring a change of pressure of about one part in 240. This measurement is very near the dead-zone limitations of many static pressure gages; consequently, the error of measurement would be large.

Differential-pressure measurement is therefore suited to level measurement in closed vessels. Any of the differential-pressure meters such as the mercury manometer meter, the bell meter, the bellows meter, the ring meter or the pneumatic-balance pressure meter may be used. These meters measure variations in liquid level as low as 1 in. and as high as 100 ft, even up to static pressures of 10,000 psi.

The use of the manometer is indicated in Fig. 9-7. The relation between head and float displacement is

$$h = \left[\frac{\gamma_m}{\gamma_l}\left(1 + \frac{A_2}{A_1}\right) - 1\right] d - h_0 \tag{9-9}$$

where γ_m = density of manometer fluid
A_2 = area of float chamber
A_1 = area of range tube
d = float displacement
γ_l = liquid weight density

By selecting the value of area A_1, the desired head can be properly related to any given float displacement. By suitable cali-

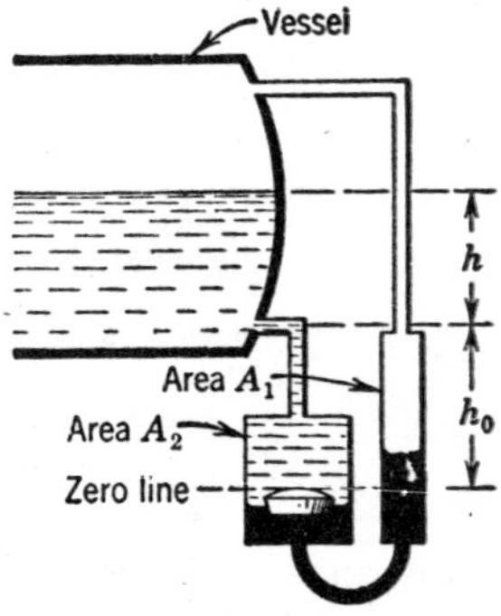

FIG. 9-7. Differential-Pressure Manometer for Measuring Liquid Level

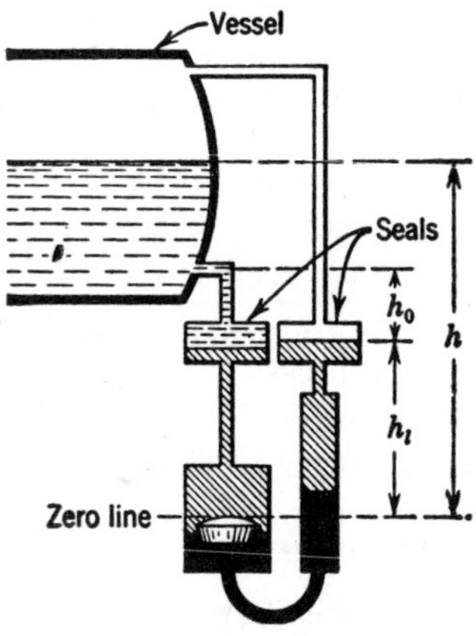

FIG. 9-8. Use of Liquid Seals with a Manometer

bration shift, the instrument can be corrected for the head h_0 from the instrument to the bottom tap. This may be accomplished through a mechanical shift of the pen or pointer of the manometer instrument or simply by adding more mercury to the manometer until the desired suppression is attained.

In many cases the liquid in the vessel cannot be allowed to enter the meter. When the liquid is corrosive, contains suspended solids, or is highly volatile, it is necessary to use the sealing methods previously described for pressure gages. Those often employed are the liquid seal and the purge system. The use of the liquid seal is shown in Fig. 9-8. The effect of the zero head $(h_l + h_0)$ can be corrected by a calibration shift in the meter.

The differential-pressure measurement of liquid level is theoretically as accurate as the meter itself. However, the following factors introduce variable errors:

1. The zero head between the meter and the lowest pressure tap.
2. The head of condensed liquids in the vessel pressure tap leg.
3. The head in each leg caused by the sealing methods.
4. The head of heavy vapor over the liquid surface in the vessel.
5. The variation of densities of sealing liquid and condensed liquid.

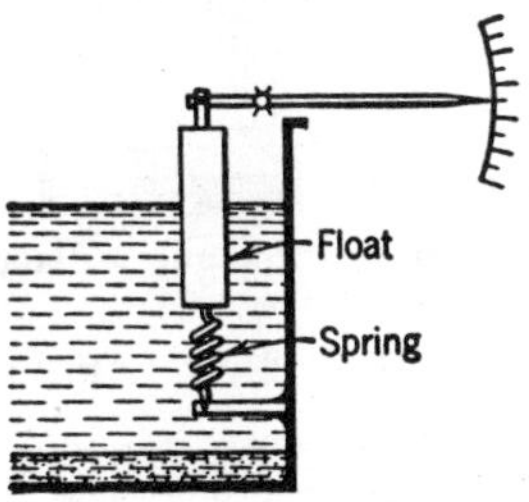

FIG. 9–9. Schematic Arrangement—Displacement-Float Liquid-Level Gage

By careful planning these errors can at least be reduced to a minimum.

The displacement-float-type liquid-level gages are also in general use, not only for pressure vessels, but also for open vessels. Their range is usually from about 14 to 180 in. change in level, and they may be used up to about 600 psi vessel pressure. One type is shown schematically in Fig. 9–9. A long float of constant area is suspended in the liquid in which the level is measured. The weight of liquid displaced causes an upward force on the float, which is counterbalanced by the force exerted by the spring. The movement of the float is carried by the shaft to an indicating pointer or can be made to operate a pneumatic transmitting means. The displacement-float gage usually employs a torque tube, as described for differential-pressure meters, in place of the shaft and spring shown in Fig. 9–9. This eliminates the necessity for sealed shafts and bearings.

The pneumatic-balance displacement-float gage is shown schematically in Fig. 9–10. The Republic Flowmeters Company and the Moore Products Company liquid-level transmitters operate on these principles. The force caused by the float is balanced by the pneumatic servosystem, as has already been described in the similar pressure gage and thermometer. The range of this kind of gage may be as low as 2 in. and as high as the length of the float permits. Liquid head up to 60 ft has been measured in this manner. The pneumatic-balance level gage can be used on pressure vessels since the float can be sealed in the pressure chamber by a sealing bellows or by a sealing

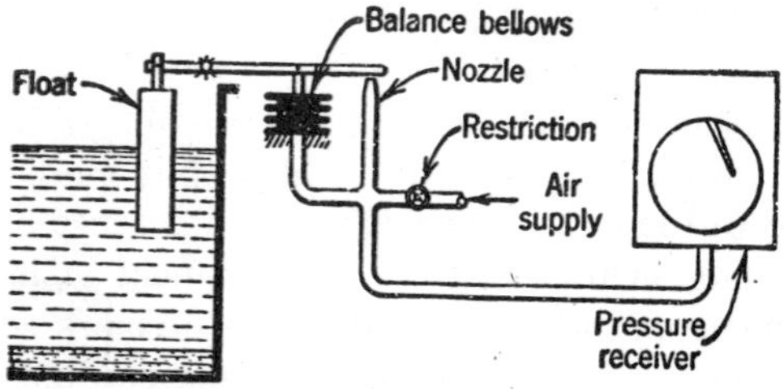

FIG. 9-10. Pneumatic-Balance Displacement-Float Liquid-Level Meter

diaphragm, thus eliminating the necessity for pressure-tight shafts.

The accuracy of the displacement-float gage is probably better than that of any other method as long as the specific gravity of the liquid is not low. The deposition of solids on the float is a source of error.

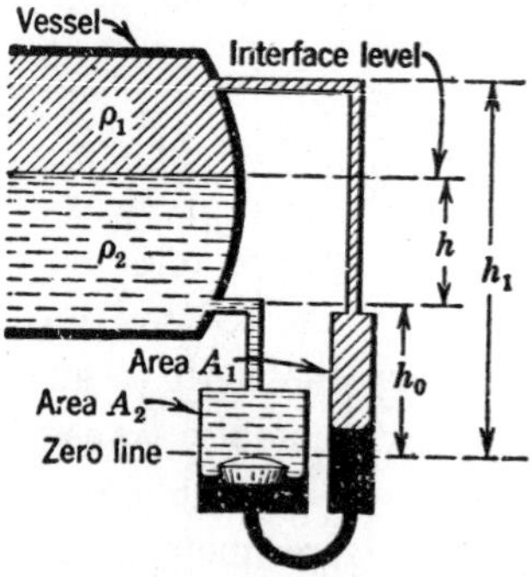

FIG. 9-11. Measurement of Interface Level

Measurement of Interface Level

In the measurement of liquid level described in the previous sections of this chapter, it has been assumed that the surface of the liquid level was bounded by a gas or a vapor of negligible weight density. However, the surface level between two immiscible liquids of differing specific gravity or between a liquid and a gas or vapor of appreciable density may also be measured. This is illustrated in Fig. 9-11. The relation between float displacement and head of interface level, if it is assumed that the less dense liquid always covers the upper tap, is

$$h = \left[\left(\frac{\gamma_m - \gamma_1}{\gamma_2 - \gamma_1}\right)\frac{A_2}{A_1} + \left(\frac{\gamma_m - \gamma_2}{\gamma_2 - \gamma_1}\right)\right] d - h_0 \qquad (9\text{-}10)$$

where γ_m = weight density of manometer liquid
γ_1 = weight density of lighter fluid
γ_2 = weight density of heavier fluid
A_2 = area of float chamber
A_1 = area of range tube
d = float displacement

The calibration of the meter depends on the difference in densities of the two fluids. Obviously, the measurement of interface level is more difficult as the density difference becomes small.

Almost any of the liquid-level gages may be used for interface level, but the differential-pressure meter, the free float, and the displacement-float types are most commonly employed. With the differential-pressure meter or the displacement-float gage it is not necessary that the interface level be clearly defined, but it may extend over a height of several inches. The level gage will measure an average position of the interface.

Density Measurement

The measurement of density or specific gravity may be accomplished by two methods: (1) the pressure at the bottom of a column of liquid (or gas) is proportional to the density; (2) the weight of a given volume of liquid (or gas) is proportional to density.

The measurement of liquid density using the pressure method is shown in Fig. 9–12. The head h_1 is held constant in the measuring tank or vessel by an overflow arrangement. The head h_0 in the reference chamber is constant, providing the reference liquid is nonvolatile. Therefore the differential pressure measured by the manometer is that due to the difference in densities of the two liquids. The system is calibrated to accommodate the different heads in the tank and reference chamber by adding sufficient mercury to the manometer to bring the float to the zero position when the densities of the two liquids are equal. Density can be continuously measured if the flow of liquid through the measuring tank is continuous. Note that the pressure in each chamber is measured by means of the bubbler system previously described.

Temperature corrections are necessary with this method of measurement. The manometer is calibrated in terms of density

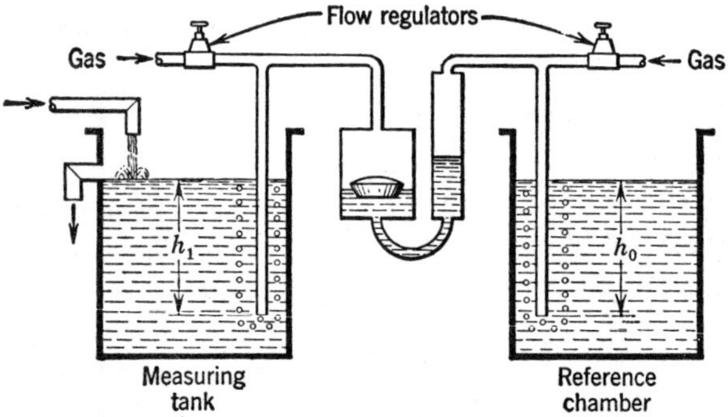

FIG. 9–12. Liquid-Level Method of Measuring Specific Gravity or Density

(or specific gravity) at a particular temperature. If the temperature of the flowing liquid varies, then it is necessary to measure this temperature separately and manually correct the manometer readings. To avoid the necessity for temperature corrections, a thermostatic heater can be used to maintain constant the temperature of the flowing liquid.

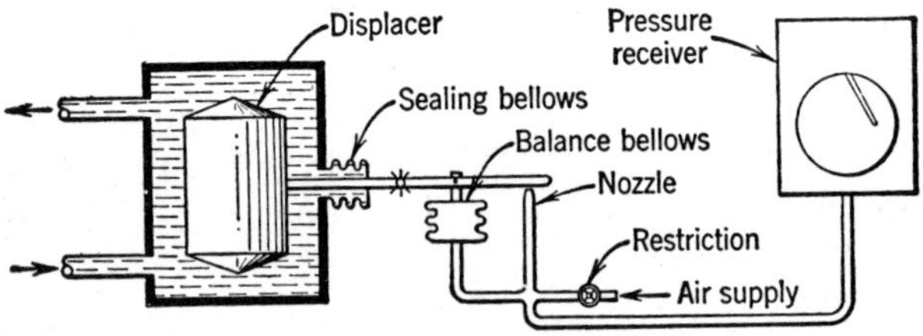

FIG. 9–13. Displacement Meter for Measuring Specific Gravity—Republic Flowmeters Company

The measurement of liquid density, using the weight method, is shown in Fig. 9–13, representing The Republic Flowmeters Company liquid-density gage. The liquid flows continuously in and out of the displacer chamber at a rate of about 3 gpm. The upward force on the balance beam is due to the displaced volume of liquid. This force is balanced by the pneumatic servomechanism previously described. As a result, the pressure transmitted to the receiver is proportional to the density of the liquid. This equipment can be used for liquids having specific gravities 0.5 and higher. The densities of corrosive liquids can be measured, providing suitable materials of construction are selected. Also, the densities of liquids under pressure may be measured. Temperature corrections are necessary, however, if the temperature of the flowing liquid changes. A thermostatic heater may be necessary to hold the temperature of the flowing liquid constant.

The hydrometer method can be used for continuously measuring specific gravity of liquids, as shown in Fig. 9–14. The level of liquid is maintained constant at the overflow tube. The hydrometer sinks or rises in the liquid as the specific gravity varies (the action of the hydrometer depends on buoyancy effect of liquid displaced). The lower end of the hydrometer supports an armature in an inductance coil, and the specific gravity can be remotely indicated. Usually the specific-gravity instrument is arranged to indicate or record the temperature of the liquid as well as the value of specific gravity, so that correction for temperature variation of specific gravity can be made.

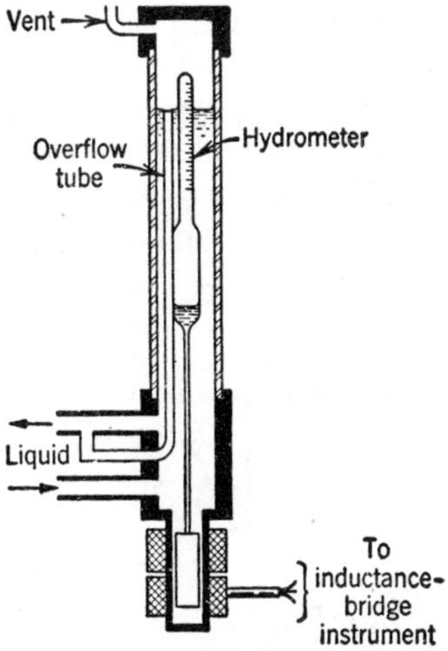

FIG. 9–14. Hydrometer Specific Gravity Meter

Specific-gravity meters for gases generally operate on the same principles as meters used for liquids. One type utilizes the weight of a known volume of gas contained within a float. The float is mounted on a balance beam so that variations in density of the gas cause the balance beam to deflect and operate a recording device. Automatic compensation for ambient temperature and barometric pressure is made. Since the forces involved are much smaller, the balance is necessarily more delicate than with liquids. Another common method of measuring specific gravity of gases is to compare the weight of a column of the gas with the weight of a column of air of the same height. A thermostatic coil is used to compensate for temperature variations.

Level Measurement by Weighing

The simplest and most obvious way to measure the level of a liquid or a dry material is to weigh the contents of the entire vessel. When the area of the vessel is known, the level can be determined. Actually, this is often done for dry materials where other methods fail. However, it is expensive when large and heavy equipment is involved.

Weighing methods are most accurate when total weight of material is the important factor, rather than density, level, or volume. In addition, it is about the only reliable method for continuously determining the amount of dry materials in a vessel or tank.

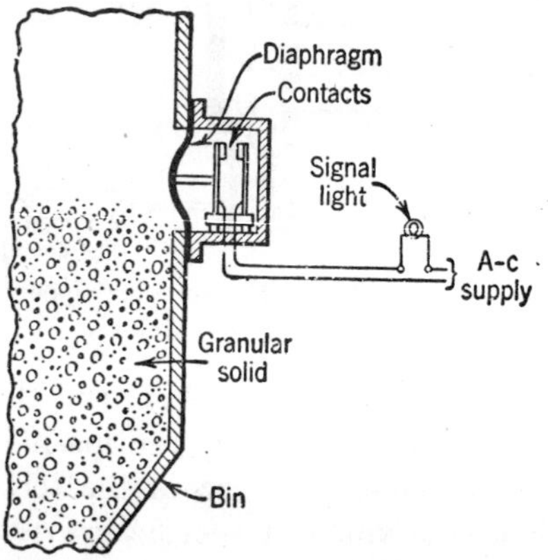

FIG. 9–15. Level Indicator for Dry Materials—Bin-Dicator Company

Level of Dry Materials

Oftentimes it is merely sufficient to know whether or not a bin or vessel has a specified level of material. In this case a level indicator of the signaling type can be employed.

The use of a signaling indicator, the device of the Bin-Dicator Company, is shown in Fig. 9–15. This device operates from a light, flexible diaphragm which mechanically positions a switch. The weight of the dry material in the bin acts against the diaphragm when the material in the bin reaches a desired level. These devices are satisfactory for all dry bulk materials of sufficient weight density.

Probe or contact-type electric indicators may also be used. These may operate from the dielectric effect or from the electric conductance of the material. In most cases it is necessary to select different types, depending on the characteristics of the material.

Problems

9–1. A cylindrical vessel has plane ends, diameter d, and length l. When the vessel is vertical (axis of cylinder vertical) and contains water, find an expression for:

(*a*) Volume of water contained versus height of liquid level from bottom.
(*b*) Weight contained versus pressure (pounds per square inch gage) at bottom of vessel.

9–2. When the cylindrical vessel of problem 9–1 is horizontal, find an expression for:

(*a*) Volume contained versus height of liquid level from bottom.
(*b*) Weight contained versus pressure (pounds per square inch gage) at bottom of vessel.

9–3. It is desired to know the volume contents of a gasoline storage tank. All other factors being equal, should direct measurement or pressure measurement of level be used? Why?

9–4. It is desired to know the weight contents of a liquid-catalyst storage chamber. If weighing methods are too expensive, would you recommend a free-float level gage or a diaphragm-box and pressure-gage system? Why?

9–5. A diaphragm-box and pressure-gage system is used to measure liquid level in a cubical tank 10 ft on a side. The tank contains liquid of specific gravity 0.8 at ambient temperature. Calculate the span of the pressure gage.

9–6. A Hortonsphere (a spherical vessel of diameter d) contains a liquid of density D under pressure. An ordinary U-tube mercury manometer is used to measure the level. (*a*) Find an expression relating level height to manometer differential. (*b*) Find an expression relating weight of liquid contained to manometer differential. Assume the manometer zero level even with the bottom of the vessel.

9–7. It is desired to measure the level of water in a tank. If the level is measured with a pressure gage, and the ambient-temperature range is from 40 to 120°F, derive an expression for error in measurement due to density variation. Assume density varies linearly with temperature.

9–8. A tank of constant cross-section area is filled with liquid to height h_0. If there is an outlet at the bottom and the flow through the outlet Q is

$$Q = c\sqrt{2gh}$$

find an expression for the level in the vessel versus time.

9-9. A vessel of 60 ft height is filled by adding a second immiscible liquid (sp gr 0.6) above the first and allowing the first (sp gr 1.0) to run out the bottom. If a pressure gage is used to measure the pressure at the bottom of the tank, what is the range of the pressure gage to indicate the full motion of the interface?

10· Flowmetering

The measurement of the flow rate and flow quantity of materials is made primarily for the purpose of determining the proportions of materials introduced to a manufacturing process and the amount of materials evolved by the process. Secondarily, flow measurements are made for the purpose of cost accounting usually for steam and water services.

Measurement of flow rate and quantity is the oldest art in the science of instrumentation, and its history extends back at least to the great hydraulic and public engineering works of the Romans. Both before and since this time a long list of illustrious scientific names became associated with hydraulics and fluid mechanics; a few of these are Archimedes, Toricelli, Pascal, Euler, Bernoulli, Stokes, Reynolds, and Prandtl. These associations were brought about by the great complexity of problems in fluid flow, and, although there is a large body of theory that is useful in the study of effects, the accurate solutions to problems are, at the present time, founded on experimental data.

Flow of fluids in closed pipes can successfully be measured by a number of methods. The head meter is the most common and operates by measuring the pressure differential across a suitable restriction to flow. Area meters are very often employed, and they operate from the variation in area of the fluid stream. Velocity meters also are used to measure flow rate. Quantity meters are a different type which primarily measure the quantity of fluid passed at a given point (the time integral of flow rate).

Flow of fluids in open channels or flumes is measured by weirs and by hydraulic flumes.

Several excellent references deserve special mention, since it is neither possible nor neecssary to review all the theory and practice of fluid mechanics. *Fluid Mechanics* by R. C. Binder [1] and *Elementary Mechanics of Fluids* by H. Rouse [2] provide an excellent background of fluid statics and dynamics. The ASME book on fluid meters [3] contains a thorough discussion of theory and practice. Industrial practices and commercial flow calculations are given in complete detail in Gess and Irwin [4] and Spink.[5]

Flow of Incompressible Fluids in Pipes

If a restriction such as an orifice is introduced into a pipe line (Fig. 10–1), the relation between pressure and velocity can

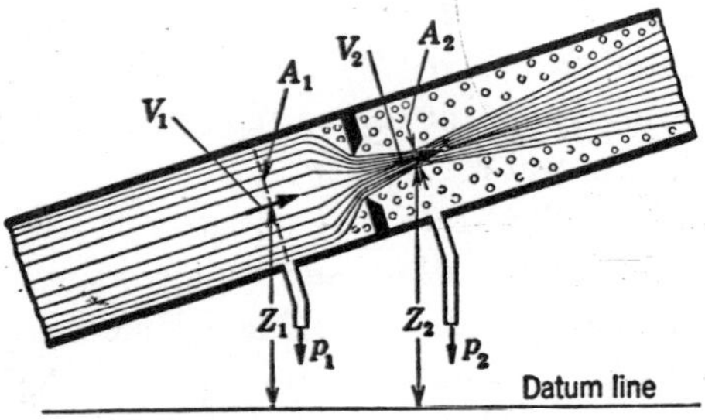

FIG. 10–1. Flow through a Restriction in a Closed Pipe

be found from the law of conservation of energy. For the flow on incompressible ideal fluids, Bernoulli's equation is used: [6]

$$\frac{p_2}{\rho} + \frac{V_{m2}^2}{2g} + Z_2 = \frac{p_1}{\rho} + \frac{V_{m1}^2}{2g} + Z_1 \qquad (10\text{–}1)$$

[1] R. C. Binder, *Fluid Mechanics,* Prentice-Hall, 1943.

[2] Hunter Rouse, *Elementary Mechanics of Fluids,* John Wiley & Sons, 1946.

[3] Fluid Meters, *Their Theory and Application,* Fourth Edition, ASME, 1937.

[4] L. Gess and R. D. Irwin, *Flowmeter Engineering Handbook,* Brown Instrument Co., Philadelphia, 1946.

[5] L. K. Spink, *Principles and Practice of Flowmetering,* Foxboro Co., 1945.

[6] Symbols used conform, in general, to those given by the ASME Special Research Committee on Fluid Meters.

where p = static pressure, absolute
V_m = fluid stream velocity
Z = elevation of center line of pipe
ρ = fluid weight density
g = acceleration due to gravity

For fluid-flow measurement the difference in level $(Z_2 - Z_1)$ is very small, even when the pipe is vertical, since the length of flow measuring restrictions, such as orifices, is very small. Therefore,

$$V_{m2}^2 - V_{m1}^2 = \frac{2g}{\rho}(p_1 - p_2) \tag{10-2}$$

Measuring the flow of fluids involves determining the rate of flow, and so, from the equation of continuity,

$$q = A_2 V_{m2} = A_1 V_{m1} \tag{10-3}$$

where A = area of flow stream
q = flow rate

However, the area A_2 in Fig. 10–1 is not an area that is explicitly known. Furthermore, physical considerations indicate that the vena contracta, the point of minimum area of the flow stream, is located at a position along the pipe, depending on the flow rate. With any practical measuring device the pressure taps for determining the upstream and downstream pressure must be fixed at their selected positions. Consequently, it is necessary to measure the pressure at the points indicated and to base area and velocity measurements directly at the restriction. For this reason equation 10–3 is modified by adding a coefficient of contraction and a coefficient of velocity. Actually these coefficients may be represented by one coefficient. Combining equations 10–2 and 10–3 gives

$$q = K_O A_2 \sqrt{\frac{2g}{\rho}(p_1 - p_2)} \tag{10-4}$$

where $K_O = C/\sqrt{1 - \beta^4}$ = flow coefficient
C = discharge coefficient
$\beta = \sqrt{A_2/A_1}$ = diameter ratio = d/D
d = diameter of orifice
D = inside diameter of pipe

The factor $\sqrt{1 - \beta^4}$ is termed the velocity of approach factor, since it is 1.0 if the upstream velocity V_{m1} is zero. To reduce the previous equation to a more useful form, it is often written

$$q = \frac{\pi K_O \beta^2 D^2}{4} \sqrt{\frac{2g}{\rho}(p_1 - p_2)} \tag{10-5}$$

where q and $(p_1 - p_2)$ are the variable quantities in flow measurement. This equation is based on flow of incompressible ideal fluids.

Actual fluids are not ideal but viscous. Furthermore, the flow of fluids in industrial practice is almost always turbulent. The exact analysis of the general problem of turbulent flow has never been made; consequently, it is necessary to rely on the law of dynamical similarity that, if the flows of two fluids around geometrically similar bodies are also geometrically similar, then the two flows are dynamically similar if the corresponding fluid forces on the two bodies are proportional at any instant. Now, fluid flow through orifices is characterized by inertia forces and viscous forces, and these are related by the Reynolds number:

$$R_d = \frac{\rho V_m d}{\mu} \tag{10-6}$$

where R_d = Reynolds number, dimensionless
V_m = average velocity of fluid stream
d = orifice diameter
μ = absolute viscosity

To use the symbols of the flow problem we may write the Reynolds number as

$$R_D = \frac{4\rho q}{\pi \mu \beta D} \tag{10-7}$$

If therefore the fluid flow is known for one set of conditions providing a given Reynolds number, the flow may be found at many other sets of conditions giving the same Reynolds number. The relationship of the flow coefficient K to Reynolds number and the diameter ratio of the orifice is given in Tables A–21, A–22, and A–23 (appendix).

Turbulent flow occurs at the Reynolds number above 2000 to 2200. As stated previously, the range of velocities, densities,

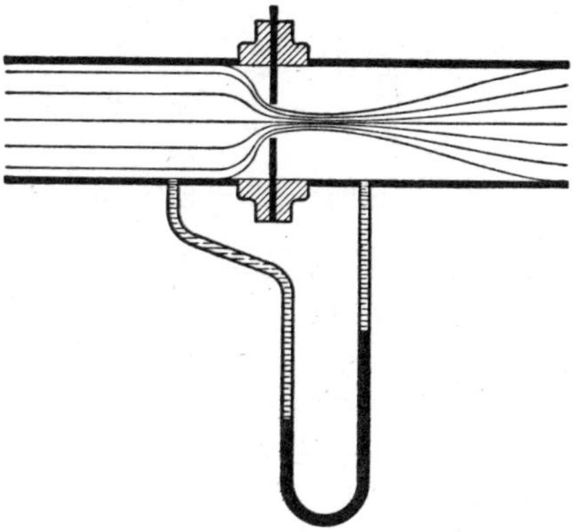

FIG. 10–2. Orifice and Pressure-Differential Measurement

pipe diameters, and viscosities met with in commercial practice is such as to result in a Reynolds number between about 10,000 and 1,000,000, and the flow is therefore in the range of turbulent flow. There is an important exception to this rule encountered in the flow of heavy or viscous oils. In such cases the flow may be laminar with a Reynolds number less than 2000 or else turbulent but with a Reynolds number between 2200 and 10,000. In either case the flow coefficients are not known (see Tables A–21, A–22, and A–23, appendix), and it is necessary to use a smaller pipe diameter to increase the Reynolds number to a point where the coefficient is known.

Measuring fluid flow with an orifice and differential-pressure manometer, as shown in Fig. 10–2, requires that the effect of the fluid over the manometer liquid be taken into account. Furthermore, the pressure differential at the orifice is usually expressed in liquid-column height. Then,

$$p_1 - p_2 = (\gamma_m - \gamma_f)h \tag{10-8}$$

where h = differential at restriction, feet liquid column height
γ_m = weight density of manometer fluid
γ_f = weight density of fluid over the manometer fluid

The flow rate is ordinarily expressed in terms of given base conditions, usually 60°F, and so the ratio of densities should be included in the final equation.

The flow equation for incompressible fluids (liquids) is

$$q = \frac{\pi}{4} K_O \beta^2 D^2 \frac{\sqrt{(\gamma_m - \gamma_f)\rho}}{\rho_b} \sqrt{2gh} \tag{10-9}$$

where q = flow rate at given base density (ρ_b), cubic feet per second
K_O = flow coefficient, Tables A–21, A–22, A–23
β = diameter ratio (d/D)
d = orifice diameter, feet
D = inside pipe diameter, feet
γ_m = density of manometer fluid, pounds per cubic foot
γ_f = density of fluid over manometer, pounds per cubic foot
ρ = density of flowing fluid, pounds per cubic foot
ρ_b = density of flowing fluid at given base condition, pounds per cubic foot
h = manometer differential, feet
μ = absolute viscosity of flowing fluid, pounds per foot-second

The only empirical fact required in the use of the foregoing equations is the flow coefficient K. Data on density of mercury and water are given in Table A–18 (appendix). More complete data on densities and viscosities of fluids can be found in the references mentioned at the beginning of this chapter.

Flow of Compressible Fluids in Pipes

The relation between pressure and velocity for flow of a compressible fluid through an orifice can be found from the law of conservation of energy as employed in thermodynamics. Assuming no heat flow to or from the fluid (adiabatic flow) and no external work done on or by the fluid, and neglecting the very small datum level difference ($Z_1 - Z_2$), we have

$$p_2v_2 + \frac{V_{m2}^2}{2g} + JE_2 = p_1v_1 + \frac{V_{m1}^2}{2g} + JE_1 \qquad (10\text{–}10)$$

where E = internal molecular energy of fluid
J = work equivalent of heat
v = specific volume of fluid

Employing the definition of enthalpy H gives

$$V_{m2}^2 - V_{m1}^2 = 2gJ(H_1 - H_2) \qquad (10\text{–}11)$$

For an ideal gas and if specific heats are constant,

$$H_1 - H_2 = \frac{KR}{J(K-1)} T_1 \left[1 - \left(\frac{p_2}{p_1}\right)^{(K-1)/K}\right] \qquad (10\text{–}12)$$

where K = ratio of specific heats = C_p/C_v
R = gas constant for a given gas
T = absolute temperature

From the equation of continuity (conservation of mass),

$$w = \frac{A_2V_{m2}}{v_2} = \frac{A_1V_{m1}}{v_1} \qquad (10\text{–}13)$$

where w is the weight rate of flow, pounds per second. Combining the foregoing equations and simplifying, we have the relation for isentropic flow of ideal gases,

$$w = \frac{\pi C\beta^2D^2}{4}\sqrt{\frac{2gKp_1}{(K-1)v_1}}\sqrt{\frac{\left(\frac{p_2}{p_1}\right)^{2/K} - \left(\frac{p_2}{p_1}\right)^{(K+1)/K}}{1 - \beta^4\left(\frac{p_2}{p_1}\right)^{2/K}}} \qquad (10\text{–}14)$$

where C is the discharge coefficient.

It is at once evident that this equation must be simplified if any simple means is to be used for flow measurement. Specifically we would prefer to base the flow measurement on the pressure differential $p_1 - p_2$ rather than on the pressure ratio

p_2/p_1. This allows the use of the data on flow coefficient and Reynolds number. By employing a series expansion[7] by the binomial theorem and dropping terms of second and higher order, we may write

$$\left(\frac{p_2}{p_1}\right)^{2/K} \cong 1 - \frac{2}{K}\left(1 - \frac{p_2}{p_1}\right) \tag{10–15}$$

By performing a similar expansion in a binomial series it may also be shown that

$$\left(\frac{p_2}{p_1}\right)^{(K+1)/K} \cong 1 - \left(\frac{K+1}{K}\right)\left(1 - \frac{p_2}{p_1}\right) \tag{10–16}$$

Substituting these relations in equation 10–14 and simplifying yields

$$w = \frac{\pi C \beta^2 D^2}{4} \sqrt{\frac{2g(p_1 - p_2)}{v_1\left[1 - \beta^4\left(\frac{p_2}{p_1}\right)^{2/K}\right]}} \tag{10–17}$$

The velocity of approach factor $[1 - \beta^4(p_2/p_1)^{2/K}]$ can be included in the value of flow coefficient K. The flow-coefficient tables show that the value of the flow coefficient is not influenced to any appreciable extent by a reasonable variation of the pressure ratio. Therefore, the velocity of approach factor can be taken as $(1 - \beta^4)$, and equation 10–17 reduces to

$$w = \frac{\pi \beta^2 D^2}{4} \frac{C}{\sqrt{1 - \beta^4}} \sqrt{\frac{2g(p_1 - p_2)}{v_1}} \tag{10–18}$$

This equation, however, is the simple hydraulic equation for incompressible fluids. It applies to flow of compressible fluids only when the change in density or specific volume is small. It is necessary to establish the range of validity of equation 10–18. This can be done by comparing the compressible flow rate to the incompressible flow rate:

[7] By writing $p_2/p_1 = 1 - x$.

$$\phi = \frac{w \text{ (compressible flow)}}{w \text{ (incompressible flow)}}$$

where ϕ is termed the rational expansion factor. It is indicative of the error in the simple equation, and

$$\phi = \sqrt{\frac{(1-\beta^4)\left[\left(\frac{p_2}{p_1}\right)^{2/K} - \left(\frac{p_2}{p_1}\right)^{(K+1)/K}\right]}{\left(\frac{K-1}{K}\right)\left(1-\frac{p_2}{p_1}\right)\left[1-\beta^4\left(\frac{p_2}{p_1}\right)^{2/K}\right]}} \qquad (10\text{–}19)$$

The value of the rational expansion factor and the error in equation 10–18 are given in Table 10–1.

TABLE 10–1. VALUE OF RATIONAL EXPANSION FACTOR ϕ
($\beta = 0.5$ and $K = 1.4$)

$\frac{p_2}{p_1}$	$\frac{p_1 - p_2}{p_1}$	ϕ	% Error in Equation 10–18 $(1-\phi)100$
1.000	0.000	1.000	0.000
0.990	0.010	0.994	0.6
0.980	0.020	0.9884	1.16
0.960	0.040	0.9765	2.35
0.940	0.060	0.9646	3.54
0.920	0.080	0.9526	4.74
0.900	0.100	0.9405	5.95
0.800	0.200	0.878	12.2
0.700	0.300	0.812	18.8
0.600	0.400	0.744	25.6

In industrial practice, the pressure ratio p_2/p_1 is often greater than 99 per cent, and the simple flow equation 10–18 may be used. When the pressure ratio is less than 99 per cent, there is too great an error in the calculation, and it is desirable to correct the equation without having to employ the rational expansion factor of equation 10–19. An empirical expansion factor has been determined for orifices using flange taps or vena-contracta taps

$$Y = 1 - [0.41 + 0.35\beta^4]\left[\frac{p_1 - p_2}{p_1}\right]\left[\frac{1}{K}\right] \qquad (10\text{–}20)$$

The expansion factor for orifices using pipe taps is [8]

$$Y = 1 - [0.333 + 1.145(\beta^2 + 0.7\beta^5 + 12\beta^{13})]\left[\frac{p_1 - p_2}{p_1}\right]\left[\frac{1}{K}\right] \tag{10-21}$$

The expansion factor Y is dimensionless.

When the gas contains appreciable moisture, a further correction is required to account correctly for the density of the vapor. The moisture factor M is

$$M = 1 + \frac{p_V}{p}\left(\frac{G_V}{G} - 1\right) \tag{10-22}$$

where p_V = vapor pressure
G_V = specific gravity of vapor referred to air at the same conditions of pressure and temperature

The specific volume of the gas may be found from

$$v = \frac{yRT}{p} \tag{10-23}$$

where y = compressibility factor
R = gas constant (2.70 in this equation)

The compressibility factor is found from compressibility charts [9] and is defined simply as a number that corrects the ideal-gas law to actual pressure, volume, and temperature data. The value of the compressibility factor is approximately 1.0 for many ordinary conditions of pressure and temperature.

The flow equation for gases is

$$q = \frac{\pi}{4} K_O\beta^2D^2Y\frac{v_b}{M_b}\sqrt{\frac{\gamma_m M_1}{v_1}}\sqrt{2gh} \tag{10-24}$$

[8] See ASME Fluid Meters, *loc. cit.* The empirical expansion factors should be used only in the limits $0.8 \leqq p_2/p_1 \leqq 1.0$ for which they are accurate to within ±0.5 per cent for all gases and vapors except steam For steam they are accurate to within ±1.0 per cent.

[9] See Gess, *loc. cit.*

where q = flow rate at given base temperature T_b and pressure p_b, cubic feet per second
K_O = flow coefficient, Tables A–21, A–22, A–23
β = diameter ratio, d/D
d = orifice diameter, feet
D = inside pipe diameter, feet
γ_m = density of manometer fluid, pounds per cubic feet
g = acceleration due to gravity, feet per second per second
h = manometer differential, feet
v_1 = specific volume of gas at upstream conditions, cubic feet per pound, equation 10–23
v_b = specific volume of gas at base conditions, cubic feet per pound, equation 10–23
M_1 = moisture factor at upstream conditions, equation 10–22
M_b = moisture factor at base conditions, equation 10–22
Y = expansion factor, equation 10–20 or equation 10–21
μ = absolute viscosity of flowing fluid, pounds per foot-second

The only experimental data required for gas-flow calculations are values for the flow coefficient K and the compressibility factor X. The flow coefficients are given in Tables A–21, A–22, and A–23 (appendix). The Reynolds number for flowing gases is

$$R_D = \frac{4}{\pi}\frac{q}{\mu v_1 \beta D} \tag{10–25}$$

The compressibility factor (often denoted by μ or Z) may be found from any compressibility chart. Values for density, specific volume, specific gravity, viscosity, specific heat, and gas constant may be found in a number of handbooks.

For the flow of a vapor such as steam, the only modification required in equation 10–24 is the inclusion of the density of the liquid over the manometer liquid. The equation for flow of a vapor is given by

$$q = \frac{\pi}{4} K_O \beta^2 D^2 Y \frac{v_b}{M_b} \sqrt{\frac{M_1(\gamma_m - \gamma_f)}{v_1}} \sqrt{2gh} \tag{10–26}$$

where γ_f is the density, pounds per cubic foot, of the condensed vapor over the manometer liquid; the other values have been given previously.

For all flow equations for orifices, that is, those for liquid, gas, and vapor, one further correction is required if the temperature of the flowing fluid is high. Thermal expansion of the metal of which the orifice plate is constructed increases the orifice diameter. The diameter ratio β remains approximately constant. The correction for thermal expansion can be made by multiplying the flow rate q by [10]

$$1 + 0.000015t$$

where t is the temperature at the orifice plate in degrees Fahrenheit. Ordinarily, this correction is small.

A limitation to the maximum flow rate of gases exists when the pressure ratio p_2/p_1 becomes small. This can be investigated by setting

$$\frac{\partial w}{\partial p_r} = 0 \tag{10-27}$$

where $p_r = p_2/p_1$. Performing this operation on equation 10–14 and assuming that the specific heat ratio is not a function of pressure ratio yields

$$(K+1)\left(\frac{p_2}{p_1}\right)^{(K-1)/K} - (K-1)\beta^4\left(\frac{p_2}{p_1}\right)^{(K+1)/K} = 2 \tag{10-28}$$

A simplification of the solution to this equation results if the second term is neglected in comparison to the other two. Then,

$$\frac{p_2}{p_1} \cong \left(\frac{2}{K+1}\right)^{K/(K-1)} \tag{10-29}$$

This pressure ratio is called the critical pressure ratio, because it is the condition for maximum flow of gas through an orifice or other such restriction to flow. For air the critical pressure ratio is approximately 0.528.

Of greater significance is the fact that, when the pressure ratio is less than the critical value, the flow rate does not depend

[10] See Gess and Irwin, *loc. cit.*

on the pressure difference but only on the upstream pressure p_1. In industrial practice this condition is rarely encountered when measuring fluid flow in pipe lines. On the other hand, when the fluid is discharged through an orifice or nozzle into the atmosphere, the upstream pressure is often great enough so that the pressure ratio is less than critical. In the latter case the flow rate must be computed by equation 10–14 in which the critical pressure ratio is used.

Orifice Installations

The restriction to flow in the pipe line most often takes the form of a thin-plate square-edge orifice. Figure 10–3 illustrates three types.

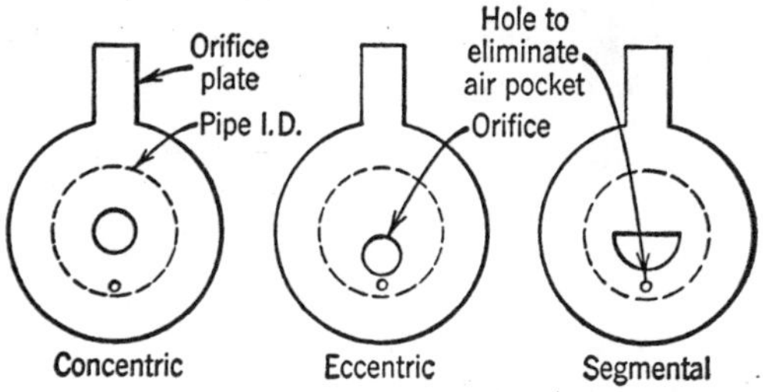

FIG. 10–3. Orifice Plate Styles

The *concentric orifice* is by far the most widely used. The *segmental* and *eccentric orifices* are used for measuring flow of fluids containing solids. In both, the orifice plate is located so that the bottom of the hole is nearly flush with the bottom inside of the pipe, as shown in Fig. 10–3. The segmental and eccentric orifices require special calibration since the standard flow coefficients are usable only for standard thin-plate, concentric orifices.

The concentric orifice plate is made of flat metal sheet with a circular hole, and it is installed in the pipe line with the hole or orifice concentric to the pipe. Orifice plates are made from steel, stainless steel, Monel, phosphor bronze, or almost any metal that will withstand the corrosive effects of the fluid. Its thickness is only sufficient to withstand the buckling forces caused by the pressure differential. The circular hole or orifice is carefully made with 90°, square, sharp edge upstream because this type can be manufactured more uniformly than one with round edges.

Wear and abrasion of this sharp edge greatly affect the accuracy of the orifice-flow measurement. It is necessary to err on the side of safety in selecting an orifice-plate material that will withstand the erosive effects caused by the fluid. In some cases it is advisable to replace the orifice frequently to maintain accuracy. In this respect a venturi tube may be worth consideration.

The orifice with flange taps (Fig. 10–4) is constructed so that the taps for measuring pressure differential are an integral part

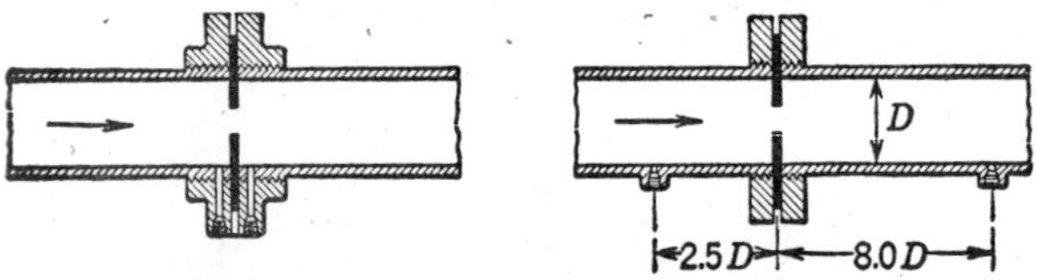

Flange taps Pipe taps

FIG. 10–4. Orifice with Flange or Pipe Taps

of the orifice assembly. The taps are usually located 1 in. either side of the orifice plate. This arrangement has the advantages that the orifice assembly is easily replaceable, no alterations in the pipe are required, and the pressure taps are accurately located.

The orifice with pipe taps is shown in Fig. 10–4. The pipe taps are made directly in the side of the pipe: the upstream tap located $2\frac{1}{2}$ pipe diameters from the orifice, and the downstream tap located 8 pipe diameters from the orifice. With this type, only the permanent pressure differential across the orifice is utilized; this differential being smaller for a given flow than with either flange taps, or vena-contracta taps. Pipe taps are commonly used in measuring the flow of gases.

The orifice with vena-contracta taps (Fig. 10–5) is arranged so that the downstream pressure tap is located a variable distance from the orifice, depending on pipe and orifice size. The taps are made directly in the pipe with the upstream tap one pipe diameter from the orifice and the downstream tap at the vena contracta. The location of the downstream tap is indicated in Fig. 10–5. The vena-contracta tap arrangement has the advantages that standard flanges may be used for the orifice plate, and the pressure differential is a maximum for a given flow.

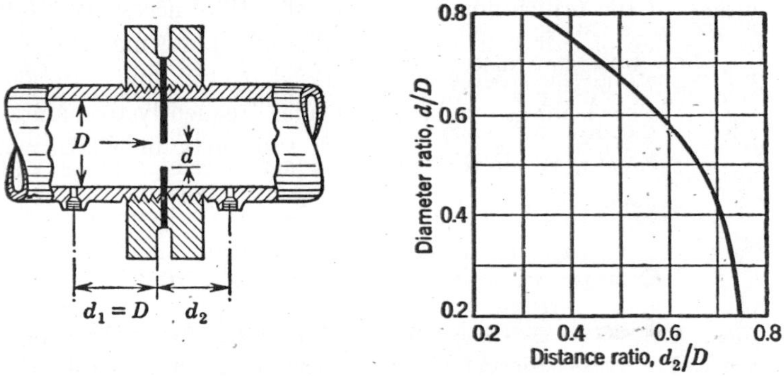

FIG. 10–5. Orifice with Vena-Contracta Taps

Flow disturbances in the pipe line adjacent to the orifice greatly alter the value of the flow coefficient K. The flow coefficients were determined experimentally in tests on straight runs

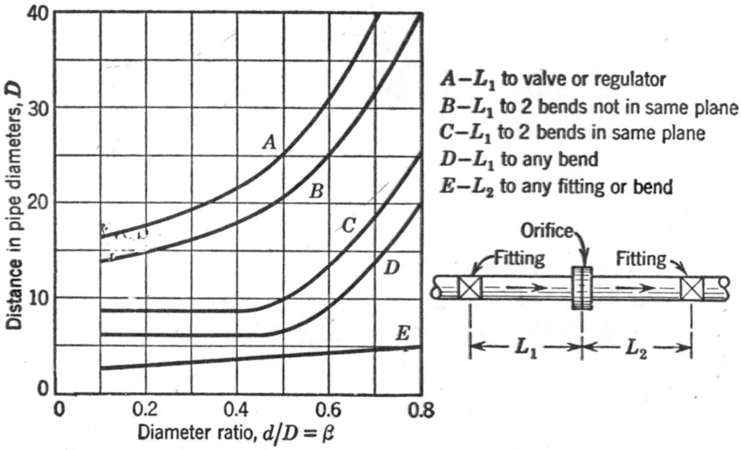

FIG. 10–6. Location of Orifice Adjacent to Valves, Fittings, and Bends

of smooth pipe. Pipe fittings of any kind cannot be used near an orifice. There should be no fittings closer than 5 pipe diameters from the orifice on the downstream side. There should be no fittings closer than about 20 pipe diameters upstream. These figures may vary somewhat depending on the diameter ratio β and the type of fittings as indicated in Fig. 10–6. By employing straightening vanes, a bundle of smaller tubes welded inside the pipe, these upstream distances can be reduced somewhat. Any other flow disturbances which cause a change in velocity distribution in the pipe will cause an error. These are, for example, excessive roughness on the inside surface of small pipes and heavy sediment on the bottom inside surface of the pipe.

Example of Orifice Liquid-Flow Calculation

Given: A 2.280-in. orifice is installed in a 3-in. standard pipe. The orifice is arranged for vena-contracta taps, and a mercury manometer reads 10.80 in. Hg at 80°F. The temperature of the flowing water is 200°F. Find the rate of flow in gallons per minute based on 60°F.

$$D = \text{standard 3-in. pipe} = 3.068 \text{ in.} = 0.256 \text{ ft}$$

$$\beta = 2.280/3.068 = 0.743$$

$$\gamma_m = 845 \text{ (Hg at 80°F) lb per cu ft}$$

$$\gamma_f = 62.2 \text{ (water at 80°F) lb per cu ft}$$

$$\rho = 60.1 \text{ (water at 200°F) lb per cu ft}$$

$$\rho_b = 62.4 \text{ (water at 60°F) lb per cu ft}$$

$$h = 0.900 \text{ ft}$$

$$\mu \cong 0.00021 \text{ (water at 200°F)}$$

Calculation:

$$q = \frac{\pi}{4} K_O (0.743)^2 (0.256)^2 \frac{\sqrt{(845 - 62.2)60.1}}{62.4} \sqrt{2g(0.9)}$$

$$q = 0.747 K_O$$

$$R_D = \frac{4}{\pi} \frac{60.1 \times q}{0.00021 \times 0.743 \times 0.256}$$

$$R_D = 1{,}900{,}000q$$

Note that calculations of Reynolds number to two significant figures is sufficient. It is necessary now to estimate the Reynolds number before the flow coefficient can be selected. Assume that the Reynolds number is 1,000,000. The flow coefficient is found from Table 9–1 at corresponding values of β and R_D.

$$K_O \cong 0.728$$

The flow is therefore

$$q = 0.747 \times 0.728 = 0.544 \text{ cfs}$$

Next check the Reynolds number to determine whether a correction of K_O is necessary.

$$R_D = 1{,}020{,}000$$

Since the Reynolds number is very close to the estimate, a recalculation is unnecessary. There are 7.48 gallons in a cubic foot, and the flow is

$$q = 244 \text{ gpm} \qquad \textit{Ans.}$$

Example of Orifice Gas-Flow Calculation

Given: A 3.00-in. orifice with flange taps installed in a 10-in. standard pipe. Air containing a negligible amount of moisture flows through the orifice at the rate of 40,000 cu ft per hr at standard conditions of 14.70 psi pressure and 60°F temperature. The actual temperature of the air is 200°F, and the actual upstream pressure is 100 psig. Find the differential pressure in inches of mercury at 70°F.

$$D \text{ (standard 10-in. pipe)} = 10.02 \text{ in.} = 0.835 \text{ ft}$$

$$\beta = 3.00/10.02 = 0.300$$

$$\gamma_m = 845 \text{ (mercury at 70°F) lb per cu ft}$$

$$v_1 = \frac{yRT}{p} = \frac{1.0 \times 53.3 \times 660}{115 \times 144} = 21.3 \text{ cu ft per lb}$$

$$v_b = \frac{y_b R T_b}{p_b} = \frac{1.0 \times 53.3 \times 520}{14.7 \times 144} = 13.1 \text{ cu ft per lb}$$

$$\mu = 0.0000145$$

$$q = 11.1 \text{ cfs}$$

Calculation:

$$R_D = \frac{4}{\pi} \frac{11.1}{0.0000145 \times 21.3 \times 0.300 \times 0.835} = 570{,}000$$

If these values are used for R_D and β, the discharge coefficient is

$$K_O = 0.601$$

The pressure differential can be calculated after determining the expansion factor:

$$Y = 1 - [0.41 + 0.35(0.3)^4]\left[\frac{2}{100}\right]\left[\frac{1}{1.4}\right]$$

and

$$Y = 0.994$$

where it has been necessary to estimate the pressure differential as 2 psi. Therefore,

$$11.1 = \frac{\pi}{4} 0.601(0.300)^2(0.835)^2 \times 0.994 \times \frac{13.1}{1.0} \times \sqrt{\frac{845 \times 1.0}{21.3}} \sqrt{2gh}$$

and

$$h = 0.326 \text{ ft Hg} = 3.91 \text{ in. Hg} \qquad \textit{Ans.}$$

The result comes reasonably close to the estimate made previously so that the expansion factor Y does not require recalculation.

The Venturi Tube

The venturi tube is shown in Fig. 10–7. The Herschel standard design is the type most often used. The short venturi is

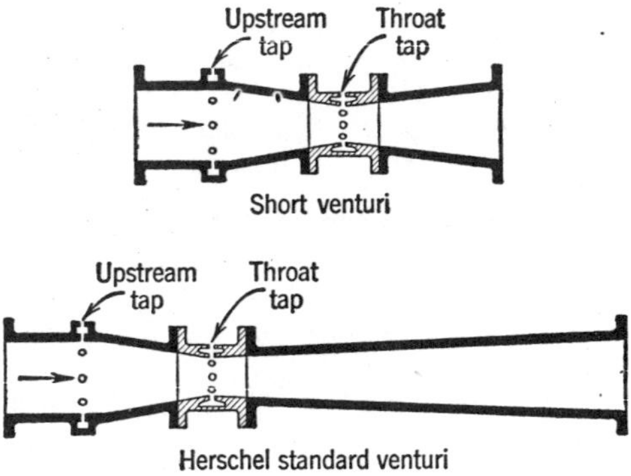

FIG. 10–7. Venturi Tubes

particularly adapted to installation in pipe lines not having long, unobstructed runs. The flow of fluid through the venturi tube establishes the pressure differential, which can then be measured and related to the flow rate. The venturi tube is usually made of cast iron or steel. In many cases the throat (the smallest section) is made a separate assembly of bronze so that it can be easily replaced. Very large venturi tubes may be made of smooth, poured concrete. The entrance section of the standard venturi has about 20° included angle and the exit section about 7° included angle. The pressure taps are made at piezometer rings so as to "average" the pressure measurement around the periphery. The diameter ratio of the venturi tube is the ratio of the throat diameter to the pipe-line diameter, usually between 0.25 and 0.50.

The advantages of the venturi tube are many, as opposed to its main single disadvantage of high cost. It has the highest accuracy of all pressure-differential flow elements when it is properly installed and calibrated, and its reproducibility is high. The permanent pressure loss in a venturi tube is usually much lower than in an orifice of the same capacity. Further, the venturi is less subject to wear and abrasion and is less likely to become clogged with sediment. The venturi tube can be employed for measurement of extremely large water-flow rates (5,000,000 gal per hr and higher).

The venturi tube should be installed in a straight run of pipe with no pipe fittings, valves, or obstructions less than 5 pipe diameters upstream. If this condition cannot be met, straightening vanes should be used. Fittings downstream have little effect on accuracy.

The flow equations for venturi tubes are the same as those for the orifice with the exception that the discharge coefficient is different. For flow of liquids,

$$q = \frac{\pi}{4} \frac{C_{VT}\beta^2 D^2}{\sqrt{1-\beta^4}} \sqrt{2g\left(\frac{\gamma_m - \gamma_f}{\rho}\right)h} \qquad (10\text{-}30)$$

$$R_D = \frac{4}{\pi} \frac{\rho q}{\mu \beta D} \qquad (10\text{-}31)$$

where q = flow rate, cubic feet per second
C_{VT} = venturi discharge coefficient
β = diameter ratio = d/D
d = venturi throat diameter, feet
D = pipe diameter, feet
γ_m = weight density of manometer fluid, pounds per cubic foot
γ_f = density of fluid over manometer fluid, pounds per cubic foot
ρ = density of flowing fluid, pounds per cubic foot
h = manometer differential, feet
μ = absolute viscosity of flowing fluid, pounds per foot-second

For the flow of gases or vapors the "hydraulic" equation can be used,

$$q = \frac{\pi}{4} \frac{C_{VT}\beta^2 D^2}{\sqrt{1-\beta^4}} \frac{\phi v_b}{M_b} \sqrt{\frac{M_1(\gamma_m - \gamma_f)}{v_1}} \sqrt{2gh} \tag{10-32}$$

and

$$R_D = \frac{4}{\pi} \frac{q}{\mu v_1 \beta D} \tag{10-33}$$

where ϕ = the rational expansion factor
v_1 = specific volume of gas at upstream conditions, cubic feet per pound, equation 10–23
v_b = specific volume of gas at base conditions, cubic feet per pound, equation 10–23
M_1 = moisture factor at upstream conditions, equation 10–22
M_b = moisture factor at base conditions, equation 10–22

The definitions of other symbols have been given previously. When the flow of gases is measured, the liquid density γ_f is generally negligible.

The venturi discharge coefficients C_{VT} are known[11] for the Herschel standard design made by Builders Iron Foundry and Simplex Valve and Meter Company. For unusual sizes of the standard venturi and for the short venturi tube the discharge coefficient must be obtained from the manufacturer or by calibration test. Normally the discharge coefficient is greater than 0.90 and less than 1.0.

The venturi nozzle is a short venturi tube which has the advantages of fitting more easily into pipe lines. The entrance and exit sections are shortened and rounded as shown in Fig. 10–8. Their flow coefficients are given by the manufacturer, or they must be determined by test.

The flow nozzle is a further variation of the venturi in which the exit section is omitted so that it resembles an orifice with a well-rounded upstream edge as shown in Fig. 10–9. The upstream (high-pressure) tap is usually made about one pipe diameter from the entrance to the nozzle. The downstream (low-

[11] See ASME Fluid Meters, *loc. cit.*

pressure) tap is usually made in the pipe directly opposite the straight portion of the nozzle. The flow coefficients of flow noz-

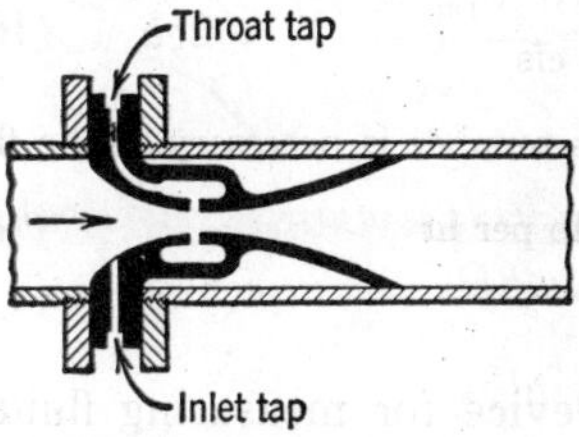

FIG. 10–8. Venturi Nozzle

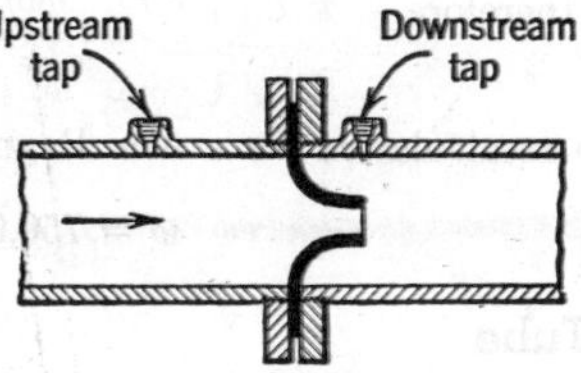

FIG. 10–9. Flow Nozzle

zles are known to some extent, but it is usually best to rely on the manufacturer.

Example of Venturi-Flow Calculation

Given: An 8 × 4-in. venturi tube is installed in an extra heavy 8-in. pipe. The inside diameter of the pipe is 7.625 in., and the throat diameter of the venturi is 4.023 in. Water flows through the venturi at 40°F. The differential pressure is 20 in. Hg at 80°F. Find the flow in pounds per hour at 40°F.

$$D = 7.625 \text{ in.} = 0.635 \text{ ft}$$

$$\beta = 4.023/7.625 = 0.528$$

$$\gamma_m = 845 \text{ (Hg at 80°F), pounds per cubic foot}$$

$$\gamma_f = 62.2 \text{ (water at 80°F), pounds per cubic foot}$$

$$\rho = 62.4 \text{ (water at 40°F), pounds per cubic foot}$$

$$h = 1.67 \text{ ft}$$

$$\mu \cong 0.0010 \text{ (water at 40°F)}$$

Calculation:

$$q = \frac{\pi}{4} \frac{C_{VT}(0.528)^2 \times (0.635)^2}{\sqrt{1 - (0.528)^4}} \sqrt{2g \frac{(845 - 62.2)}{62.4} 1.67}$$

and

$$q = 3.40 C_{VT}$$

$$R_D = \frac{4 \times 62.4 \times q}{\pi \times 0.001 \times 0.528 \times 0.635}$$

$$R_D = 237{,}000q$$

To determine C_{VT} we use the table on page 100 of ASME Fluid Meters. Venturi coefficients are usually close to 1.0. If the ideal flow, 3.40 cfs, is used, the Reynolds number is 806,000, and the table gives $C_{VT} = 0.982$. The flow is therefore

$$q = 3.34 \text{ cfs}$$

and no recalculation for a new Reynolds number is necessary. The flow is

$$w = 750{,}000 \text{ lb per hr} \qquad \textit{Ans.}$$

Pitot Tube

The pitot tube is primarily a device for measuring fluid velocity. The principles employed in the pitot tube are shown in Fig. 10–10. A flow stream in a pipe line has a given velocity V and a given static pressure p_1 in a stream tube directly in line with a solid body which is held stationary in the pipe. As the fluid in the stream tube approaches the body, the fluid particles are decelerated until, at a point directly in front of the body, the velocity is zero. This point is called the stagnation point. Accompanying the deceleration is an increase in pressure from p_1 to p_2. The process taking place in the fluid is one of converting the "velocity" head to an additional static head. If therefore the pressure difference $(p_2 - p_1)$ can be measured, the velocity of the fluid in the stream tube may be found. Note that this measurement has little practical relation to the velocity in other adjacent stream tubes.

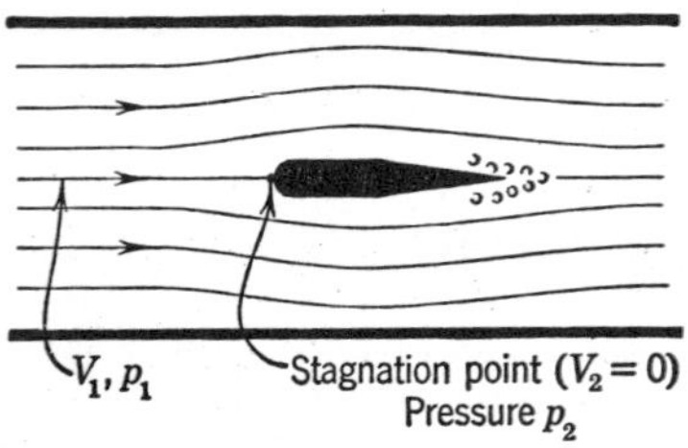

FIG. 10–10. Flow around a Solid Body

The sketch of the pitot tube in Fig. 10–11 shows a type often used for flow of liquids. It consists simply of a tube supported in the pipe with the impact opening of about ⅛ to ¼ in. diameter arranged to point directly toward the oncoming fluid. This is called the impact opening and is used to measure the stagnation pressure. The static pressure is measured through the ordinary pressure tap or through a piezometer ring in the side of the pipe line.

A combined pitot tube is shown in Fig. 10–12. The static-pressure openings are made directly in the pitot-tube assembly, the stagnation pressure being measured through the inside tube and the static pressure through the outside tube. The static-pressure measurement made at the pitot tube may be slightly in error because of eddies around the pitot-tube assembly.

The main advantages of the pitot tubes are that installation is relatively simple and they are readily adaptable for flow measurements made in very large pipes or ducts. They can be employed in either open-channel streams or pipes. The pressure

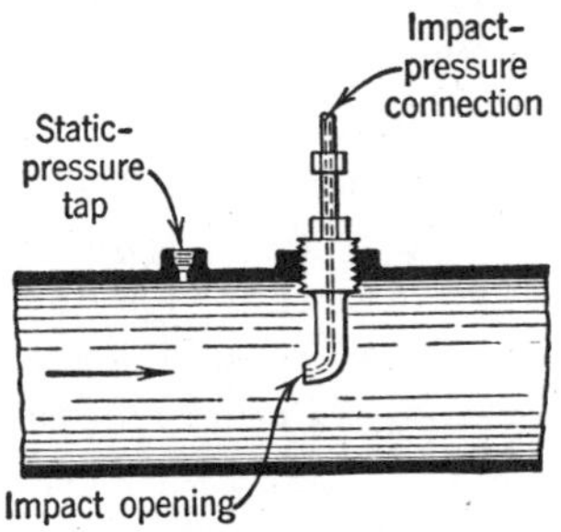

FIG. 10–11. Pitot Tube—Short Single-Opening Type

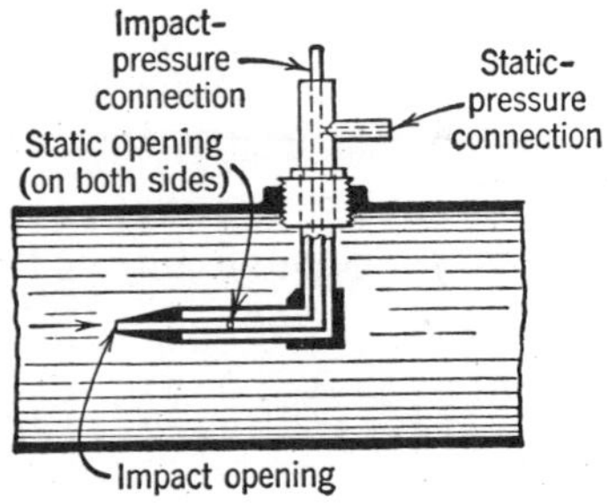

FIG. 10–12. Combined Pitot Tube

loss caused by the pitot tube is, of course, very small. The pitot tube is about the only means by which a velocity traverse of a large pipe or duct can be made. The main disadvantage of the pitot tube is that it cannot be used in fluids containing solid particles.

The energy equation for the flow of liquid may be written, by referring to Fig. 10–10,

$$\frac{p_2}{\rho} = \frac{p_1}{\rho} + \frac{V_1^2}{2g} \tag{10–34}$$

where any difference in height to a datum line $(Z_2 - Z_1)$ is ignored and where

p_2 = stagnation pressure, absolute
p_1 = static pressure in flowing fluid, absolute
ρ = weight density of flowing fluid
V_1 = velocity of flowing fluid in a stream tube in line with the axis of the impact opening

Solving for velocity and introducing a velocity coefficient and the manometer head, we find

$$V_1 = C_{PT} \sqrt{\frac{\gamma_m - \gamma_f}{\rho}} \sqrt{2gh} \tag{10-35}$$

where V_1 = velocity of flowing fluid, feet per second
C_{PT} = velocity coefficient
γ_m = density of manometer fluid, pounds per cubic foot
γ_f = density of fluid over the manometer fluid, pounds per cubic foot
ρ = density of flowing fluid, pounds per cubic foot
h = manometer differential, feet

The energy equation for the flow of gas may be written

$$p_2 v_2 + JE_2 = p_1 v_1 + \frac{V_1^2}{2g} + JE_1 \tag{10-36}$$

where v = specific volume
E = internal (molecular) energy

Assuming that the ratio of specific heats is constant and that the process is adiabatic and reversible, there results,

$$V_1 = \sqrt{\frac{2gp_1v_1K}{K-1}} \sqrt{\left(\frac{p_2}{p_1}\right)^{(K-1)/K} - 1} \tag{10-37}$$

where K = ratio of specific heats = c_P/c_v. For industrial use it is desirable to measure the pressure differential $(p_2 - p_1)$ rather than the pressure ratio (p_2/p_1). Employing an expansion by the binomial theorem, we may write

$$\left(\frac{p_2}{p_1}\right)^{(K-1)/K} \cong 1 - \left(\frac{K-1}{K}\right)\left(1 - \frac{p_2}{p_1}\right) \tag{10-38}$$

where terms of second and higher order are discarded. Substituting in equation 10–37 gives

$$V_1 = \sqrt{2gv_1(p_2 - p_1)} \tag{10-39}$$

For gases flowing at low velocity and high static pressure[12] the error in using equation 10–38 is small. Fortunately, in nearly all industrial applications, the gas velocity V_1 is about 10 to 40 fps, and the gas pressure is about 65 to 165 psia, and equation 10–39 may be employed with good accuracy.[13] Introducing the manometer head and a flow coefficient,

$$V_1 = C_{PT}\sqrt{(\gamma_m - \gamma_f)v_1}\sqrt{2gh} \tag{10–40}$$

where v_1 is the specific volume of the flowing gas, cubic feet per pound. If the gas contains an appreciable amount of moisture, the specific volume v_1 requires correction by employing the moisture factor as given in equation 10–22. The pitot tube should be used cautiously in measuring the velocity of flowing vapor, because condensation of liquid in the impact tube would destroy the calibration.

The velocity coefficient of the pitot tube is varied by the details of construction, and no standardization of pitot-tube construction has been made. Consequently, the velocity coefficient must be either determined by test, or obtained from the manufacturer. It is, however, safe to assume that the velocity coefficient for the pitot tube with a short impact-opening extension (Fig. 10–11) will be between 0.84 and 0.88, and between 0.98 and 1.02 for a pitot tube with a long impact-opening extension (Fig. 10–12). It is assumed that the impact opening of the pitot tube faces directly into the flow stream without more than a 3° angle between the axis of the opening and the axis of the pipe.

The ordinary pitot tube has no provision for changing its range of operation; that is, the velocity and the pressure differential are the only variables employed in calibration. Thus,

[12] The error in this formula caused by discarding terms higher than second is, if the ratio of specific heats is taken as 1.4,

$\frac{p_2 - p_1}{p_1}$	0	0.1	0.2	0.4	0.6
Error %	..	+1.4	+3.1	+5.5	+15

[13] For a more accurate form of the equation at high velocity and low pressure see S. A. Moss, "Flow of Air and Other Gases with Special Reference to Small Pressure Differences," *Am. Machinist* (Sept. 20, 1906).

for a small velocity of flowing fluid, the pressure differential is so small that its measurement becomes difficult with practical industrial instruments. For this reason, a variation of the pitot tube shown in Fig. 10–13 is often used. The impact opening faces directly upstream, and the "static" opening faces directly downstream. The pressure differential is greater than in the ordinary pitot tube because "suction" head is added to the difference of stagnation pressure and static pressure. Another arrangement for obtaining greater differential pressure at low flow is to construct a small venturi-shaped shield around the impact opening so as to increase the local velocity of flow. Such pitot tubes must be calibrated by test, because there is little reason to assume that the simple hydraulic formula is applicable.

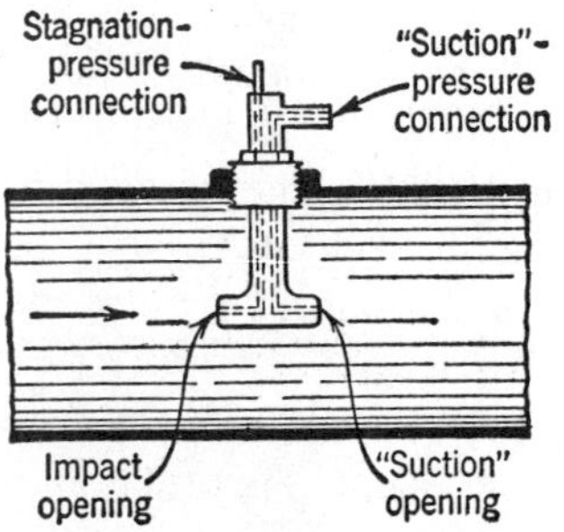

FIG. 10–13. Multiplying Pitot Tube

The flow rate in the pipe or duct cannot be calculated from velocity unless the velocity distribution is known. However, the pitot tube can be calibrated for actual flow rate in the pipe if the velocity distribution is time-invariant:

$$q = K_{PT} \frac{\pi D^2}{4} V_1 \tag{10–41}$$

where q = rate of fluid flow, cubic feet per second
D = inside pipe diameter, feet
K_{PT} = flow coefficient

For normal distribution of velocity in a pipe, the pitot tube placed at the center of the pipe measures the maximum velocity. The mean velocity is approximately 0.83 of the maximum so that flow coefficient K_{PT} is about 0.83. It is usually necessary to determine the flow coefficient by means of a velocity traverse of the pipe or duct.

Head Flowmeters

The head flowmeter consists of a primary element such as an orifice, venturi, or pitot tube used with a differential-pressure meter to measure the differential head caused by the flowing fluid at the primary element. The differential-pressure meter may be any of the various meters such as the enlarged-leg mercury manometer, the bell gage, the bellows gage, the diaphragm gage, the ring gage, or the tilting U-tube gage previously described in the chapter on pressure measurement. By far the most common is the enlarged-leg mercury manometer.

The differential-pressure meter and the primary element require careful connection and installation. It must be remembered that the meter is used for the purpose of measuring differential pressure, not flow, in the range of an inch or two of mercury head or even less. Any extraneous or false head introduced by the connecting piping causes a serious error. The proper connections for several examples are given in Fig. 10–14. The factors in selecting proper piping arrangement are, considering only the important ones:

1. The piping arrangement must be absolutely free of leaks.
2. The connecting lines must be clean and free from obstructions (¼- to 1-in. sizes are most common). Use as few fittings as possible.
3. The connecting lines must pitch 2 in. per ft to prevent gas pockets and aid drainage.
4. The connecting lines should not be over 50 ft long, preferably less.
5. The connecting lines must be maintained at a temperature between 32 and 110°F.
6. The differential-pressure meter should be installed below the level of the primary element if possible.
7. Drain legs should be installed at the meter when dirt and sediment exist in the connecting lines.
8. Condenser chambers used for steam lines must be level.
9. Sealing chambers used for corrosive fluids must be level and should be vented.

A short study of these requirements together with reference to Fig. 10–14 will illustrate the necessity for each in terms of accuracy of measurement.

Square-root compensation is often employed with differential-pressure meters in order to provide an indicating scale or chart with linear calibration spacings. Because flow rate through an orifice is proportional to the square root of pressure differential,

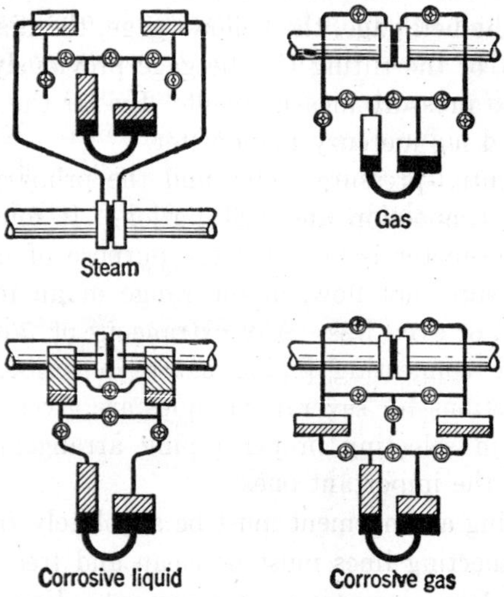

FIG. 10–14. Installation of Orifice Head Meters

the chart or scale calibration is nonlinear with close spacing at low flow, as shown in Fig. 10–15. Methods of square-root compensation are as varied as the meters themselves, and they

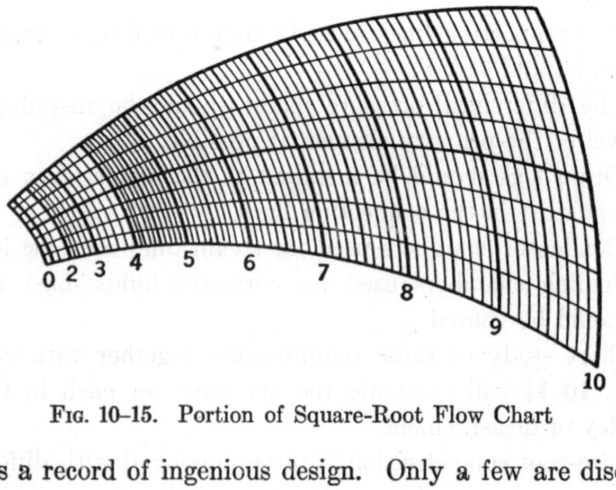

FIG. 10–15. Portion of Square-Root Flow Chart

serve as a record of ingenious design. Only a few are discussed here.

The LeDoux bell meter of the Bailey Meter Company is sketched in Fig. 10–16. Here square-root compensation is achieved by controlling the amount of mercury displaced from the inside of the bell of a bell-type differential-pressure meter. By shaping the inside of the bell to parabolic form, the flow calibration of the meter can be made linear. The electrical-resistance-type differential-pressure meter of the Republic Flowmeters Company has square-root compensation through suitable vertical spacing of the contact rods. At low differential pressure

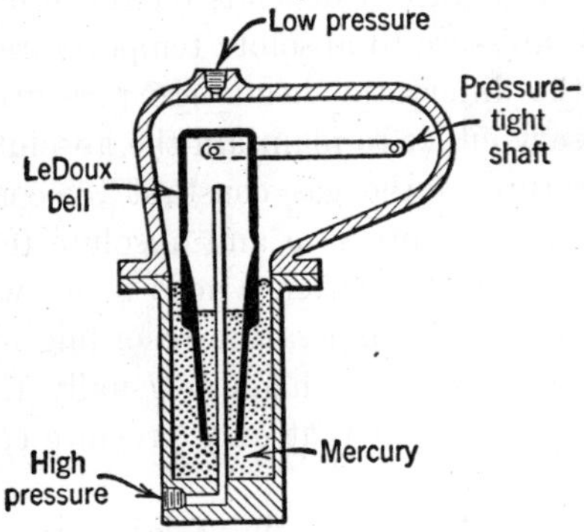

FIG. 10–16. LeDoux Bell Meter—Bailey Meter Company

the contact rods are spaced more closely. The ring manometer or tilting U-tube manometer may have square-root compensation by a suitable cam arrangement which simply magnifies the lower portion of the scale to linear magnitude.

The advantage of square-root compensation in an orifice head meter is simply that the scale is made more readable, especially to unskilled operators. On the other hand, the use of square-root compensation does not improve accuracy of flow calibration because the first 1 per cent of pressure differential causes the first 10 per cent flow. No amount of manipulation will improve the accuracy of measurement of the actual differential pressure.

Density variations of the flowing fluid will, of course, cause the meter to be in error since density appears directly in the fluid-flow relations. In measuring flow of liquids, density, and therefore temperature, since density varies considerably with temperature, must be maintained substantially constant if density errors are to be avoided.

Pressure compensation in orifice-head meters is usually required for gas-flow measurement, because the rate of flow of a gas through an orifice depends on the absolute static pressure as well as the differential pressure. Simplifying equation 10–24 and substituting for the density,

$$q = (\text{a constant}) \times Y \sqrt{\frac{p_1 h}{T_1}} \qquad (10\text{–}42)$$

we find that the flow rate q depends on the square root of the ratio of absolute pressure to absolute temperature as well as the square root of the differential head. In measuring the flow of gases it is necessary either to maintain the absolute pressure and absolute temperature of the gas constant by control means or to measure the absolute pressure and absolute temperature and correct for these variables. Often a head meter will be arranged to include pressure- and temperature-recording elements to enable manual corrections to be made. Usually the temperature remains sufficiently constant that only pressure compensation or correction is required.

Viscosity variations of the flowing fluid cause slight errors in flow measurement with head meters. Since viscosity varies considerably with temperature, the temperature of the flowing fluid must be substantially constant. Actually, maintaining a constant temperature of flowing fluid increases accuracy of measurement because viscosity and density both are more constant. Since considerable variation of Reynolds number is required to alter the flow coefficient of an orifice K (See Table A–21, appendix), variation of viscosity must be appreciable to cause appreciable error.

Pulsating flow is a problem peculiar to flow measurement with head meters. Pulsations or cyclic variation of flow rate are often observed in flow lines connected to reciprocating pumps and other such reciprocating machinery. A head meter, because of the square-root relation of head and flow rate, does not register the correct value of average flow but tends to read high when square-root compensation is not provided. Obviously this is due to the fact that the average of the square root of a series of readings is not the same as the square root of the average of the same series. The head meter attempts to record an average since its sluggishness caused by unavoidable lag prevents it from instantaneously indicating the full amplitude of a pulsating flow.

The best method to avoid pulsation errors is to eliminate the pulsation. Often this is easier said than done. Pulsation can be reduced by providing sufficient fluid capacitance or fluid resistance ahead of the orifice or venturi to attenuate the amplitude of pulsation to reasonable magnitude. Also, operating the orifice at a higher pressure differential and reducing the pipe size to obtain the desired diameter ratio are helpful.

Integration of flow rate to establish a quantity unit is often required, generally for the purpose of cost accounting of the flow of materials to various plants and processing units. Based on

$$Q = \int_{t_1}^{t_2} q(t)\, dt \qquad (10\text{–}43)$$

Q is the total quantity of flow of fluid past an orifice in the time interval t_1, t_2. The flow rate q is generally a random function of time.

Integration can be accomplished manually by employing a special planimeter for circular recording charts. These special planimeters can be obtained from the manufacturers of recording head meters. The planimeter can be used quite easily and if handled skillfully produces sufficient accuracy for cost-accounting purposes. Different-style planimeters are used for the charts of head meter with and without square-root compensation.

Numerical integration is often used by summing the average flow rate over short-time intervals directly from the flow chart. Then,

$$Q \cong \sum_{i=1}^{i=n} q(\zeta_i)(\Delta t_i) \qquad (10\text{–}44)$$

where the time interval Δt_i is generally 15 min and the average flow rate $q(\zeta_i)$ is estimated from the chart. The time value ζ_i is taken anywhere in the time interval Δt_i. For gas-flow measurements it is necessary to correct for absolute pressure and temperature, using

$$Q \cong \sum_{i=1}^{i=n} \lambda \left[\frac{p(\zeta_i) \times h(\zeta_i)}{T(\zeta_i)} \right]^{1/2} (\Delta t_i) \qquad (10\text{–}45)$$

where λ is a multiplying number representing all the constants in the gas-flow equation. The average absolute pressure, average absolute temperature, and average differential or head in each time interval Δt_i are estimated from the chart. This process gives satisfactory accuracy only when the flow rate changes gradually.

Automatic integrators may be used with orifice head meters and the meters are supplied with counters for indicating total

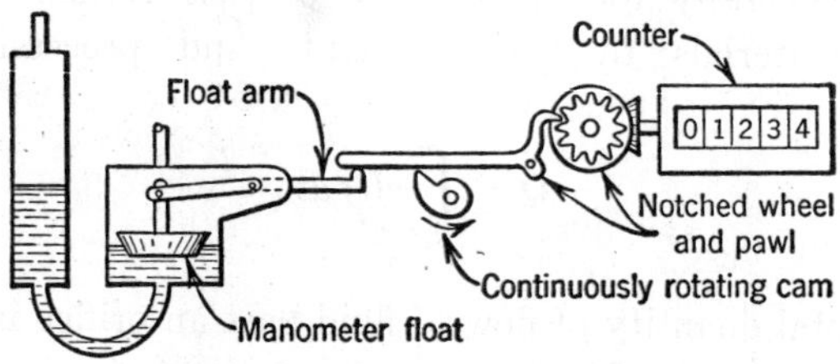

Fig. 10–17. Schematic Arrangement of Flowmeter Integrator

flow units. An integrator of the wheel-and-cam type is shown schematically in Fig. 10–17. A synchronous-motor drives the cam continuously. The meter float positions the integrator arm near the cam: high off the cam at low flow and close to the cam at high flow. When the cam comes in contact with the integrator arm, the arm is made to rotate and drive a mechanical counter through gears. The amount of rotation given the counter depends on the height of positioning the arm, the position of the float, the differential pressure, and, therefore, the flow rate. Mechanical integrators for head meters without square-root compensation must include this compensation in the integrator because the total flow is proportional to the time integral of the square root of pressure differential. In gas-flow measurement, it is also necessary to integrate the absolute pressure. However, this can be done only for flow rate and pressures that do not vary widely. Obviously the product of the integrals of the separate factors is not equal to the integral of the product, even when the absolute temperature is considered constant. Mechanical integrators usually operate on a 15-sec cycle; that is, the summation intervals occur every 15 sec.

Area Flowmeters

The area meter operates on the principle that the variation in area of the flow stream required to produce a constant pres-

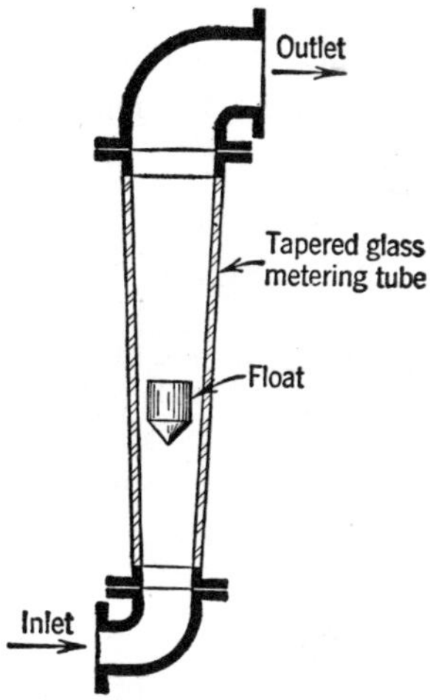

FIG. 10–18. Area Meter (rotameter)

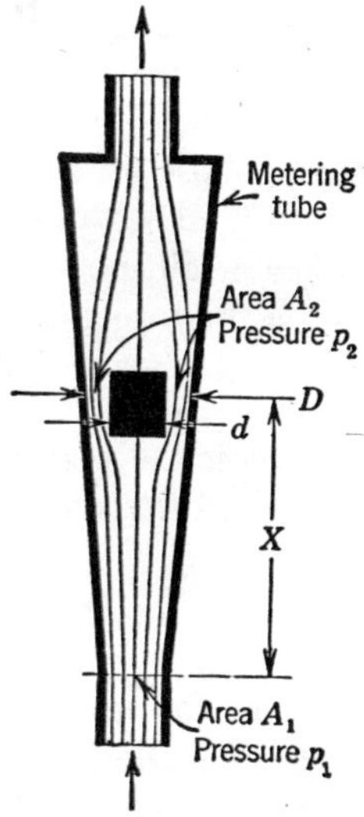

FIG. 10–19. Schematic Diagram of an Area Meter

sure differential at a restriction to fluid flow is proportional to the flow rate. The rotameter illustrated in Fig. 10–18 is one type of area meter. The flowing fluid enters the bottom, passes upward through the metering tube and around the float, and exits at the top. The metering tube is hollow, usually made of glass, and has tapered sides, so that the cross-section area at the top is greater than at the bottom. Often the metering tube has protruding vertical ridges at the inside diameter, for guiding the float in its travel up and down the tube. The tube must be installed vertically; that is, it must be plumb to within about 2 geometrical degrees, because the float must center itself in the fluid stream. At a high flow rate the float assumes a position toward the top of the metering tube, and at a low flow rate the float positions itself lower in the tube. The float (or bob) is in static equilibrium at a given flow rate because of gravity and buoyant forces, which are discussed in greater detail in the following analysis. The flow rate is read by noting the position of the float against a calibrated scale etched on the glass.

The energy equation for flow of liquids is

$$\frac{p_2}{\rho} + \frac{V_{m2}^2}{2g} + Z_2 = \frac{p_1}{\rho} + \frac{V_{m1}^2}{2g} + Z_1 \tag{10-1}$$

where p = static pressure, absolute
V_m = fluid stream mean velocity
Z = height above datum line
ρ = fluid weight density
g = acceleration due to gravity

If we proceed as before, the difference in height $(Z_2 - Z_1)$ is negligible, and we have

$$V_{m2}^2 - V_{m1}^2 = \frac{2g}{\rho}(p_1 - p_2) \tag{10-2}$$

Considering the rotameter in Fig. 10–19, we will assume that the pressure acting upward on the bottom of the float is the stagnation pressure. It is also assumed that the pressure acting downward on the float is the static pressure p_2. These assumptions are questionable, and they are made strictly for the purpose of demonstrating the principles of operation. Thus, for static equilibrium of the float at any given position,

$$A_f\left(p_1 + \frac{V_{m1}^2\rho}{2g}\right) + V_f\rho = A_f p_2 + V_f\rho_f \tag{10-46}$$

where A_f = effective area of float
V_f = volume of float
ρ_f = weight density of float
ρ = weight density of flowing fluid

The continuity relation is also required,

$$q = V_{m1}A_1 = C_c V_{m2} A_2 \tag{10-47}$$

where q = rate of flow of fluid
A_1 = area of inlet to tapered tube
A_2 = annular area between float and tube
C_c = coefficient of contraction

The coefficient of contraction is required because it is preferred to base the calculation of area A_2 on float and tube diameters, whereas the area of flow past the float is smaller owing to contraction. Combining equations 10–2, 10–46, and 10–47, to eliminate p_2, p_1, V_{m2}, and V_{m1}, we have

$$q = C_c A_2 \sqrt{\frac{2gV_f}{A_f}\left(\frac{\rho_f}{\rho} - 1\right)} \qquad (10\text{–}48)$$

and this is the basic equation of the rotameter for flowing liquids.[14] The coefficient C_c is generally about 0.60 to 0.80 but varies with different designs.

For flow of gases through a rotameter a similar analysis can be made. However, in doing so it is necessary to assume another relationship for the forces acting on the float. This analysis is not attempted here because the simple equations for liquids serve to demonstrate the principles of operation.

[14] This relation can also be derived from a consideration of total drag (pressure) force on the float. Suppose that the forces acting upward on the float are the drag force D and the buoyant force. The force acting downward is the weight of the float. Then,

$$V_f\rho + D = V_f\rho_f$$

or

$$D = V_f(\rho_f - \rho)$$

The drag force is in the form,

$$D = C_d A_f \frac{\rho}{2g} V_m^2$$

where V_m is the fluid mean velocity and C_d is a drag coefficient. Substituting and combining yields

$$V_m = \frac{1}{C_d}\sqrt{\frac{2gV_f}{A_f}\left(\frac{\rho_f}{\rho} - 1\right)}$$

Compare this result to equation 10–48.

For area meters with straight-sided metering tubes,

$$d = d_f + aX \tag{10–49}$$

where d = inside diameter of tube
d_f = diameter of tube at inlet
a = a scale length factor
X = height of float over the tube inlet

and where for simplicity it is assumed that the diameter of the float is the same as the diameter of the tube inlet. Equation 10–48 can then be written

$$q = a\pi C_e \left(X + \frac{a}{2d_f} X^2\right) \sqrt{\frac{gV_f d_f}{2A_f}\left(\frac{\rho_f}{\rho} - 1\right)} \tag{10–50}$$

Although this equation indicates a quadratic relation between flow and float position X, the calibration is very nearly linear, because, for practical values of the constants a and d_f, the linear term in X predominates. By suitably shaping the metering tube a linear scale characteristic can be obtained.

Partial compensation of the area meter for changes in density of the flowing liquid can be provided by a suitable choice of float material. For flow rate expressed in weight units, equation 10–48 may be simplified to

$$w = C_1 A_2 \sqrt{(\rho_f - \rho)\rho} \tag{10–51}$$

If the flow rate is to be independent of density,

$$\frac{\partial w}{\partial \rho} = 0$$

and, by performing these operations, there results

$$\rho_f = 2\rho \tag{10–52}$$

Thus, if the density of the float material is made twice the density of the flowing liquid, compensation is almost exact at all flows. For example, with the proper float material a change of liquid density of about 4 per cent results in a 0.1 per cent error in weight flow rate. For flow rate expressed in volume units, density compensation requires that the density of the float be made as great as possible. Selecting a dense float material such as silver, lead, or tantalum provides partial compensation. For example, with a very dense float a change of 4 per cent in density of flowing liquid results in a 2 per cent error in volumetric flow rate.

Viscosity compensation in area meters can be provided by suitably shaping the float. The reason for changes in viscosity affecting the calibration of the area meter is that the viscous drag acting on the float disturbs the static-equilibrium position

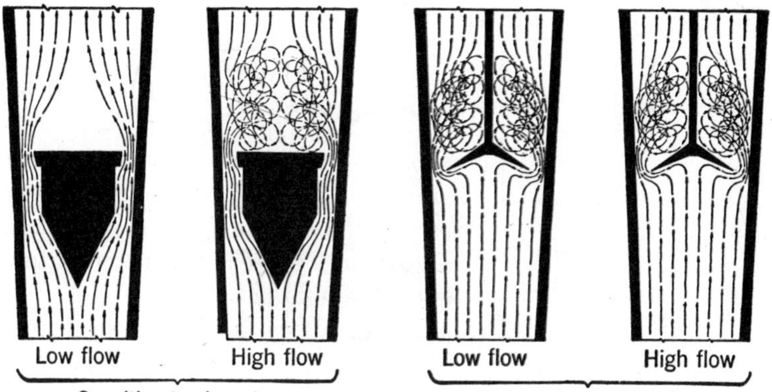

Fig. 10–20. Viscosity Compensation by Float Shape (*Reproduced by permission of Fischer & Porter Company*)

of the float for a given flow rate. The force of viscous drag will naturally vary with fluid viscosity. If, then, the float can be shaped so as to provide a constant viscous drag at all flow rates, the operation of the area meter is made independent of viscosity. Figure 10–20 illustrates this effect. With the ordinary plumb-bob-shape float the turbulence and viscous drag are greatly different at different flow rates. Thus the upward force of viscous drag varies at different flow rates. With the inverted-cone-shape float the turbulence and viscous drag are nearly alike at all flows except at extremely small Reynolds numbers. Therefore, the upward force of viscous drag is approximately constant, and the meter operation is insensitive to viscosity variations. Actually several patented arrangements of this kind are in common industrial use for measurement of flow of liquids having variable viscosity.

There are many different styles of rotameters, all suited to different applications. Since the rotameter contains all the necessary metering elements in one assembly, there is negligible effect due to any type of valves, fittings, and so on placed immediately before or after the meter. Ordinarily the simple indicating rotameter is constructed with a glass metering tube, and the flow is read by noting the location of the float against a scale etched on the glass. Armored rotameters are used for high-pressure service.

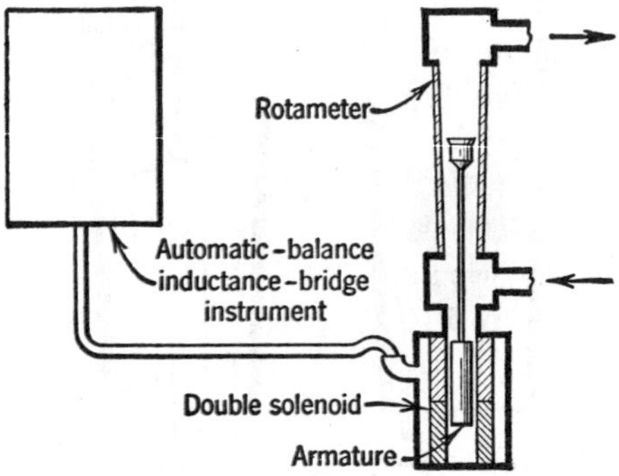

FIG. 10–21. A Recording Rotameter

Rotameters for indicating and recording purposes employ various methods for transmitting the motion of the float. These methods and their use are very similar to the methods employed in differential-pressure meters. Figure 10–21 illustrates one type using an inductance-bridge system. A nonmagnetic tube at the bottom of the rotameter allows an iron armature to be suspended from the float. A double solenoid mounted on the outside of the nonmagnetic tube is connected into the inductance bridge of an automatic-balance instrument similar to a potentiometer in many physical respects. The instrument can then be calibrated in terms of rotameter float position. Other types employ a magnetic-clutch and pneumatic-transmission means or a self-balancing inductance-bridge system.

The rotameter has many advantages in industrial service. The pressure loss at the meter is nearly constant, and it is small. The meter can handle almost any corrosive fluid encountered in industrial processing plants. It can be compensated for changes in fluid density and viscosity (therefore temperature) when it is used in metering certain fluids. A rotameter has quite good accuracy, particularly at low flows. Its main disadvantage is that it is not so rugged as head meters because of the glass metering tube. Recently, metal and plastic metering tubes have been employed so that the rotameter is as rugged as any other flow-measuring device.

The *cylinder-and-piston-type* area meter (The Brown Instrument Company) is shown schematically in Fig. 10–22. The operation is in many respects similar to that of a rotameter; the difference is that the area for fluid flow is provided by a series of reamed holes in the walls of the cylinder. These holes are spaced helically around the cylinder in rows, so that the variation in area for various heights of the piston is continuous. The pressure differential is constant, because the weight of the piston is constant. By properly spacing the holes in the cylinder the calibration of the instrument for flow rate is made linear. An area meter of similar principles is made by the Bailey Meter Company, the helical row of holes being replaced by two vertical rectangular slots. A similar meter is the Hays-Cochrane area meter, which utilizes a flat-bottom float in a tapered throat installed in a vertical body. Generally these area meters employ the inductance-bridge system for remote indication and recording of flow rate.

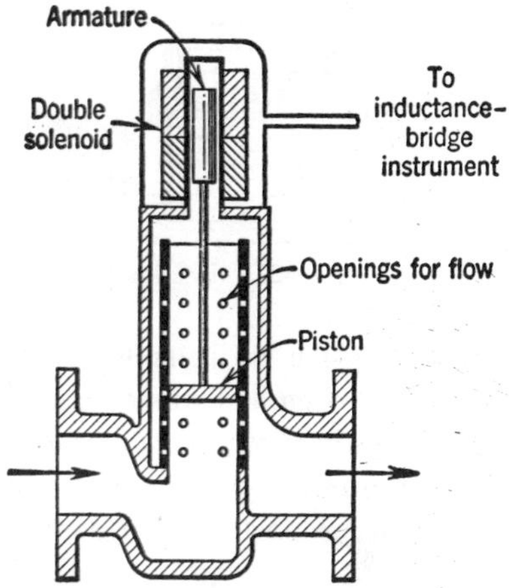

Fig. 10–22. Cylinder-and-Piston-Type Area Meter—The Brown Instrument Company

The cylinder-and-piston area meter is most often used for measuring flow of fuel oils, Bunker-C fuel oil, tar, chemical liquors, and other such high-viscosity fluids. Its use is advantageous particularly when a mercury manometer or bellows meter with orifice cannot be used, because of corrosion problems, possibility of plugging orifice connecting lines, or lack of adequate information on flow coefficients at very low Reynolds numbers. Also, in shipboard use its installation is simpler.

Open-Channel Meters

Head-area meters such as weirs, nozzles, and flumes are used in open channels for liquid-flow measurement. The current meter is used in large and open streams. Open-channel meters are employed in central power stations, hydraulic power-generating stations, municipal waterworks, sewage-disposal units, and

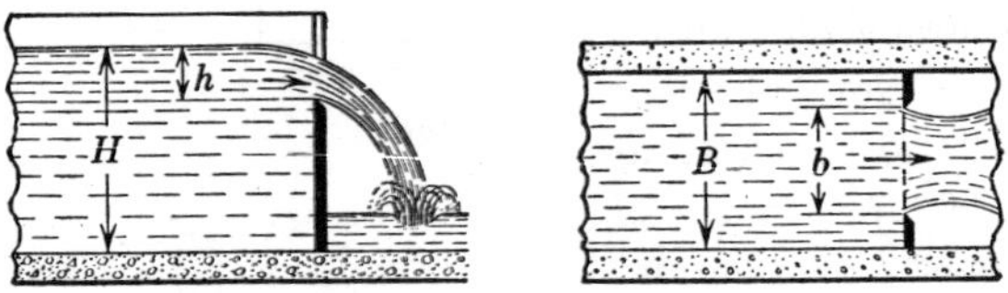

FIG. 10–23. Rectangular Weir with Side Contraction

water-irrigation projects for measuring the flow of water, sewage sludge, chemical wastes, and other semifluids.

The familiar *rectangular-notch weir* shown in Fig. 10–23 is most often employed for measuring large flow rates. Since the velocity is proportional to depth at the weir,

$$V = \sqrt{2gZ} \tag{10–53}$$

where V = velocity at given lamina
Z = depth from surface to lamina

the total flow is the product of velocity times area,

$$q = \int_0^h \sqrt{2gZ} \cdot b\, dz \tag{10–54}$$

and

$$q = \tfrac{2}{3} b \sqrt{2gh^3} \tag{10–55}$$

where q = flow rate
b = width of rectangular notch
h = head over weir crest

Since the actual flow is less than the ideal because of contraction of the jet,

$$q = C_{RW}\tfrac{2}{3}b\sqrt{2gh^3} \tag{10–56}$$

where C_{RW} is the flow coefficient. The weir shown in Fig. 10–23 has side contraction in that the width of the weir notch b is less than the channel width. Weirs without side contraction have a weir-notch width the same as the channel width and are generally considered more accurate because of greater certainty in the accuracy of width b. Needless to say, the flow coefficient C_{RW} is different for each.

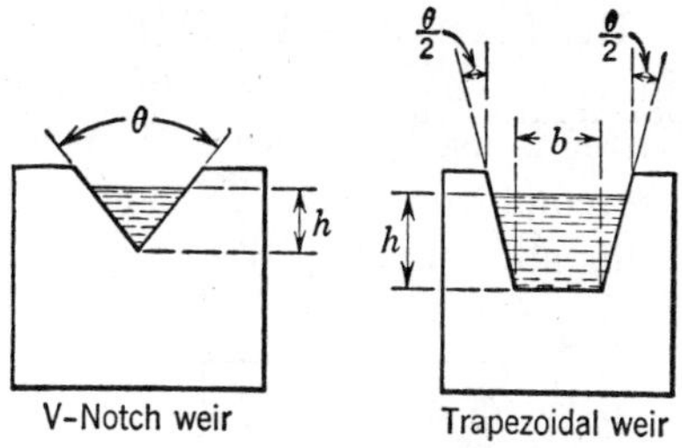

FIG. 10–24. V-Notch and Trapezoidal Weirs

The *V-notch* or *Thomson weir* is used with the apex of the triangle downward as in Fig. 10–24 and is generally more useful in measuring small flow rates and where good accuracy is desired. Its flow relation can be developed as before,

$$q = C_{VW}\left(\frac{8}{15}\tan\frac{\theta}{2}\right)\sqrt{2gh^5} \tag{10–57}$$

where C_{VW} is the flow coefficient.

The *trapezoidal-notch* or *Cipoletti weir* is shown in Fig. 10–24. In this weir the slope of the sides is such as to compensate fcr side contraction, and the included angle θ is made $2\tan^{-1}(\tfrac{1}{4})$. If side contraction is compensated, then,

$$q = C_{TW}\tfrac{2}{3}b\sqrt{2gh^3} \tag{10–58}$$

where C_{TW} is the flow coefficient.

Weirs for industrial use are made by individual contractors and manufacturers; consequently, their specifications are not standardized. Generally a weir is constructed on the job for a particular application. The discharge coefficient may be determined by calibration on the job by a weigh test or sometimes by calibration to another flow-measuring device, or the flow coefficients given in the literature by various experimenters can be employed.[15] Approximate discharge coefficients are given in Table A–24 (appendix). Unfortunately the flow coefficients of most weirs vary with head.

In constructing a weir, several limitations on dimensions of head, width, and depth must be observed for reasons of accuracy. These are given in Table A–24 (appendix). In addition, a weir should be preceded by a straight, uniform cross-section channel of sufficient length to insure uniform velocity distribution. Baffle plates and screens extending across the channel are often necessary to reduce turbulence. The upstream edge of the weir plate should be square. The downstream edge of the weir plate should be beveled, so that the stream will break free of the edge and not fall down the face of the weir plate. Any wetness of the open face of the weir plate may be taken as indicative of considerable error. In a rectangular weir without side contraction the downstream face of the weir plate must be aerated in order to provide breakaway flow.

Hydraulic flumes are used in open streams where a sufficient fall or head cannot be obtained to use a weir. Ordinarily too flumes are used for greater flow rates than can be accommodated by a weir.

The *Parshall flume* shown in Fig. 10–25 is a common type. The Palmer-Bowlus flume has a similar arrangement. These types are used in throat sizes from about 6 in. to 8 ft for flows ranging from 1 to 70 million gallons per day. The head is measured at a single point about one third the distance downstream from the inlet in the entrance section. The head may vary from 6 to 30 in. The discharge is determined by an empirical formula or from test data.[16]

Open nozzles are used for measuring open flow of raw sewage, raw and digested sludge, and industrial waste in partially filled pipes and channels. The *Kennison nozzle* (Builders-Providence) is sketched in Fig. 10–26. The unique cross-section shape of the nozzle produces a nearly linear relationship between head and

[15] There is a great body of literature available on weirs. See ASME Fluid Meters.

[16] R. L. Parshall, "The Improved Venturi Flume," *Trans. ASCE,* **89** (1926).

flow. The capacity of the nozzle for inlet diameters from 6 to 36 in. is 0.275 to 18.2 million gallons per day.

The *Parabolic flume* (Simplex Valve and Meter Company) is a similar open nozzle having a gradual transition from a circular to a vertical parabolic cross section. This nozzle is open at the top for cleaning, and the flow area of the nozzle has no obstruc-

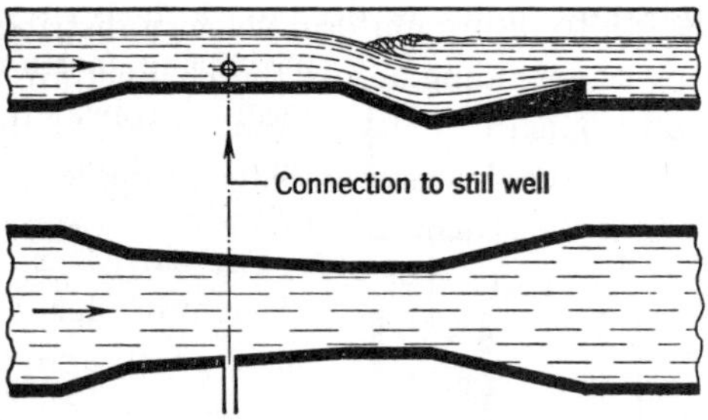

Fig. 10–25. Parshall Flume

tions. The capacity of the nozzle for inlet diameters from 6 to 36 in. is 0.208 to 19.07 million gallons per day.

The head measurement for weirs, flumes, and open nozzles is preferably performed in a still well, which is an open standpipe connected by a pressure line to the piezometer opening in the

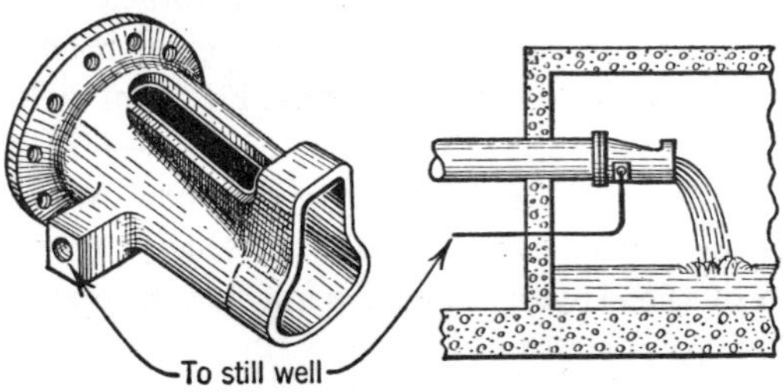

Fig. 10–26. Kennison Nozzle—Builders-Providence, Inc.

side or bottom of the weir, flume, or nozzle. Figure 10–27 illustrates the arrangement of a still well. The advantages of using a still well are that solids are prevented from interfering with the head measurement and waves and stream currents do not directly impinge on the instrument float. Sometimes the still well is installed directly in the flow channel. The head should be measured at a point located at least 3 times the head $3h$ upstream from the weir or nozzle opening.

Head is measured by any suitable liquid-level measuring device. The float-and-cable instrument is used almost exclusively. Since the float-and-cable instrument measures head directly, the calibration may be made directly in terms of volume flow units, independent of liquid density. If calibration in terms of weight-flow units is desired, a buoyancy-type level measuring instrument should receive consideration, since then the calibration is independent of density.

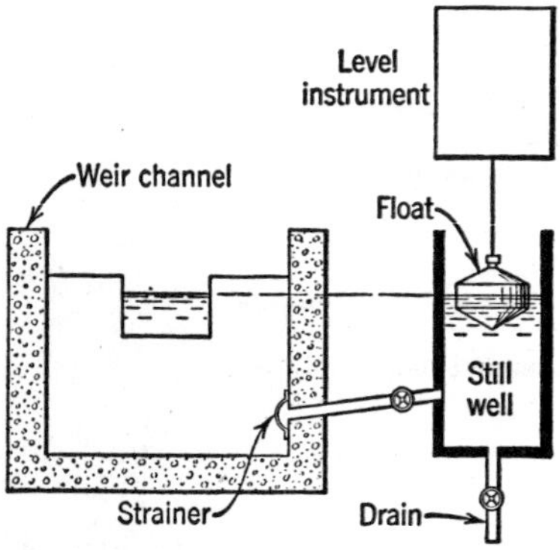

FIG. 10–27. Still Well for Head Measurement in Open-Channel Flow

The *current meter* consists of a propeller turning free on a shaft and arranged to operate a revolution counter. Generally the shaft operates a tachometer so that a recording instrument can be used for indicating flow rate and flow quantity. Ideally the velocity of the fluid stream at the point of measurement is proportional to the speed of rotation of the propeller. Actually friction exists in the propeller bearings, and these meters do not indicate very small stream velocities. A current meter can be calibrated by a towing test in still water. In order to calibrate the current meter for flow rate of a whole stream, it is often necessary to make a velocity traverse of the cross section of the stream. The stream flow rate is then calculated, usually graphically.

Velocity Meters

Many meters for measuring flow operate by determining the velocity in one part of the flow stream and are calibrated to indicate either the flow rate or the flow quantity for the whole stream. Such meters are the propeller meter and the thermal meter. It is of interest to note that the pitot-tube head meter is actually a velocity meter.

The *propeller meter*, normally used for measuring flow of liquids, is indicated in Fig. 10–28. Other types of propeller meters employ a turbine-type impeller with the geometric axis of the impeller perpendicular to the axis of the pipe.

The propeller or impeller of the meter is continuously rotated by the motion of the stream and is connected by gearing to a register calibrated to read in total flow quantity. Often the meter is arranged to transmit the readings by electrical means to a remote receiver, where both total flow and flow rate may be indicated.

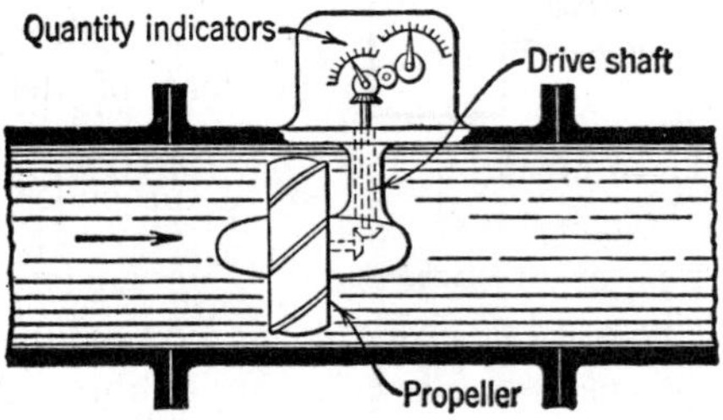

Fig. 10–28. Propeller Meter

Propeller speed would be linearly related to velocity of flow if the propeller and gearing in the meter were completely frictionless. Because of unavoidable friction there is a small stream velocity below which the propeller is not turned. The propeller meter operates best, therefore, when the stream velocity is continuously above the velocity threshold.

The accuracy of the propeller meter is generally about two per cent. The main factors influencing calibration are the velocity distribution in the pipe, and the density and viscosity (therefore temperature) of flowing fluid. Since the propeller meter infers the quantity of flow from the mass velocity of a part of the stream, usually 80 to 98 per cent of total, the velocity distribution is assumed to remain unchanged from the given distribution used during calibration. For this reason many propeller meters contain built-in straightening vanes. The density of the flowing fluid affects the impulse–momentum relationship at the propeller and therefore the calibration. Furthermore, the viscosity of the flowing fluid determines the propeller drag and affects calibration. It is best to calibrate the meter under actual conditions of density and viscosity, if these conditions are considerably different from those for which the meter was originally calibrated.

The advantages of the propeller meter are that it has almost no upper limit to capacity, it causes very little pressure loss, and it can be used for liquids containing abrasive materials. If the liquid being measured contains debris or heavy foreign material, a screen can be installed ahead of the meter. Many of these meters are arranged so that they subtract on reverse flows, a feature not usually found in other meters.

The *cup anemometer* employs virtually the same principles as the propeller meter. The familiar cup anemometer as used at weather stations for indicating wind speed can be made to indicate air speeds from about 150 up to 10,000 fpm. Generally the electric-tachometer system is employed to transmit readings to the desired point of indication. Cup anemometers are not accurate when measuring flow rate or flow quantity because of the large number of variables affecting calibration.

The *deflecting-vane anemometer* is a mechanical instrument, which operates by allowing a jet of air or gas to impinge directly on a pivoted vane. The vane is mounted on jewel bearings, may be damped magnetically, and carries a pointer for indicating stream velocity. The deflection of vane and pointer is opposed by the force of a hairspring much as in millivoltmeters. This meter requires special calibration for indicating flow rate. It is relatively inexpensive and is widely used in measuring duct velocities in ventilating and air-conditioning work.

The *hot-wire anemometer* consists of a small resistance wire inserted in the fluid stream and heated by electric current. The heat loss from the wire depends on the mass velocity and specific heat of the fluid. There are three methods of determining velocity of flow. First, the temperature of the wire can be maintained constant by adjusting the current flow to the wire. The velocity of flow is then proportional to current through the resistance wire. Second, the temperature difference in the flowing fluid before and after the resistance element can be maintained constant by adjusting the current flow to the element. This temperature difference is usually 1 or 2°F. The velocity of flow is then proportional to the current to the element. Third, the current to the resistance wire can be maintained constant, and the temperature, and therefore the resistance, of the wire is measured. An indicating milliammeter or an indicating Wheatstone-bridge instrument is used with the hot-wire anemometer. All hot-wire anemometers require special calibration in flow rate or flow quantity measurement.

Quantity Meters

Quantity meters operate by passing the fluid being measured through the meter in separate and distinct increments by filling and emptying containers of known weight or volumetric capacity. The quantity of flow is determined simply by totalizing the number of increments.

There are two basic kinds of quantity meters: those that weigh discrete increments of fluid, and those that pass discrete volumes of fluid. Weighing meters for measuring liquid flow are not commonly used. However, the true weighing meter operates on the beam-balance principle and determines weight quantity of flow. Tilting trap meters (Fig. 10–29) operate by gravity, so that the trap is automatically upset when the center of gravity passes over the stable equilibrium point. They are mechanically arranged for continuous flow by employing two tilting traps or by temporarily diverting the liquid to a holding chamber during the dumping operation.

The quantity meter operating on a volumetric basis (called a volumetric meter) is found in several types. Most of them resemble hydraulic piston pumps, gear pumps, or vane pumps, which when operated "backward" become motors. Actually volumetric-fluid meters are nothing more than fluid motors operating with minimum friction from the pressure differential across the motor.

In all volumetric meters the flow rate is given by

$$q = Q_i n \tag{10–59}$$

where q = flow rate in volume units at flowing condition
Q_i = fluid volume increment
n = frequency of passing increments of fluid

The weight flow rate is given by

$$w = \rho Q_i n \tag{10–60}$$

where ρ = weight density of flowing fluid. In both cases the flow rate is given in terms of flowing conditions, and, if reference to standard conditions of temperature and pressure is desired, suitable corrections must be made. Ordinarily, the flow of liquids is given in gallons per unit time and the flow of gases in cubic feet per unit time.

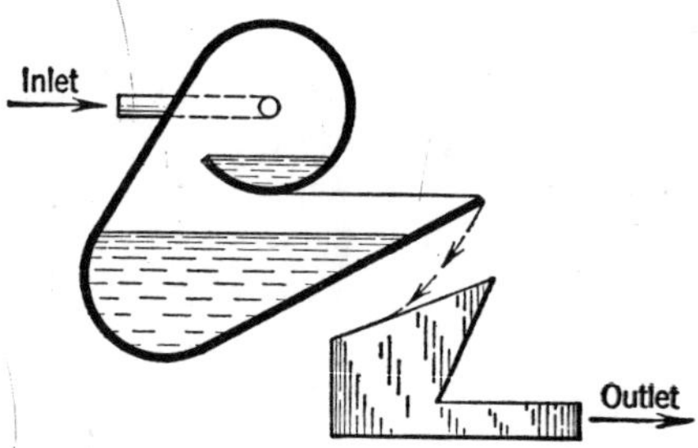

FIG. 10–29. Tilting Trap Meter

The *reciprocating-piston meter* (Fig. 10–30) is suitable for measuring flow of liquids. The piston reciprocates and passes the liquid alternately through each end of the cylinder from the inlet to the outlet. The piston operates the slide valve controlling the inlet and outlet ports and also operates a counter-mechanism. Actually such meters have two to four pistons, all operating on a center crank arm. The crank arm operates a rotary valve mechanism, which opens and closes the inlet and exhaust valves at proper points in the piston stroke. The crank arm also operates a register for indicating total quantity. The pressure differential required to operate the motor at low speeds against its own friction is kept small by selecting pistons of large diameter and operating them over a short stroke. Obviously, if the meter has no friction and if it has no slip (leakage or blow-by), then volume measurement is perfect. When in good condition, a piston meter will

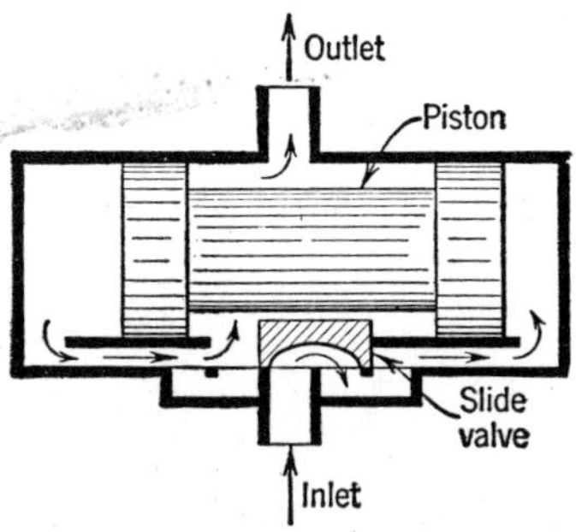

FIG. 10–30. Operating Principle of the Piston Meter

indicate low flow with reasonable accuracy. The accuracy of most piston meters is 0.2 to 0.3 per cent. The effect of fluid density and viscosity on accuracy of measuring volume flow rate is negligible. The piston meter is limited to use with noncorrosive and light (low-viscosity) liquids. It can handle flow rates of about 10 to 1000 gpm.

The *nutating-disk meter* is widely used for measuring flow quantity of liquids and is one form of the domestic water meter. The nutating-disk meter has the advantages of simplicity, compactness, and relatively low cost. Figure 10–31 indicates that the circular disk is pivoted at its geometric center and it does not rotate but "nutates" in a circular chamber with a conical roof and a conical floor. The vertical shaft of the disk generates a cone with the apex downward. Thus liquid enters the chamber and by the action of the differential pressure causes the disk to "roll" this quantity of fluid around to the outlet. The inlet and outlet chambers are separated by a partition (not shown in Fig. 10–31), which fits into a slot in the disk. This meter is suitable for almost any reasonable static pressure and temperature. The effect of density and viscosity on volume flow measurement is usually small, but not negligible, because the clearance at the disk is appreciable, and a small amount of fluid is by-passed. The static error of the nutating-disk meter is usually less than 1 per cent. This meter is used for flow rates from about 15 to 500 gpm.

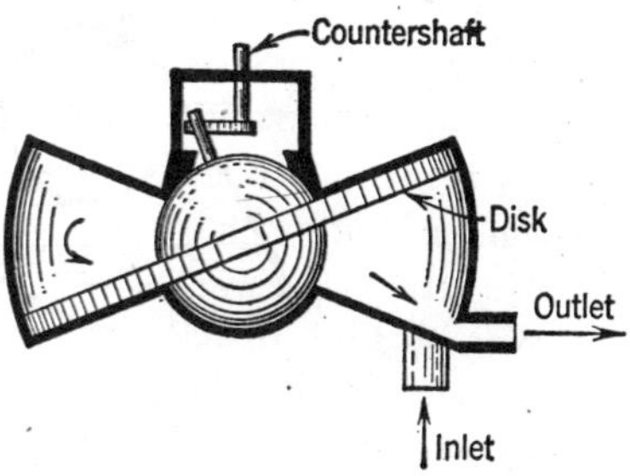

FIG. 10–31. Principle of Nutating-Disk Meter

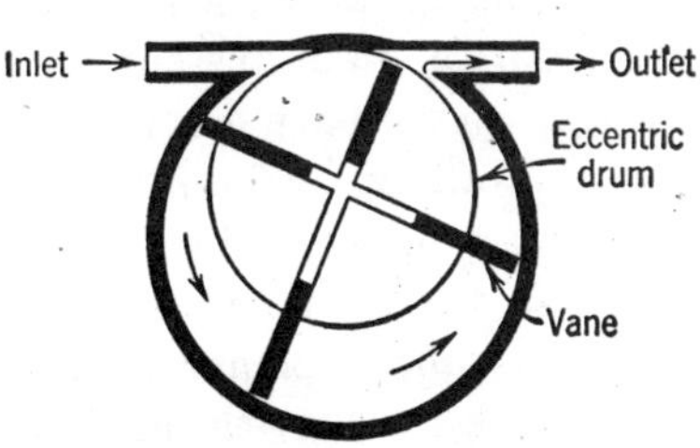

FIG. 10–32. Principle of the Rotary-Vane Meter

The *rotary-vane meter* (Fig. 10–32) is one example of a number of rotary-chamber meters. Here the drum rotates about its own center, which is eccentric to the meter body. As the drum rotates, the vanes are pushed outward by springs to form separate sealed chambers. The drum rotates by virtue of the differential pressure across the meter. A register on the drum shaft indicates total flow quantity. Since clearance is small or zero, effects of density and viscosity changes are negligible. The accuracy of the meter is therefore in the region of 0.2 to 0.3 per cent. Other types of quantity meters are the rotary-bucket meter and the rotary-piston meter. In many respects their operation is similar to that of the rotary-vane meter, and they have a comparable capacity and accuracy.

The *sealed-drum meter* used for measuring gas flow is shown in Fig. 10–33. It consists of a motor with spiral-like vanes contained in a horizontal cylindrical drum. The meter is filled with water to a height slightly over the center hub. The gas enters through the rotor hub and passes through the water to the chamber above formed by adjacent spiral vanes. When the vane emerges from the water, the gas is released to the outlet. The accuracy of this meter is about 0.5 per cent. Serious error may be caused by improper water level and by mounting the meter so that it is not level. Accurate indication of flow rate depends on the temperature and pressure of the gas, and correction of readings may be required. The water (or other sealing liquid) should be at gas temperatures and should be saturated with the gas passing through it.

The *lobed-impeller meter* has two rotors, each having lobes arranged to mesh very much like gears as shown in Fig. 10–34. The impellers rotate in close-fitting chambers. The fluid is trapped in the space between lobes and is passed from inlet to outlet. These meters may be used for either liquids or gases; the two-rotor type with two lobes on each rotor is commonly employed for gas flow.

The *bellows meter* is used only in gas-flow measurement. In construction it is simply a diaphragm pump operated as a motor and thus is quite similar to the reciprocating-piston meter. There are two bellows of oiled sheepskin alternately filled and emptied through passages controlled by slide valves, which are operated by the motion of the bellows. A register for indicating total flow is operated by the valve mechanism. Bellows meters may be used up to 500 psi static pressure. When in good condition a bellows meter is accurate to about 1 per cent. However, the calibration of the meter and, therefore, the accuracy of flow measurement depend on the static pressure of the gas, as is the case with orifice-head meters. The static pressure of the gas should therefore be separately measured and the readings of the

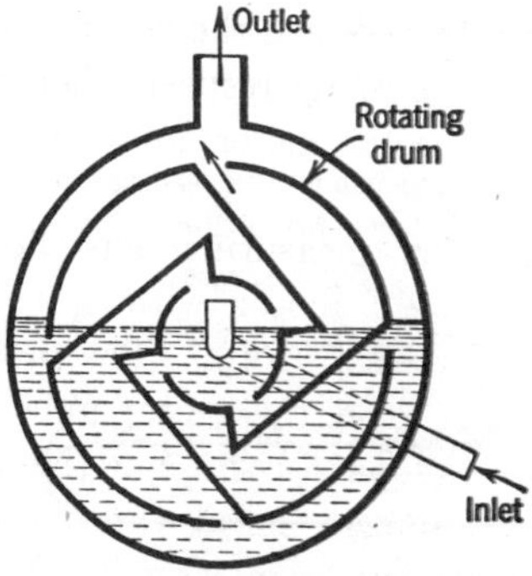

FIG. 10–33. Sealed-Drum Meter for Gas

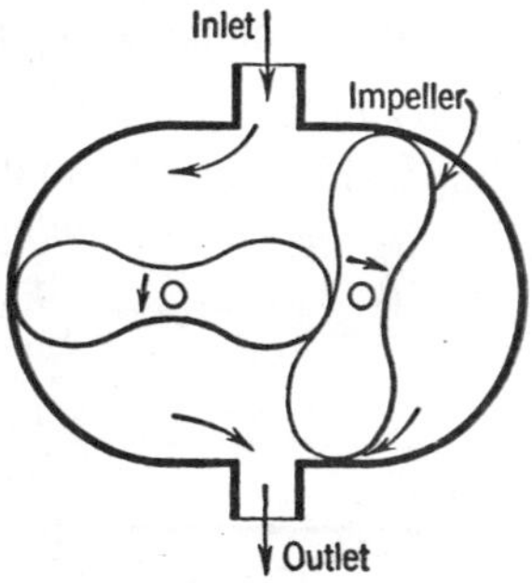

FIG. 10–34. The Lobed-Impeller Meter for Gas or Liquid

bellows meter corrected. In some meters automatic compensation for static pressure is provided.

Flow of Dry Materials

Continuous determination of flow rate of dry materials such as coal, cement, powdered chemicals, paper, and fruit is necessary in many industrial processes. Meters for measuring flow of dry materials are essentially of the weighing type, in that they determine the weight of material passing a given point.

Weighing meters are of many kinds, each developed for a particular kind of material flowing either in pipes or on open conveyers. Meters similar to the propeller meter for liquids are sometimes used for measuring flow of dry powdered or granular materials in closed pipes. Meters similar to the tilting-trap meter for liquids are used for some free-flowing dry materials. The latter meters operate on the weigh-beam principle, with suitable latches and gates for controlling the filling and dumping operations.

Conveyer-flow meters for dry materials may operate by weighing a section of the conveyer, as shown in Fig. 10–35. Here the use of a spring balance is indicated, and flow is measured in weight units such as tons per hour. If the conveyer belt is sufficiently flexible, the accuracy of flow measurement is usually within 1 per cent and sometimes 0.5 per cent. Integrators are used for determining total flow, and most such meters are compensated for conveyer-belt speed. Other conveyer meters em-

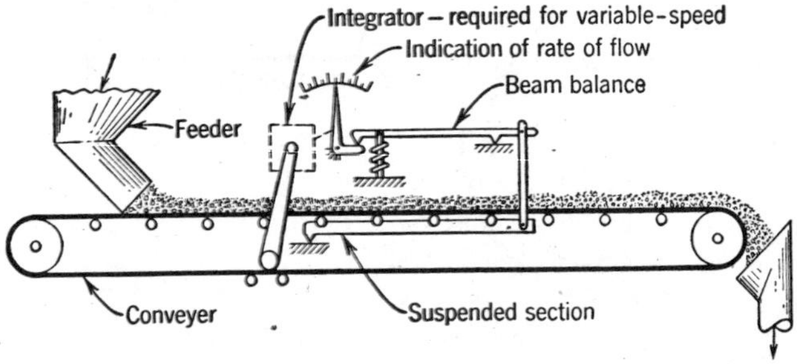

FIG. 10–35. Conveyer Weighing for Measuring Flow of Dry Materials

ploy the automatic weighing scales of either the electric or pneumatic type instead of the spring balance shown in Fig. 10–35.

Viscosity Measurement

Measurement of viscosity of fluids is difficult in continuous industrial operations, and there are few industrial instruments for this service. However, viscosity is important in many fluid operations, and it is sometimes necessary to attempt measurement.

Viscosity is a characteristic sometimes called fluid friction. Consequently, it can be measured by measuring friction or, actually, shear force. An example of this method is the Viscograph of the Brabender Corporation. Although this is not primarily an industrial instrument, it can be used for recording viscosity of liquids. Figure 10–36 shows a bowl containing the liquid in question. The bowl is rotated at constant speed by the synchronous-drive motor. A paddle wheel is supported in the bowl and is free to turn with the liquid, the rotation of the paddle wheel opposed by the spring. For low-viscosity fluids the shear or drag forces are low, and the pointer deflects a small amount. For high-viscosity fluids the forces are greater, and the pointer is deflected a large amount. Since viscosity is greatly affected by temperature, the bowl is enclosed by a housing which has thermostatic temperature control.

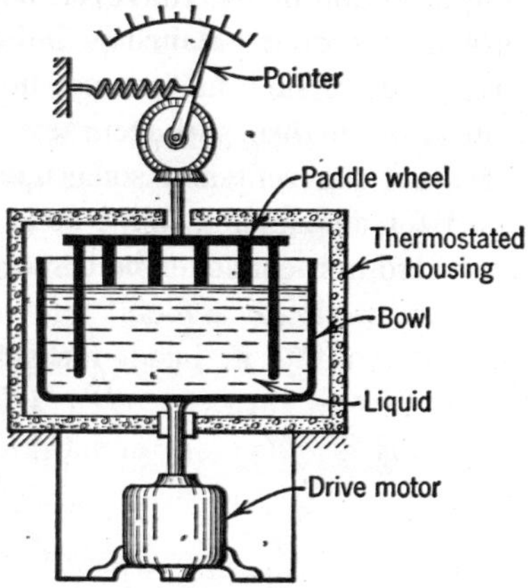

Fig. 10-36. Viscosity Meter—Brabender Corporation

The Viscosimeter of the Fischer & Porter Company is indicated in Fig. 10-37. Since flow rate of a fluid in an area meter (rotameter) is dependent on viscosity, this fact is used to measure viscosity. Briefly, a tapered tube, through which the fluid flows, contains two floats. The lower float is shaped so as to be immune to viscosity changes. The flow rate through the tube is controlled to a constant value so that the lower float remains at the flow-set-

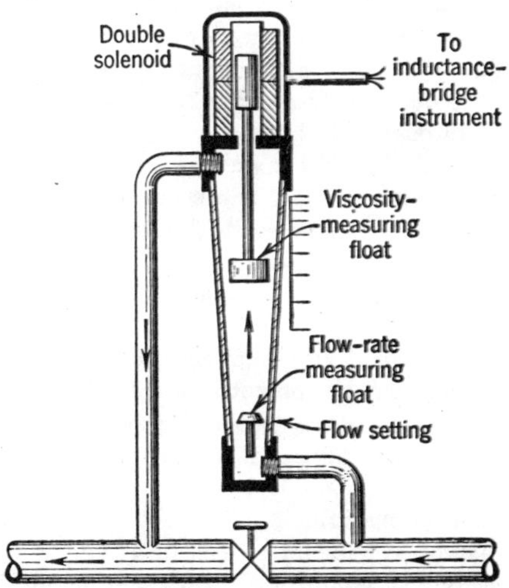

Fig. 10-37. Continuous Viscosity Meter—Fischer & Porter Company

ting mark. Since the flow is constant, the upper float, which is sensitive to viscosity changes or actually is acted on by the viscous drag forces, changes position with changing fluid viscosity. Therefore, the distance between floats is proportional to the viscosity of the fluid at the existing temperature. This instrument is made as a recorder by electrically transmitting the position of the upper float to a recorder.

Measurement of viscosity is accomplished by the Taylor Instrument Companies through measurement of the pressure differential across a smooth tube through which the fluid flows. The restriction is simply a metal tube with smooth walls about 15 in. long and about 3/8 in. inside diameter. The pressure is measured through taps at each end of the tube. The flow rate and density (or specific gravity) are maintained constant.

Problems

10–1. Water flowing in a 3-in. line (inside diameter 3.068 in.) is measured by an orifice 2.280 in. in diameter. The orifice head is measured through vena-contracta taps. The pressure differential is 10.8 in. Hg at 70°F. The temperature of the water is 300°F. Find the flow rate in gallons per minute at 70°F.

10–2. Dry air flows in a 2-in. pipe (inside diameter 2.067 in.) at 150 psig and 70°F. It is desired to meter a flow rate of 1.5 cfm at the given temperature and pressure. The manometer must employ a head of 20 in. Hg at 80°F. What orifice diameter is required?

10–3. Steam flows through a 10-in. header at 57,000 lb per hr. The inside diameter of the header is 9.76 in., and the orifice is 5.85 in. in diameter. The orifice has vena contracta taps. The steam pressure is 250 psia, and the temperature is 640°F. Using steam tables for necessary data, find the differential pressure at the manometer in terms of inches of water at 70°F. (Actually the manometer uses mercury, but the head equivalent is often given in water column.)

10–4. A large venturi is used at a pumping station for measuring water flow. The inside diameter of the conduit is 60 in., and the throat diameter of the venturi is 30 in. An automatic controller holds the differential pressure constant at 10 in. Hg at 70°F. However, in the daytime the water temperature is 80°F, and at night it is 50°F. By what percentage does the flow rate decrease at nighttime?

10–5. An orifice having a diameter of 2.00 in. is known to wear so that the diameter increases by 0.002 in. each week. It is installed in a 4-in. pipe of 4.026 in. inside diameter. If the Reynolds number is about 100,000, estimate the error after 6 weeks' operation.

10–6. A rectangular weir without side contraction has a width of 2.0 ft. The head over the weir crest is 2.0 ft. Determine the flow of water through the weir in gallons per day.

10–7. A V-notch weir is used for sewage-flow measurement. The angle of the V notch θ is 90° The flow is 60 million gallons per day. What is the head over the weir crest?

10–8. In integrating a gas flow, what is the approximate error in total quantity per day caused by not accounting for gradual linear increase of static pressure amounting to 10 per cent per day?

10–9. In a rotameter for metering gasoline of specific gravity 0.8, what should be the density of the float material if density compensation is desired?

10–10. If a rotameter is desired with linear scale calibration, derive the equation for the shape of the sides of the metering tube, and prove the calibration is linear.

10–11. Suppose that a rotameter uses a cylindrical float 1 in. high and 1 in. in diameter and of the proper density to give density compensation. If the coefficient C_c is 0.5 and the maximum inside diameter of the metering tube 2 in., what flow rate in gallons per minute can the rotameter accommodate?

10–12. In the orifice-flow problem 10–1, the temperature of the water changes from 300 to 200°F. Find the new flow rate in gallons per minute at 70°F.

10–13. Explain all the terms present in the equation 10–20 for the expansion factor Y. Is this expression empirical, or can it be derived?

10–14. Explain all the terms present in the equation 10–22 for the moisture factor M. Derive this factor.

10–15. Refer to a compressibility chart, and determine under what conditions of pressure and temperature it is safe to take the value 1.0 for the compressibility factor y. For what pressure and temperature limits for air, oxygen, hydrogen, and methane is this permissible?

10–16. From tables in the appendix plot the value of flow coefficient K versus diameter ratio β as abscissa at a Reynolds number of 100,000. Use data for flange taps and 4-in. pipe. Plot the same data, but omit the velocity of approach factor $(1/\sqrt{1 - \beta^4})$.

(a) Above what value of diameter ratio is the velocity of approach factor important?

(b) In what range of diameter ratio is variation of diameter ratio due to wear and abrasion least important?

10–17. It is desired to measure the flow rate of fuel oil in a ½-in. standard pipe. Orifice-flow coefficients are known only down to 2-in. standard pipe. What must be done to measure the flow?

10–18. Certain heavy oils, such as fuel oil and tars, and heavy liquids are characterized by high viscosity. This restricts the flow measurement of such fluids when orifice-head meters are used. Why? Investigate the Reynolds number for heavy fluids flowing in 2-in. pipes, and compare to water. To obtain satisfactory Reynolds number it is sometimes necessary to increase the velocity. Increasing temperature also helps. Why?

10–19. Derive the theoretical equation for compressible flow (equation 10–14), showing all steps and assumptions made. Which of the assumptions are the most important in that they are the most likely to be met?

10–20. Write the complete binomial expansions given in equations 10–15 and 10–16. What is the maximum value of p_2/p_1 for which these expansions are accurate to within 1 per cent?

10–21. Can the rational expansion factor ϕ in equation 10–19 be reduced to simpler form by employing a binomial expansion for the pressure-ratio terms? For what pressure-ratio p_2/p_1 values would this procedure be valid?

10–22. Compute the difference between the rational expansion factor ϕ and the empirical expansion factor for flange taps using $\beta = 0.25$, $p_2/p_1 = 0.8$. The flowing fluid is carbon dioxide at 400°F.

10–23. For moist air at 80°F, what values of relative humidity are permissible for an accuracy of 1 per cent of measuring flow rate if no correction for moisture content is made.

10–24. Derive the relation for critical pressure ratio given in equation 10–29.

10–25. Employing the exact relation for critical pressure ratio and assuming $\beta = 0.6$, find the critical pressure ratio for a 50–50 mixture by weight of hydrogen and nitrogen at 100°F.

10–26. For an orifice installed with vena-contracta taps in a 4-in. pipe line, what are the locations of the upstream and downstream taps with respect to the orifice plate for an orifice diameter of 2.0 in.?

10–27. An orifice of 4 in. diameter is installed in an 8-in. pipe line. What is the minimum allowable distance upstream to a pressure-reducing valve? What is the minimum allowable distance downstream to a 90° elbow?

10–28. Water at 150°F and 80 psi pressure flows in a 14-in. extra-heavy pipe line. The flow rate is measured with a standard Herschel venturi tube, and the manometer differential is 10 in. Hg at 40°F. Find the flow in gallons per day at 60°F.

10–29. Dry air at 300°F and 20 psi pressure flows in an 8-in. standard pipe line. The flow rate is 300 cfm at 60°F and 14.7 psi conditions and is measured by a standard venturi tube. What is the manometer differential in inches of water at 60°F?

10–30. The stream velocity in water being measured by a pitot tube is 1200 fpm. The manometer differential is 8.0 in. Hg at 80°F. What is the velocity coefficient of the pitot tube?

10–31. Air at 60 psi pressure and 100°F flows in a pipe. A pitot tube is used to measure the velocity. The manometer shows a differential pressure of 20 in. water at 60°F. What is the velocity? Assume a reasonable velocity coefficient.

10–32. For the piston-area meter of Fig. 10–22, assume that each side of the piston is acted on by the corresponding static pressures. Assume that the holes are spaced to provide a linear variation of area with height. Derive the relationship between flow and piston height.

10–33. A rectangular-notch weir with side contraction is used in measuring the flow of raw crude oil. The width of the notch is 22 in. The head is 18 in. Estimate the flow in barrels per day.

10–34. Criticize the assumptions made in writing equations 10–43 for the forces acting on a rotameter float.

11· Process Instrumentation

The modern industrial plant relies on instrumentation as a regular part of its production equipment. In a few, industrial instrumentation is simple. In the vast majority, instrumentation is more elaborate, and, in an increasing number of processes, complete instrumentation is a necessity without which the plant cannot operate efficiently.

A whole new field is gradually taking shape, and the instrumentation engineer is becoming an important factor in the engineering of newly designed manufacturing processes.

Process instrumentation, whether it is the temperature control [1] of a milk cooler, the operation control means for a distillery, or the control means for a large steel plant, is still accomplished largely by experience. As instrumentation analysis adopts more of the analytical approach, pure study will become more useful.

Recording Instruments

A recording instrument should be employed when a permanent record of the variable is desirable. A record is useful for three reasons: (1) The process operator may refer to the instrument to observe the trend of the variable as a guide to process operation; (2) the record may be useful in locating trouble on the job; (3) the record may be required for reference to past performance.

[1] It will be necessary in portions of this chapter to assume that the student has some knowledge of automatic control. This can be gained from references such as D. P. Eckman, *Principles of Industrial Process Control,* John Wiley & Sons, 1945, and G. S. Brown and D. P. Campbell, *Principles of Servomechanisms,* John Wiley & Sons, 1948.

A recording instrument should not be used when any of the foregoing functions are not required and when a simple indicator will suffice. Obviously recording instruments are more expensive and require more maintenance than indicating types. Furthermore, a recording instrument without a chart in place is not useful as an indicator.

Industrial instruments of the recording type employ two different styles of recording charts: the circular recording chart, and the strip recording chart.

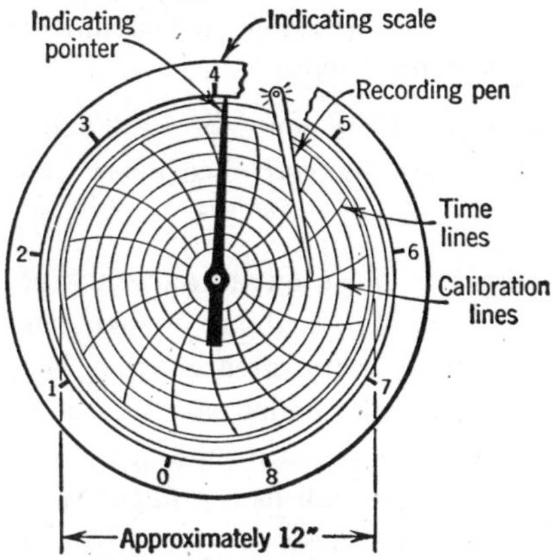

Fig. 11–1. Circular Recording Chart and Concentric Indicating Scale

A circular chart is shown in Fig. 11–1. The circular chart is used on almost every kind of industrial instrument—flow, pressure, temperature, and level. This chart is based essentially on polar coordinates with the exceptions that the "zero" is moved away from the geometric center and the time lines are segments of an arc instead of straight lines. The chart is clamped at the geometric center and rotates about the center.

The period of rotation of the circular chart is usually 8 hr, 24 hr, or 7 days, although nearly any speed of rotation can be obtained by choosing a properly geared synchronous motor. Mechanical (spring-driven) clocks can also be used where electricity is not available or not desirable.

Circular recording charts are made in basically two sizes: an 8-in. (approximate) diameter and a 12-in. (approximate) diameter. The 12-in. chart is by far the most common in industrial work. The 8-in. chart is used in recording ambient temperature and humidity and in recording temperature of heated vats and the like, where accuracy of reading is not essential.

The 12-in. circular chart has the following main advantages compared to a strip chart:

1. The entire record of one process period (4, 8, 12 hr, etc.) is available at a glance.
2. Long chart length (about 22 in. at 60 per cent calibration circumference) is available for one process period.
3. It is easily filed for future reference.

It has the following limitations:

1. Time values toward the center are difficult to read.
2. The chart must be changed once every process period (4, 8, 12 hr, etc.)

The circular chart is most advantageously used to record a mildly fluctuating variable at about 50 per cent of scale span where the process period is not greater than 24 hr.

Circular charts are printed under carefully controlled temperature and humidity conditions on a high-grade paper. Humidity variations under actual use causes the paper to expand and contract. Since the fibers in the chart paper run in one direction, this causes the chart to expand or contract to elliptical shape. The error in calibration is very small and is not corrected or compensated.

A strip recording chart is shown in Fig. 11–2. The strip chart is generally used on the potentiometer-type instrument or on the resistance thermometer. The chart is driven from the supply roll to the reroll by a synchronous motor. The speed of movement of the paper is usually given in inches per hour, 1, 2, and 4 in. per hr being very common. Speeds up to 20 in. per hr are available.

The strip recording chart used with the recording-pyrometer- and resistance-thermometer-type instruments has rectangular coordinates and a width of approximately 10 in. Strip recording charts for galvanometer-type instruments are not so common in recent years. These are usually about 6 in. in width, and the time lines consist of arc segments instead of straight lines—because the pen on such an instrument is arranged to move in an arc.

The strip chart has the following main advantages compared to the circular chart:

1. It does not require frequent changing.
2. Long process periods (24 hr or longer) are easily recorded.
3. Readability is the same at all values of scale span.

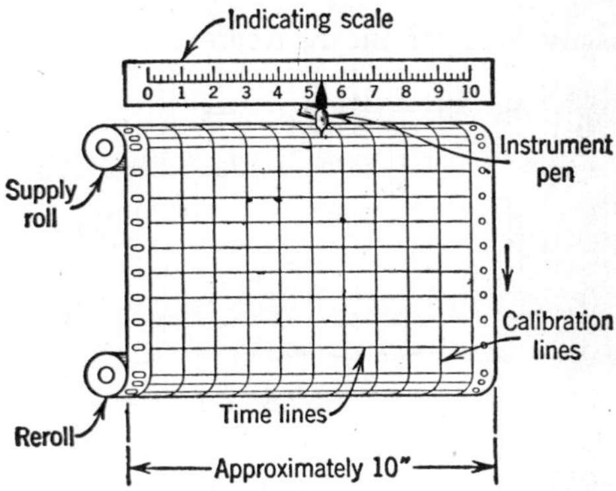

FIG. 11–2. Strip Recording Chart and Linear Indicating Scale

It has the following limitations:

1. A limited length of the chart is visible (usually about 8 in.).
2. It is difficult to file and not easily used for reference.

The strip chart is most advantageously used to record a variable changing between wide limits and to record a variable over very long process periods.

Humidity variations cause the strip chart to change width, and this is the reason for the oblong slots on the left side of the chart in Fig. 11–2. As with circular charts, the error is very small, but it is better to avoid wide changes in humidity.

Recording instruments may also include an indicating scale. The circular-chart-recording instrument of Fig. 11–1 includes an indicating scale. The strip-chart-recording instrument usually is arranged with a linear indicating scale as shown in Fig. 11–2.

Multiple-point recorders are used to record several variables on one chart. Several variables can be recorded on one chart when they do not vary greatly with time and when they are separated (not close together) in value. The variables should not vary through wide limits, because this produces a confusing array of criss-crossing lines. Furthermore, each variable should be positively identified. When one is recording several variables on a circular chart, it is well to remember that the portion of the chart from 0 to 10 or 15 per cent of scale span is not generally useful because the time scale is so short.

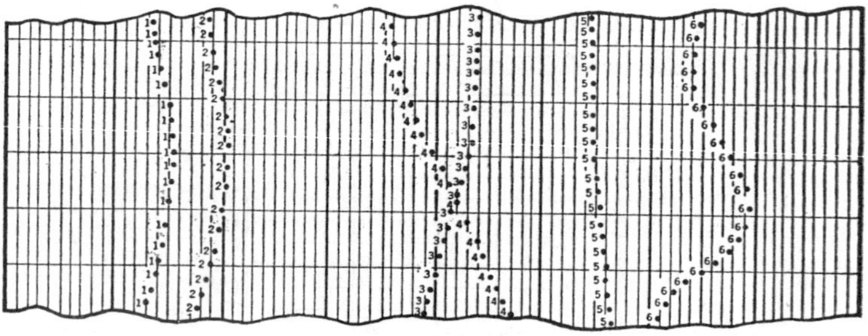

Fig. 11-3. Multirecord Recorder

Circular-chart instruments may continuously record up to three and sometimes four variables, and these may be almost any measured quantity. Such instruments usually have four separate measuring means and four separate recording pens, so that they are essentially four instruments in one, with each variable recorded by an independent means. It is possible to record up to six variables on one circular chart by employing six separate measuring means but operating one recording pen alternately from each of the measuring means.

Strip-chart instruments, generally the potentiometer or resistance-thermometer instruments, are arranged with a multiple-point switch in such a fashion that several thermocouples or resistance bulbs are alternately connecting into the measuring circuit. The instrument then "prints" by depositing a dot or a plus sign or other mark on the chart. Four, eight, and sixteen variables may be recorded in this manner, the individual records identified by means of a color or by a number or both. The printing is accomplished periodically, usually every 15 sec to 2 min, as a series of dots or dashes. A portion of a six-record chart is shown in Fig. 11–3.

Indicating and Signaling Instruments

An indicating instrument is used when only the present value of the variable has meaning and the past record is of no consequence.

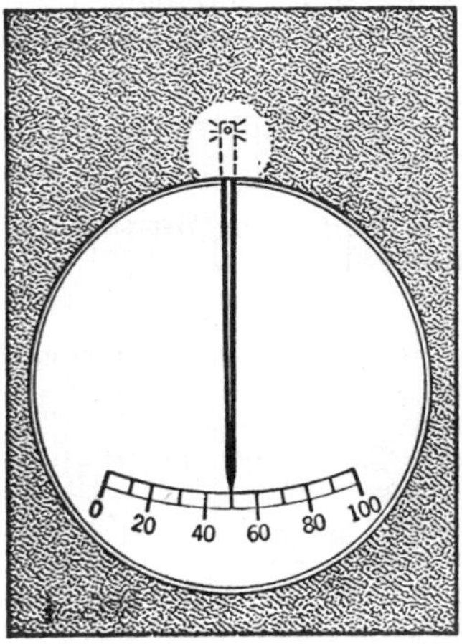

FIG. 11–4. Eccentric Scale Indicator

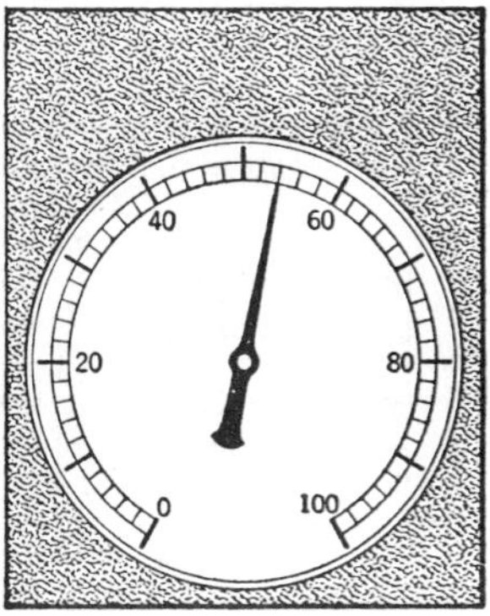

FIG. 11–5. Concentric Indicating Scale

Indicating scales are of three types: the eccentric scale, the concentric scale, and the linear scale. The eccentric scale, shown in Fig. 11–4, is commonly used on mechanical instruments such as pressure thermometers, flowmeters, and pressure gages. The scale length is usually about 8 in. but may be much smaller. The concentric scale shown in Fig. 11–1 and in Fig. 11–5 is used for nearly all kinds of instruments. The scale length is usually about 25 in. in the precision-type instrument. In simple dial-type pressure gages and thermometers the scale length may be as small as 3 in. The linear scale shown in Fig. 11–2 is commonly used only on the potentiometer–resistance-thermometer-type instrument. For such instruments an indicator is made simply by omitting the strip chart, chart drive, and pen. Galvanometer-type millivoltmeters employ a linear scale in a curved form.

The use of the various indicating instruments is generally restricted, because a free choice of scale types is not possible. For readability there can be no doubt that the concentric scale generally offers the greatest scale length in the smallest space.

Signaling instruments are very effective when it is only necessary to indicate that the variable is within certain limits. This is accomplished by electric contacts or switches suitably operated from a measuring means such as a thermometer pressure spring, liquid-level float, or thermocouple pyrometer. Signal lights or horns are used for visual or audible signal.

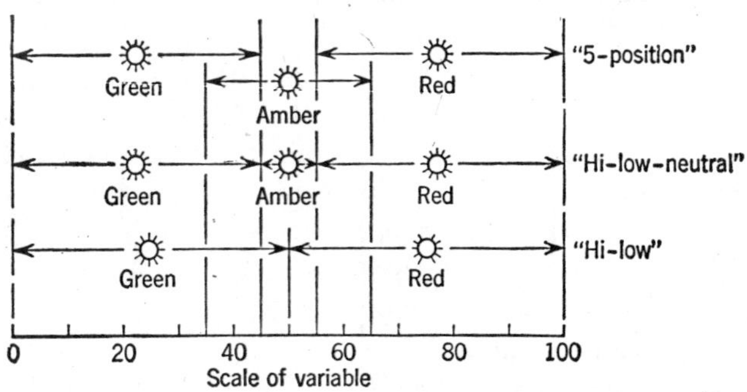

FIG. 11–6. Signaling Systems

Signaling systems should be positive; that is, a system should require an operator to look for a specific combination of signals before performing an operation. An example of poor signaling is to permit a process operation to proceed only when no signal is evident. This does not force the operation of checking a signal into the process-operation cycle, and it will likely be overlooked. Furthermore, such procedure is unsafe, because no signal may simply mean that the power is off.

The common signaling systems are shown in Fig. 11–6. These are: (1) the "hi–low" system, which is useful for indicating when a variable has passed a given point (This is sometimes used without the high or the low); (2) the "hi–low–neutral" system, which is widely used to indicate that a process operation may proceed only when the amber light is on; (3) the "5-position" system, which provides more information on the value of the variable. For example, some process operations may proceed when the amber light only is on, and some operations may proceed when the amber and either red or green lights are on.

Transmission of Instrument Readings

It is very often necessary to measure a variable at one point and to indicate its value at another point located a considerable distance from the first. The distances involved vary from a few feet up to several thousand feet.

For mechanical instruments, transmission of readings must be considered at distances greater than 50 ft, because the additional measuring lag and the possibility of ambient-temperature errors increase rapidly with distance. The manometer differential-pressure instrument (as in flowmeters) should not be used with connecting lines longer than about 50 ft. Long lines cause considerable difficulty with spurious liquid heads, and the resulting errors may be large. A pressure-spring thermometer with a capillary greater than 150 ft in length is costly, and the instrument is then subject to ambient-temperature errors. Pressure-gage connecting lines should be short, because long lines introduce considerable damping.

Pneumatic transmission has been illustrated in earlier chapters and is exemplified by the pressure thermometer with pneumatic transmission, the pneumatic differential-pressure transmitter, the pneumatic liquid-level transmitter, and the specific-gravity transmitter. Pneumatic transmission is generally operated on transmitted pressures of about 0 to 20 psig. The receiver is simply a pressure gage of 0 to 20 psi calibration. Actual calibration can then be made in terms of temperature, flow, liquid level, or whatever variable is desired.

Practically the only limitation to the distance between transmitter and receiver in pneumatic transmission is the lag caused by the resistance and volume of the connecting tubing. Generally the tubing is standard ⅜- or ¼-in. (outside-diameter) copper tubing. The lag caused by the tubing results from the resistance, and the volume and tubing up to ½ in. diameter has much less lag than tubing smaller than ¼ in. The lag of transmission is relatively small up to 500 ft length. Pneumatic transmission can be used up to 2000 ft distance, providing the lag is reduced by all possible means. To reduce lag to a minimum it is the practice to install booster pilots at intervals of 200 to 500 ft along the transmission line.

For *electric transmission* with electric instruments there is theoretically no limitation to distance between measuring element and instrument. In practice, there is the additional resistance of long lines, which may reduce the sensitiveness of measurement. In temperature measurement long connecting lines may make accurate ambient-temperature compensation difficult.

Electric transmission is illustrated by the inductance-bridge system used in the differential-pressure transmitter, the specific-gravity meter, the rotameter transmitter, the area-meter transmitter, and the viscosity meter. Electric transmission can be considered inherent in such instruments as the thermocouple pyrometer, the resistance thermometer, and other electric measuring instruments where a reasonable distance may separate the measuring element and the instrument.

In inductance-bridge systems the only practical limitation to the distance of transmission is the increasing dead zone caused by the reduced current in the bridge when line resistance is high. Distances up to 2000 ft have been covered satisfactorily. The lack of sensitivity of the instrument at long transmission distances is generally not too serious, so long as the instrument is not an automatic controller.

The electric-pulse system employs a constantly rotating cam at the transmitter to operate a switch. The percentage time on of the switch is determined by the position of the transmitter pen (or measuring element). The duration of each cycle is determined by the speed of the rotating cam (usually once every 15 sec). Thus a square-wave voltage is passed over the two connecting wires to the receiver. The receiver employs, in effect, a constantly rotating cam and an electromagnet operating the receiver pen or pointer through a clutch mechanism. The pen or pointer is positioned by the difference in duration of the pulse received from the transmitter and the pulse from the cam mechanism at the receiver. The system may operate over any two-wire system (even a telephone circuit), and synchronization is not required. The distance of transmission is therefore unlimited.

With automatic controllers an additional element is introduced into the problem of transmission by the consideration of the automatic-control circuit. When the transmission line is included in the automatic-control circuit, the measuring lag and the controlling lag can become very great. This point is of consequence in pneumatic transmission. Figure 11–7 illustrates two methods of arranging transmission when automatic control is in-

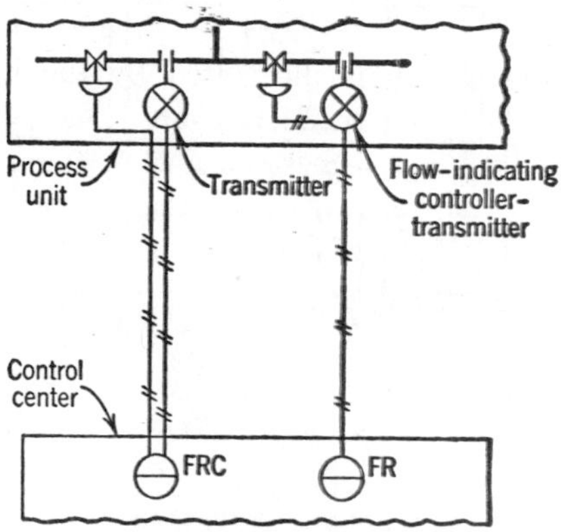

Fig. 11–7. Arrangement of Transmitter-Controllers

volved. In the circuit on the left, the transmitter is located at the pipe line and transmits the flow readings to the receiver located some distance away. The receiver is also a controller and transmits the control impulse back to the final control element. Thus the transmission-line lag is included twice in the control circuit. In the circuit on the right of Fig. 11–7, the instrument at the pipe line is a combined controller and transmitter. Automatic control is accomplished entirely at the pipe line without extending the control circuit over great distances. Such an arrangement results in much better quality of control.

In many applications there is little need for locating the controller at the control center. The industrial automatic controller is reliable in operation and generally requires no further adjustment after its initial installation. Employing a controller–transmitter allows the receiver to be made a multiple-point recorder, and several variables may be indicated on one chart.

The Control Center

Instrumentation equipment (instruments and automatic controllers) are generally grouped at one location near the processing unit or manufacturing operation which they serve. This group of instrumentation equipment may be termed a control center since

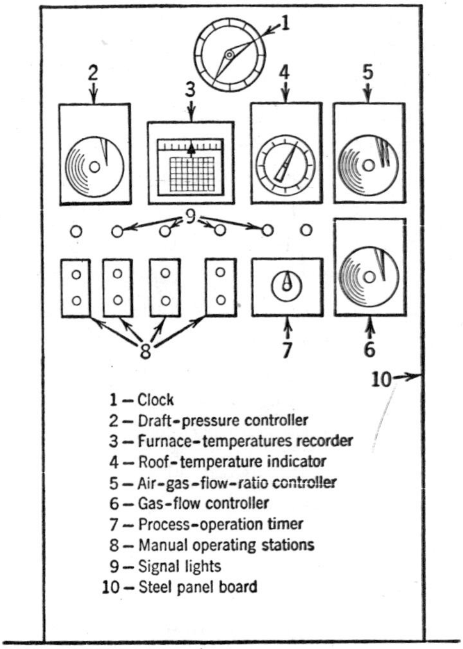

FIG. 11–8. Control Center for Open-Hearth Furnace (simplified from actual practice)

it controls or aids an operator in controlling the process operation. Actually the fundamental purpose of instrumentation is control of production quality or quantity.

The control center may include only a single instrument "hung" on the process equipment, or it may consist of hundreds of instruments and controllers housed in a separate building. The control center for an open-hearth furnace is shown in Fig. 11–8. This is a relatively simple example in which instruments and controllers for furnace-draft pressure, furnace temperatures, fuel flow, and process timing are included at one station. The center is located adjacent to the open-hearth furnace so that the operator in charge may readily maintain the desired operating conditions.

The control center should be arranged in a separate enclosure or room if this is at all feasible. A separate enclosure is the most effective way of providing clean, dry ambient conditions, which are vitally necessary in the operation of any precision equipment. In any event, the control center must have:

1. Clean dry atmosphere.
2. Relatively constant temperature and humidity.
3. No vibration.
4. Adequate light.
5. Unfailing source of electric power free of surges in voltage and frequency (when electrical measurements are used).
6. Unfailing source of clean dry air of constant pressure and required capacity (when pneumatic equipment is used).

Air conditioning at the control center is relatively common practice in satisfying the first two requirements for proper ambient conditions. In nearly all cases air conditioning pays for itself not only in the savings from instrument-equipment maintenance but also in increased efficiency of the workers and operators. In large control centers the use of color dynamics is almost a necessity in identification of equipment and connecting lines.

The control center should be located near the processing operation for close cooperation between process and control center. Because of the physical size of some plants, it is necessary to consider two types of plant and instrumentation layout: the central layout and the unit layout.

Central layout is illustrated in the hypothetical plant of Fig. 11–9 where the control center is common to all operations. The instrumentation for all plant operations (material storage, material treatment, first refining, second refining, separation, and product storage) is included in the control center. The control center then becomes the operating center of the plant. For a plant of any reasonable size the control center will probably include from 50 to 500 instruments. At the control center the instrumentation equipment is conveniently grouped on panel boards around the control-center room. This room is completely enclosed, air-conditioned, properly lighted, and clean. The obvious advantages of central layout are that the coordination of all plant operations is greatly facilitated and service and maintenance of instrumentation equipment are quickly and easily accomplished.

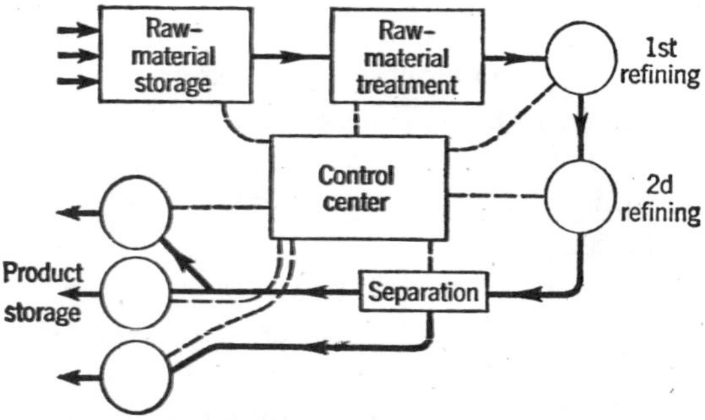

FIG. 11-9. Central Layout of Control Center and Plant

Unit layout is shown in Fig. 11-10 for another hypothetical plant. Separate control centers for each operation may be used because of the physical size of the whole plant or because the operations require no coordination. The control centers in this

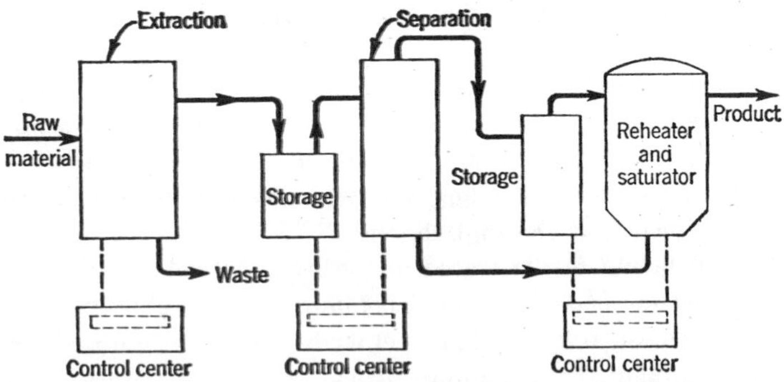

FIG. 11-10. Unit Layout of Control Center and Plant

case may be located in separate enclosures, or, as is common when the plant is inside a building, open panel boards are set up near each unit. A type of unit control center is shown in Fig. 11-11. It is often employed in steel mills and cement plants, where extra protection from dirt and damage is desirable. The house is usually constructed of concrete or cinder block with a window on one side. The window faces the process unit or operation floor, so that the process operator or supervisor may readily check on various operations without entering the enclosure. Unit layout has the advantages that the control center is close to the processing unit and coordination between processing unit and control center is improved.

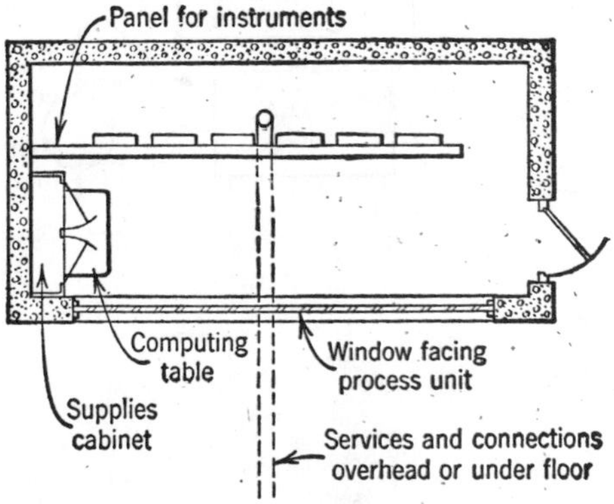

FIG. 11-11. Unit Control-Center Enclosure

Instrumentation Diagram

The purpose of an instrumentation diagram is to provide information quickly for use in process analysis, production control, specification of equipment, and preparation of equipment requisitions. The instrumentation diagram is constructed on the basis of a process flow diagram. It is intended to show the schematic layout of the process and plant equipment, together with the instrumentation equipment used with the process. The instrumentation diagram does not show the scale (size) of the equipment or the technical specifications of the equipment.

With regard to instrumentation equipment the diagram should show all major items; in some cases minor items may be omitted. It should indicate:

1. The variable being measured.
2. Whether indication, recording, or other service is required.
3. Whether control or alarm functions are required.
4. The auxiliary features of the instrument or controller.

5. The type of connecting lines.
6. The approximate location of the point of measurement and point of control.
7. Which instrumentation is to be included in the control center and which is to remain at the processing unit.

The symbols[2] for instrumentation diagrams are shown in Fig. 11–12. The specifications[2] for instrumentation diagrams are

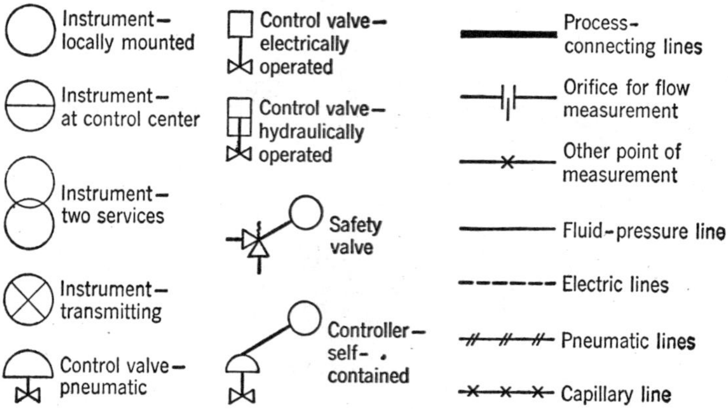

Fig. 11–12. Symbols for Instrumentation Diagrams

given in Table 11–1. The following notes pertain to the use of symbols:

1. Primary measuring elements have the same identification as the instruments to which they connect. Where an ele-

Table 11–1. Identifications on Instrumentation Diagrams

First Letter	Second Letter	Third Letter
D—Density	A—Alarm	A—Alarm
F—Flow	C—Control	C—Control
H—Hand-actuated	E—Element (primary)	V—Valve
L—Level	G—Glass (no measurement)	
M—Moisture	I—Indicating	
P—Pressure	R—Recorder	
T—Temperature	S—Safety	
	W—Well	

[2] These symbols conform in general to the tentative specifications of the Recommended Practices Committee (subcommittee RP-5) of the Instrument Society of America.

ment does not connect to any instrument, it carries separate identification. Where more than one element connects to the same instrument, they can be identified by suffix numbers after the item number.

2. Each element of any combination instrument that includes more than one service should have its own identification.

3. For remote transmission both the receiver and transmitter are identified in accordance with the combined service of the item.

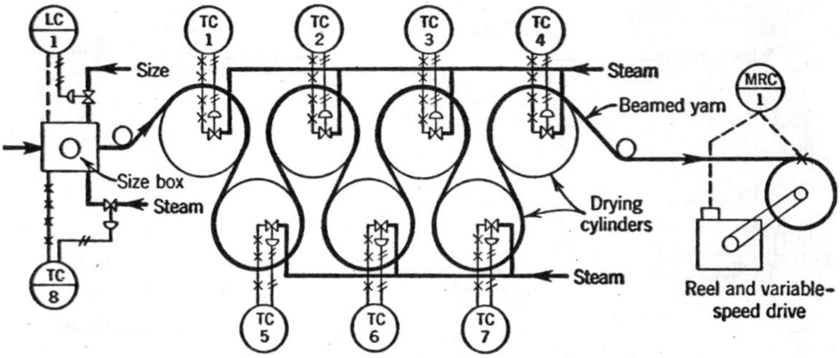

FIG. 11–13. Instrumentation Diagram (rayon slashing operation)

4. Each control valve has the same identification as the controller to which it is connected. Where more than one valve is connected to the same controller, they can be identified by suffix letters added to the number of the item.

An instrumentation diagram is given in Fig. 11–13 for a rayon-slashing operation. Briefly, the rayon cloth passes through the size box at the left of the diagram where it becomes saturated with size. (Size can be considered similar to starch.) The cloth then passes over the drying cylinders to the reel. The cloth must be dried to a closely controlled value of moisture content. The instrumentation diagram shows the following equipment. A moisture recorder–controller (MRC-1) measures the moisture of the cloth at the reel by electric means and controls electrically the variable-speed drive unit. Separate nonindicating temperature controllers (TC) are used at each drying cylinder. The controllers are pressure-spring thermometers with pneumatic control. The level in the size box is controlled by a nonindicating controller (LC-1) with electrical measuring means and pneumatic control. The temperature of the size box is controlled by a nonindicating pressure-spring thermometer with pneumatic control. The drying-cylinder temperature controllers are mounted on the unit. The moisture controller, the size-level controller and the size-temperature controller are located at the control center.

An instrumentation diagram for a more complicated process is shown in Fig. 11–14. Briefly, the material being treated passes

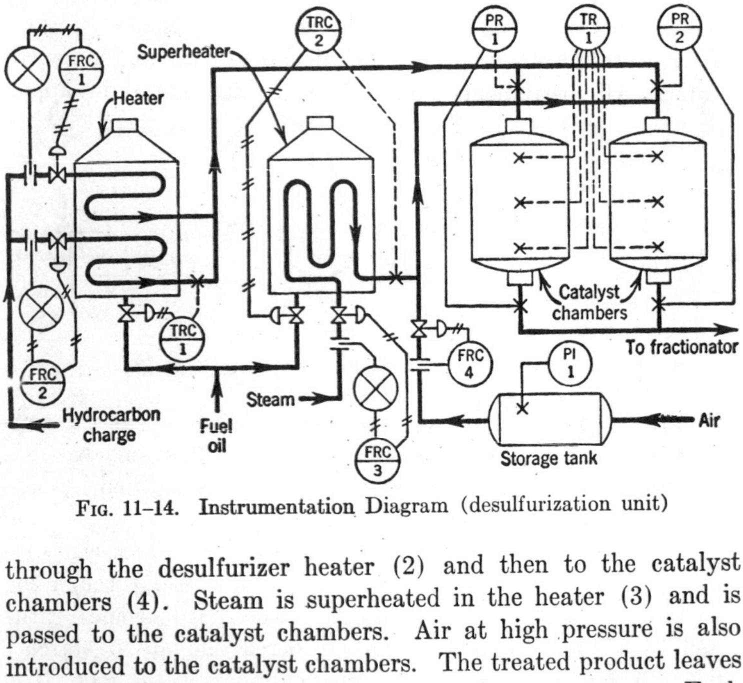

Fig. 11–14. Instrumentation Diagram (desulfurization unit)

through the desulfurizer heater (2) and then to the catalyst chambers (4). Steam is superheated in the heater (3) and is passed to the catalyst chambers. Air at high pressure is also introduced to the catalyst chambers. The treated product leaves the catalyst chambers and passes on to the next process. Each of the instruments and controllers can be identified by means of the standard symbols.

The Diagrammatic Control-Center Layout

Many large continuous-processing plants employ the central layout for the control center in order to obtain the best coordination of processing operations. The control center then consists of a large building with hundreds of instruments and automatic controllers mounted on panel boards around the walls of the building. This physical arrangement defeats its purpose by presenting such a confusing array of control equipment that coordination is difficult if not impossible to achieve. Much effort has been directed toward designing a more compact and centralized control center.

The diagrammatic control center is a means of obtaining better coordination of plant operation. This arrangement of the control center includes two sections: an operation panel and a record panel. The operation panel is made by reproducing the instrumentation diagram on a metal panel board in the control center. All instruments and controllers are placed in appropriate positions on the diagram and arranged to indicate or signal the actual magnitude of the corresponding variable at the processing unit. The record panel, arranged on a separate panel board, includes all the necessary recorders and indicators for gathering more complete data than it is necessary to provide at the operation panel.

The operation panel is illustrated in Fig. 11–15 from a schematic layout provided by Panellit Inc. and Moore Products Company. The instrumentation diagram, together with the indicators necessary for determining the conditions in the processing unit, greatly simplifies operating techniques. Abnormal conditions may be determined almost at a glance. The dial indicators are small dial-type pneumatic receivers calibrated to read in terms of the corresponding variable measured at the processing unit. Provision is made for adjusting set point and performing other necessary control manipulations directly at the operation panel. Other indicators employ color signals to show proper operating conditions.

The record panel serves to provide a record of performance and more complete data on operating conditions. This is accomplished by six-record circular-chart instruments for recording the variables indicated at the operation panel and the auxiliary variables. Multiple-point indicators are used for checking such individual variables as may require it.

The diagrammatic control center has been constructed in other forms than that of Fig. 11–15. However, the principle employed

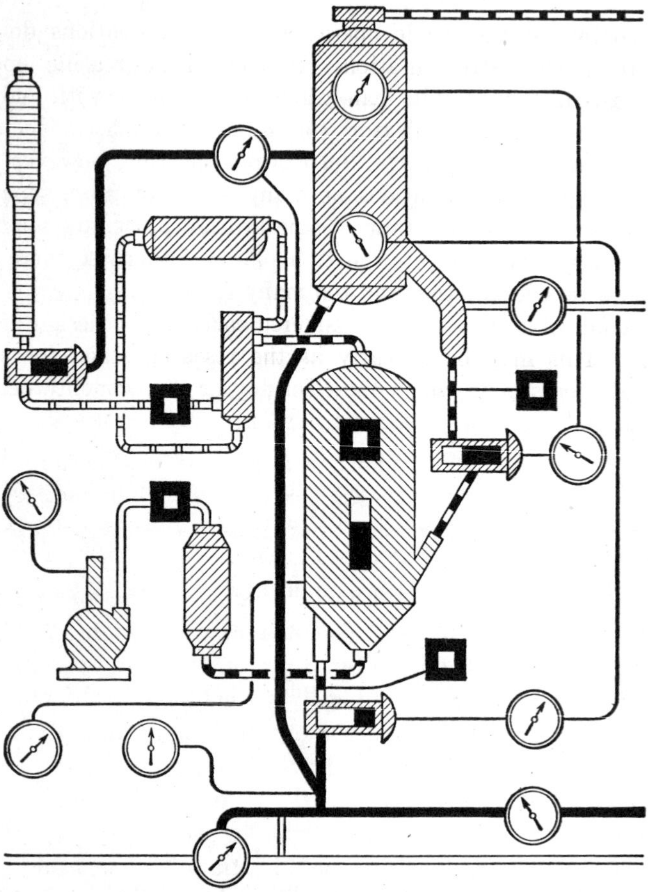

FIG. 11-15. Diagrammatic Control Center (*Reproduced by permission of Panellit Inc. and Moore Products Company*)

in all such control centers is to provide simple operating procedures.

Process Analysis

In applying instrumentation to a process it is first necessary to determine what variables in the process operation are of significance. In other words, it is necessary to decide what is to be measured and what it means. Measuring instruments are useful in providing information on which can be based the guidance and control of the whole operation. Some operations do not require much instrumentation. In some heat-treating operations satisfactory results can often be obtained with only a simple temperature indicator (often an experienced operator can do very well by eye!). Such is not the rule, however. In modern industrial operations as many as 30 or more separate variables may be important. Some orderly procedure must be used to apply instrumentation to a plant. Otherwise there will be too little coordination and too many mistakes.

Instrumentation must be considered early in the design of the plant. This may be as early as the research stage for some critical processes in which the efficiency or the economy depend on the accuracy of performance of the instrumentation. For example, if all the operations of a continuous process can be closely controlled, then provision for temporary storage between operations is not necessary. Research and development on the process operations then need to be concentrated on the performance of the instrumentation equipment.

A procedure for analyzing a process and applying instrumentation can be given as follows:

1. Divide the plant function into the smallest operation elements or operation units.

2. List all variables (temperature, pressure, flow rate, composition, etc.) that may affect each process operation element.

3. Add to this list ambient temperature, ambient humidity, barometric pressure, sun and wind conditions.

4. Divide this list of variables into four groups:

   (*a*) Variables to be controlled automatically.

   (*b*) Variables to be measured continuously.

   (*c*) Variables to be measured periodically.

   (*d*) Variables neither measured nor controlled.

5. For each variable of the automatic control group, select:

   (*a*) A method of measurement that will provide data most indicative of the desired process performance.

   (*b*) A location for the controller—at the control center or on the unit.

   (*c*) A style of controller—nonindicating, indicating, or recording.

(*d*) A mode of control that will provide the desired performance in view of the process dynamics.

6. For each variable of the continuous measurement group, decide:

(*a*) What method of measurement will provide data most indicative of the desired process performance.

(*b*) Whether signaling, indication, or recording is most desirable.

(*c*) Whether the signal device, indicator, or recorder is to be located at the control center or at the unit.

7. For each variable of the periodic measurement group, decide on a method of performing the measurement and the frequency with which the measurement must be made.

8. Construct an instrumentation diagram.

This outline is, of course, not complete in detail. Before arriving at the last step every detail of the installation and accessory equipment of each instrument and controller will have been considered. The example at the end of this chapter is designed to illustrate the procedure of applying instrumentation.

Instrumentation in the Modern Plant

The supervision of instrumentation within the organization of modern plants is the responsibility of the factory manager or works manager. The operation or use of the instrumentation equipment is usually separated from the service or maintenance of the instrumentation equipment.

The instrumentation equipment in a plant is used as a part of the regular production equipment. As such, the operating departments of the plant are responsible for its efficient operation. The servicing and maintenance of instrumentation equipment is the responsibility of the maintenance center of the plant.

The organization chart in Fig. 11–16 is typical of many average-size plants. The instrument department, as it is generally called, consists of all the personnel involved in the servicing and maintenance of instrumentation equipment. Direct authority for this function is delegated by the factory manager through a plant engineer. Usually the purchasing agent, the personnel director, and the manager of engineering act in an advisory capacity directly with the instrument department. These relations have become necessary because of the complex and technical nature of instrumentation equipment.

There is generally an instrumentation engineer associated with the engineering department. His responsibility is to guide the selection of instrumentation equipment for new manufacturing and processing operations. The instrument engineer therefore

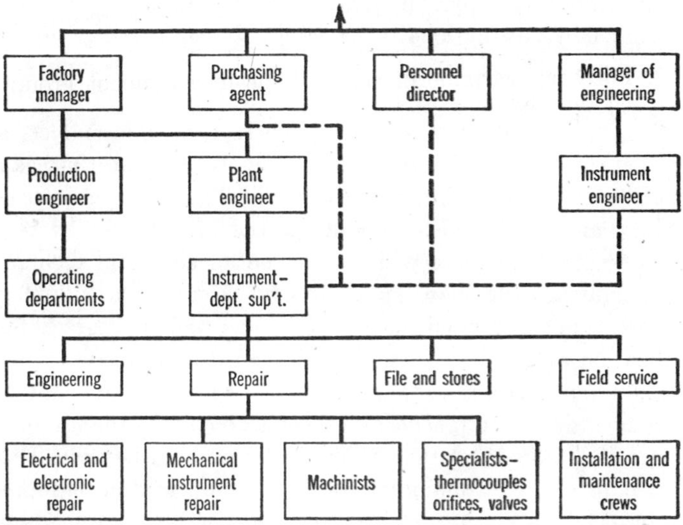

Fig. 11–16. Location of Instrumentation Functions in Plant Organization

is primarily concerned with the process design, the process layout, and the control center via the instrumentation diagram. Generally the instrumentation engineer consults with the instrument department and with the operating departments in the specification of new instrumentation equipment. This is very desirable from the standpoint of obtaining better cooperation and understanding in new operating and maintenance procedures.

In the purchase and specifications for instrumentation equipment the instrument department acts directly with the purchasing agent. In actual practice, it usually results that the instrument department becomes responsible for final selection of the instrumentation equipment, because, as has been mentioned previously, the equipment is so technical in nature.

The director of personnel cooperates directly with the instrument department in obtaining the technical personnel and the specialized and skilled workers.

The instrument department is generally divided into sections, the number of which depends on the magnitude of production operations and the amount of instrumentation equipment in use. There is usually an engineering section, a repair section, a file and stores section, and a field-service section.

The engineering section is called on for detailed orifice calculations; flow calculations; electric-circuit engineering; design of instrument air, water, oil, and electricity power supplies; and all such functions of engineering on the job.

The repair section takes on all the repair, maintenance, calibration, and testing of instrumentation equipment that can be brought into the maintenance center. This section is usually divided along lines corresponding to the kind of skilled technicians available. There will usually be an electrical technician; a mechanical technician; specialists in valves, orifices, and electronics; and the skilled machinists. Too high a degree of specialization at this level is not considered desirable.

The file and stores section is necessary, first, to maintain complete records of equipment, and second, to control the use of supplies. It is important that a numbering and card system be employed in checking maintenance on instrumentation equipment. An orderly system insures that periodic service is properly performed and provides ready information on localized failures of equipment.

The field-service section is responsible for repairs, service, maintenance, and installation of equipment at the processing plant or unit. Field service is required when local failure, that is, any abnormal operation of the instrumentation equipment, occurs. Prompt and immediate field service is necessary to prevent costly process shutdown, and all functions of the field-service section must be directed toward this end.

The instrument department may be located at the control-center building if the central control-center layout is used. This is desirable so that maintenance can be performed quickly. There is generally better cooperation between operating and servicing departments with this arrangement. Furthermore, better working conditions are usually available at the control center than elsewhere. In the unit control-center layout the instrument department should be located conveniently in relation to the plant and processing operations.

The instrument department is vitally concerned with the testing and calibration standards for instrument and automatic-control equipment. The standards for temperature, pressure, and voltage are the most important.

Mercury-in-glass thermometers (laboratory type), thermocouples, and resistance-thermometer bulbs are all used for reference standards of temperature. These may be obtained with an error curve certified by the manufacturer. It is desirable to have these standards certified by the National Bureau of Standards.

The reference standard for pressure is generally a series of dead-weight testers with an error curve for each. In addition, liquid-column manometers, when carefully used, may serve as reference standards for pressure.

Voltage standards in millivolt ranges are used in calibrating potentiometer-, millivoltmeter-, and resistance-thermometer-type instruments. Laboratory precision potentiometers, together with resistance standards, can then be used for voltage, resistance, and current measurements.

Illustrative Problem in Instrumentation

The soaking pit in the steel mill is used for bringing ingots up to a temperature of about 2300°F. They are then withdrawn

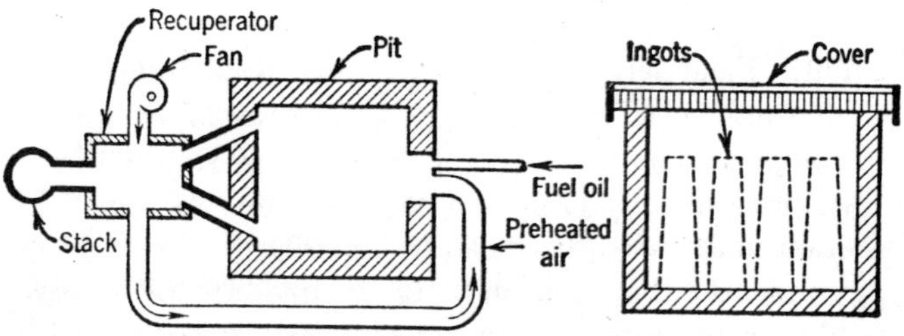

Fig. 11–17. Direct-Fired Soaking Pit

by a crane while heated and taken to rolling and forming operations. The soaking pit in Fig. 11–17 is sunk into the ground and is covered by a steel and ceramic brick cover. The cover is lifted and replaced by a crane during charging and drawing operations. The pit is fired by fuel oil, and the combustion gases pass out through a recuperator where waste heat is extracted.

The operation requires heating the ingots to the required temperature, uniform throughout the ingot, in as short a time as possible without excessive scaling or burning of the metal.

Making a list of variables affecting the over-all operation of the process, we have:

1. Ingot temperature.
2. Amount of metal in pit.
3. Pit temperatures.
4. Pit-temperature distribution.
5. Wall and cover temperatures.
6. Preheated-air temperature.
7. Oil-flow rate.
8. Oil-flow quantity.
9. Heating value of fuel oil.
10. Pressure of fuel oil.
11. Ratio of oil to air for combustion.
12. Fan speed.
13. Stack draft.
14. Stack temperature.

to which are added:

15. Ambient temperature.
16. Ambient humidity.
17. Barometric pressure.
18. Sun and wind conditions.

Before subdividing this list, it is necessary to consider the temperature control. It is desired to maintain the ingot at constant temperature. In accordance with the rule of direct measurement, we should therefore measure the surface temperature of the hottest ingot during heating so as to avoid scaling and burning. After the ingots are up to temperature, the center temperature of the coolest ingot should be measured so as to insure that all ingots have reached the minimum temperature for forming.

It is readily apparent that this procedure is physically impractical. Therefore it is necessary to measure the pit temperature, which will be an average of wall, roof, ingot-surface, and pit-atmosphere temperatures.

Next it is necessary to decide how to effect temperature control. The pit temperature can be controlled by varying the fuel-oil flow rate, the heating value of fuel oil, the fan speed, or the stack draft. A short consideration of each of these manipulated variables shows that most effective control is obtained by varying fuel-oil flow rate.

The following group of variables will not be measured: 1, 2, 5, 9, 10, 12, 15, 16, 17, and 18. It will be the job of the automatic control to overcome changes in all these variables. Actually all the ambient conditions affect the operation of this process. The student should consider the reasons why.

The following group of variables will be intermittent measured only when necessary: 6 and 14.

The following group of variables will be continuously measured: 4, 7, and 8. Oil-flow quantity is measured because in a heating operation of this magnitude the amount of fuel oil used must be accounted in job costs.

The following group of variables will be measured and controlled: 3, 11, and 13. The oil–air ratio must be controlled for efficient combustion. The stack draft will be controlled to reduce waste heat up the stack. The stack draft is controlled most easily by maintaining a positive *static* pressure within the pit.

All instruments and controllers will be located in an enclosed control center because of dust conditions, highly fluctuating ambient temperatures, and possibility of mechanical damage in a soaking-pit building.

As shown on the instrumentation diagram in Fig. 11–18, the following instrumentation equipment is selected:

6 and 14. These temperatures may be required infrequently to aid in burner and damper adjustments and to insure efficient combustion. Use an indicating millivoltmeter with thermocouples (TI-1) and a manual selector switch (TI-1a).

4. Pit-temperature distribution can be measured by several thermocouples or radiation receivers located around the pit. Actually it is sufficient to measure the temperatures at the exits

from the pit to the recuperator. An even distribution of temperature here indicates uniformity in the pit. This will require a two-record potentiometer with thermocouples (TR-1).

7 and 8. Oil-flow rate and quantity can be measured either by a quantity meter or an area meter. A recording area meter with integrator is preferred, because the fuel-oil supply is not usually clean in steel plants (often tars are added to the fuel oil). A recorder is used, because a reference is desired for de-

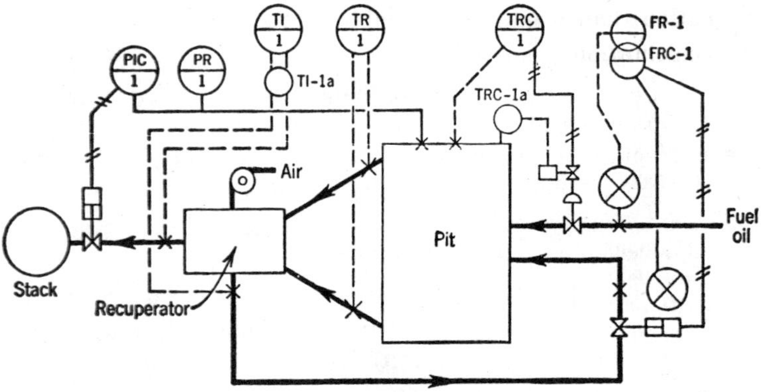

Fig. 11-18. Instrumentation Diagram—Illustrative Problem on Direct-Fired Soaking Pit

termining rates used during heating and for setting burners (FR-1 and FRC-1).

3. Pit temperature can be measured with either a thermocouple or a radiation receiver. Although the choice is quite open, we shall select a radiation unit with target tube to avoid the necessity of replacing thermocouples at this temperature (2300°F). The controller will be a single-point recording potentiometer (automatic balance) (TRC-1). Pneumatic control is selected because of the ease of accomplishing shutoff. This occurs when the cover is removed to draw an ingot. It is necessary to shut off the fuel at this time. This is accomplished by arranging a limit switch at the cover (TRC-1a) to close a solenoid valve in the air line to a diaphragm valve in the fuel-oil line. This automatically closes the diaphragm valve since it opens by air pressure and closes by a spring.

11. The fuel–air ratio does not need to be either indicated or recorded. The oil-flow rate is measured by an orifice, by the burner nozzle-back pressure, or by an area meter. The ratio of the two quantities is measured automatically, and a butterfly valve in the air line is controlled.

13. The static pressure in the pit should be indicated by a draft gage and controlled by positioning a stack damper (PIC-1 and PR-1).

Actually in this problem we have used considerably more information than would be available to the beginning student of instrumentation.

Problems

11–1. Make a list of instruments and primary elements having good over-all operation in the following ranges:

(*a*) –100°F.
(*b*) 0–100°F.
(*c*) 100–600°F.
(*d*) 1000–1400°F.
(*e*) 2000–2800°F.

11–2. In a temperature-measurement application there is a question of accuracy. In the following examples, briefly describe methods for checking accuracy, assuming no physical limitations:

(*a*) Mercury thermometer measuring temperature in a closed vessel containing hot liquid at 300°F.

(*b*) Iron–Constantan thermocouple with potentiometer pyrometer measuring temperature in an air-annealing furnace at 1200°F.

(*c*) Radiation pyrometer sighted on a brick wall at 2000°F.

11–3. A temperature-measuring instrument is to be used in a gas plant having an explosive atmosphere. The temperature is 325°F. What instrument should be selected?

11–4. Instruments are often installed in open country, as in pipe-line installations. Make a sketch showing how a manometer flowmeter could be protected from low and high ambient temperatures. Electricity is available.

11–5. A temperature of 150°F is to be measured in a gas holder located 1 mile from the point where a recorder is desired. Select an instrument for this service.

11–6. It is desired to measure the flow quantity in a 6-in. gas line with a line pressure of 20 psig. No electricity or compressed air is available. Select an instrument for this service.

11–7. What flowmeters are satisfactory for measuring flow rate of liquids containing suspended solids?

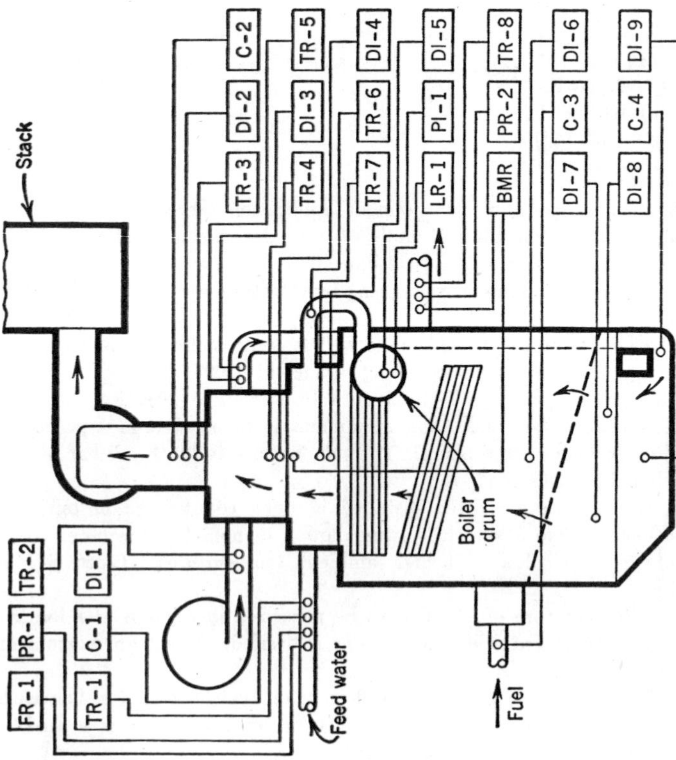

- TR-1 Feed water to economizer
- TR-2 Air entering preheater
- TR-3 Flue gas leaving preheater
- TR-4 Flue gas leaving economizer
- TR-5 Air leaving preheater
- TR-6 Feed water leaving economizer
- TR-7 Flue gas entering economizer
- TR-8 Steam in header
- PR-1 Feed water
- PR-2 Steam in header
- PI-1 Boiler drum
- DI-1 Forced draft
- DI-2 Preheater-outlet draft
- DI-3 Preheater-outlet pressure
- DI-4 Economizer-outlet draft
- DI-5 Economizer-inlet draft
- DI-6 Furnace draft
- DI-7 Fore-compartment draft
- DI-8 Aft-compartment draft
- DI-9 Windbox pressure
- FR-1 Feed water
- LR-1 Boiler-drum water level
- C-1 Feed-flow control from LR-1
- C-2 Draft control from PI-1 + BMR
- C-3 Fuel-flow control from PI-1 or BMR
- C-4 Draft control from DI-6
- BMR Boiler meter (steam-air flow)

FIG. 11-19. Instrumentation for a Steam Power Plant

11–8. What types of flowmeters operate most satisfactorily when density, viscosity, or temperature of the flowing fluid fluctuates?

11–9. Instrument the process, and draw an instrumentation diagram for an 8-room house with an oil-fired steam-heating system. Assume a control center is placed in a hallway.

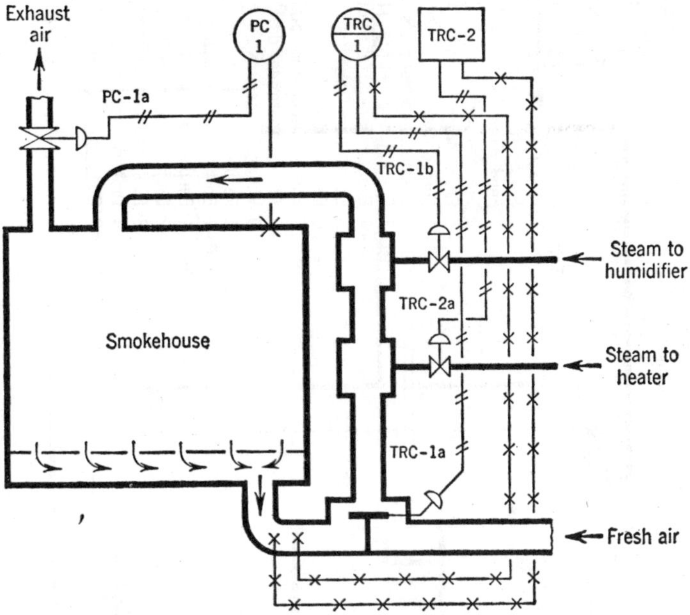

PC-1 Static pressure
TRC-1 Wet-bulb temperature
TRC-2 Dry-bulb temperature
TRC-1b Diaphragm valve
TRC-2a Diaphragm valve
TRC-1a Damper control
PC-1a Damper control

FIG. 11–20. Instrumentation of Air-Conditioned Smokehouse for Smoking Meat

11–10. Instrument the process, and draw an instrumentation diagram for a medium-size steam plant with oil firing. The plant produces steam at 200 psig for heating and processing uses, and the demand is fairly steady.

11–11. Instrument the process, and draw an instrumentation diagram for an air-conditioned film-storage room 8 ft by 8 ft by 30 ft long. The air must be changed once in 10 min. The relative humidity is to be 40 per cent and the temperature 60°F.

11–12. Study the instrumentation arrangement for the steam power plant in Fig. 11–19. (*a*) For a small power plant, the budget does not allow all the instruments shown. Which 10 instruments could be dispensed with? (*b*) Select the make, style, and model of each of the instruments

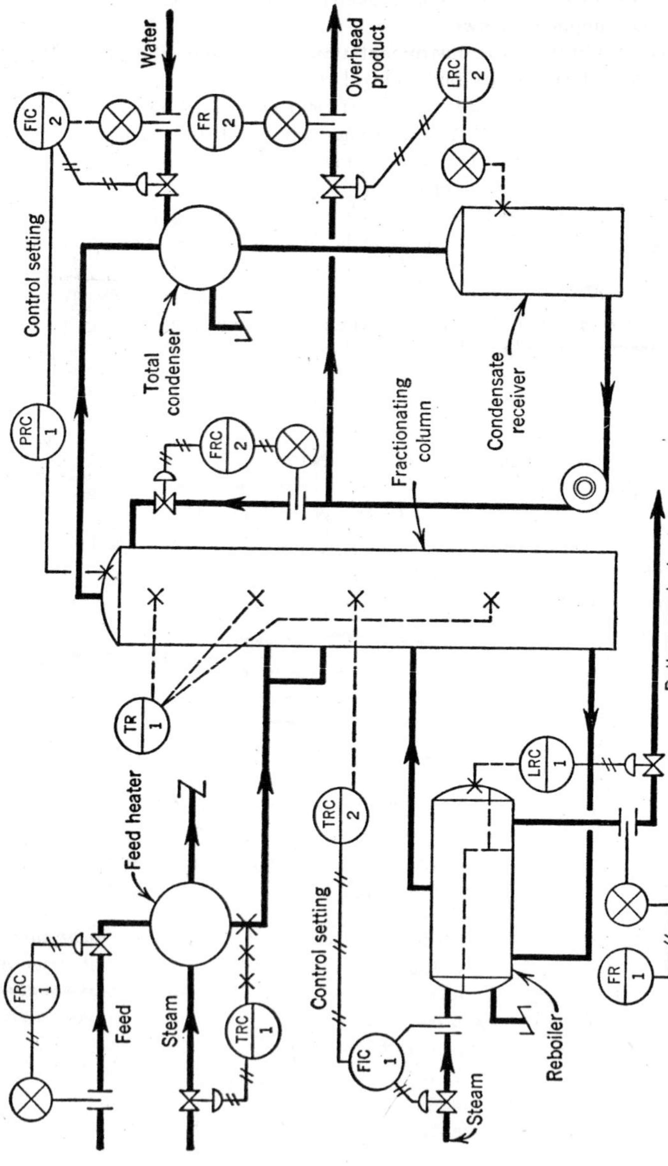

Fig. 11-21. Instrumentation for Narrow-Range Fractioning Column

shown. (*c*) For the four controllers shown, select the controller action (mode) for each. (*d*) Make a sketch of a panel board for the instrumentation equipment shown.

11–13. Criticize the instrumentation of the smokehouse in Fig. 11–20. (*a*) Why does the dry-bulb control the heater, and the wet-bulb the fresh air damper and humidifier? (*b*) If rigid control of moisture content were desired, suggest another measuring instrument to replace the two temperature controllers. (*c*) Why is static-pressure control necessary? Should static pressure or absolute pressure be controlled?

11–14. Study the instrumentation diagram of the fractionating column in Fig. 11–21. (*a*) If the column is used to separate normal propane and isopropane, can Gibbs' phase rule be used to determine whether both pressure control and temperature control are necessary? (*b*) If pure component separation is attempted, would a continuous-recording spectrometer be useful? (*c*) Liquid level in the condensate tank is controlled by a valve in the overhead product line. Does self-regulation in the process help this controller? (*d*) Pressure control is accomplished by adjusting the set point of the cooling-water-flow controller at the condenser. Pressure control through thermal effect is often poor. Why? Suggest a different arrangement for pressure control.

11–15. Write down as many specifications as you can for the instruments shown in Fig. 11–14, using only data shown by the instrumentation diagram.

Appendix

Index of Tables

TABLE A-1. FIXED POINTS OF THE INTERNATIONAL TEMPERATURE SCALE *

Fundamental and Primary Points

Equilibrium at Standard Pressure of	°C	°F
Oxygen, liquid and vapor	−182.970	−297.346
Ice and air-saturated vapor	0	32
Water, liquid and vapor	100	212
Sulfur, liquid and vapor	444.600	838.280
Silver, solid and liquid	960.8	1761
Gold, solid and liquid	1063.0	1945.4

Secondary Fixed Points

Point of	°C	°F
Carbon dioxide, solid–vapor equilibrium	−78.5	−109
Mercury, freezing	−38.87	−37.97
Naphthalene, liquid–vapor equilibrium	218.0	424.4
Tin, freezing	231.9	449.4
Benzophenone, liquid–vapor equilibrium	305.9	582.6
Cadmium, freezing	320.9	609.6
Lead, freezing	327.3	621.1
Mercury, liquid–vapor equilibrium	356.58	673.85
Zinc, freezing	419.5	787.1
Antimony, freezing	630.5	1167
Aluminum, freezing	660.1	1220
Copper, freezing in reducing atmosphere	1083	1981
Nickel, freezing	1453	2647
Palladium, freezing	1552	2826
Platinum, freezing	1769	3216
Rhodium, freezing	1960	3560
Iridium, freezing	2443	4429
Tungsten, melting	3380	6116

* *Natl. Bur. Standards J. Research*, **42**, 209 (Mar. 1949).

Table A–2. Temperature Conversion, °C to °F to °C *

−459.4 to −120			−120 to −80			−80 to −40			−40 to 0		
°C		°F	°C		°F	°C		°F	°C		°F
−273	**−459.4**		−84.4	**−120**	−184.0	−62.2	**−80**	−112.0	−40.0	**−40**	−40.0
−268	**−450**		−83.9	**−119**	−182.2	−61.7	**−79**	−110.2	−39.4	**−39**	−38.2
−262	**−440**		−83.3	**−118**	−180.4	−61.1	**−78**	−108.4	−38.9	**−38**	−36.4
−257	**−430**		−82.8	**−117**	−178.6	−60.6	**−77**	−106.6	−38.3	**−37**	−34.6
−251	**−420**		−82.2	**−116**	−176.8	−60.0	**−76**	−104.8	−37.8	**−36**	−32.8
−246	**−410**		−81.7	**−115**	−175.0	−59.4	**−75**	−103.0	−37.2	**−35**	−31.0
−240	**−400**		−81.1	**−114**	−173.2	−58.9	**−74**	−101.2	−36.7	**−34**	−29.2
−234	**−390**		−80.6	**−113**	−171.4	−58.3	**−73**	−99.4	−36.1	**−33**	−27.4
−229	**−380**		−80.0	**−112**	−169.6	−57.8	**−72**	−97.6	−35.6	**−32**	−25.6
−223	**−370**		−79.4	**−111**	−167.8	−57.2	**−71**	−95.8	−35.0	**−31**	−23.8
−218	**−360**		−78.9	**−110**	−166.0	−56.7	**−70**	−94.0	−34.4	**−30**	−22.0
−212	**−350**		−78.3	**−109**	−164.2	−56.1	**−69**	−92.2	−33.9	**−29**	−20.2
−207	**−340**		−77.8	**−108**	−162.4	−55.6	**−68**	−90.4	−33.3	**−28**	−18.4
−201	**−330**		−77.2	**−107**	−160.6	−55.0	**−67**	−88.6	−32.8	**−27**	−16.6
−196	**−320**		−76.7	**−106**	−158.8	−54.4	**−66**	−86.8	−32.2	**−26**	−14.8
−190	**−310**		−76.1	**−105**	−157.0	−53.9	**−65**	−85.0	−31.7	**−25**	−13.0
−184	**−300**		−75.6	**−104**	−155.2	−53.3	**−64**	−83.2	−31.1	**−24**	−11.2
−179	**−290**		−75.0	**−103**	−153.4	−52.7	**−63**	−81.4	−30.6	**−23**	−9.4
−173	**−280**		−74.4	**−102**	−151.6	−52.2	**−62**	−79.6	−30.0	**−22**	−7.6
−169	**−273**	−459.4	−73.8	**−101**	−149.8	−51.7	**−61**	−77.8	−29.4	**−21**	−5.8
−168	**−270**	−454	−73.3	**−100**	−148.0	−51.1	**−60**	−76.0	−28.9	**−20**	−4.0
−162	**−260**	−436	−72.8	**−99**	−146.2	−50.6	**−59**	−74.2	−28.3	**−19**	−2.2
−157	**−250**	−418	−72.2	**−98**	−144.4	−50.0	**−58**	−72.4	−27.8	**−18**	−0.4
−151	**−240**	−400	−71.7	**−97**	−142.6	−49.4	**−57**	−70.6	−27.2	**−17**	+1.4
−146	**−230**	−382	−71.1	**−96**	−140.8	−48.8	**−56**	−68.8	−26.7	**−16**	3.2
−140	**−220**	−364	−70.6	**−95**	−139.0	−48.3	**−55**	−67.0	−26.1	**−15**	5.0
−134	**−210**	−346	−70.0	**−94**	−137.2	−47.7	**−54**	−65.2	−25.6	**−14**	6.8
−129	**−200**	−328	−69.4	**−93**	−135.4	−47.2	**−53**	−63.4	−25.0	**−13**	8.6
−123	**−190**	−310	−68.9	**−92**	−133.6	−46.6	**−52**	−61.6	−24.4	**−12**	10.4
−118	**−180**	−292	−68.3	**−91**	−131.8	−46.1	**−51**	−59.8	−23.9	**−11**	12.2
−112	**−170**	−274	−67.8	**−90**	−130.0	−45.6	**−50**	−58.0	−23.3	**−10**	14.0
−107	**−160**	−256	−67.2	**−89**	−128.0	−45.0	**−49**	−56.2	−22.8	**−9**	15.8
−101	**−150**	−238	−66.7	**−88**	−126.4	−44.4	**−48**	−54.4	−22.2	**−8**	17.6
−95.6	**−140**	−220	−66.1	**−87**	−124.6	−43.8	**−47**	−52.6	−21.7	**−7**	19.4
−90.0	**−130**	−202	−65.6	**−86**	−122.8	−43.3	**−46**	−50.8	−21.1	**−6**	21.2
−84.4	**−120**	−184	−65.0	**−85**	−121.0	−42.8	**−45**	−49.9	−20.6	**−5**	23.0
			−64.4	**−84**	−119.2	−42.2	**−44**	−47.2	−20.0	**−4**	24.8
			−63.8	**−83**	−117.4	−41.7	**−43**	−45.4	−19.4	**−3**	26.6
			−63.3	**−82**	−115.6	−41.1	**−42**	−43.6	−18.9	**−2**	28.4
			−62.7	**−81**	−113.8	−40.6	**−41**	−41.8	−18.3	**−1**	30.2
			−62.2	**−80**	−112.0	−40.0	**−40**	−40.0	−17.8	**0**	32.0

Interpolation Factors

0.56	**1**	1.8	2.22	**4**	7.2	3.89	**7**	12.6	5.56	**10**	18.0
1.11	**2**	3.6	2.78	**5**	9.0	4.44	**8**	14.4			
1.67	**3**	5.4	3.33	**6**	10.8	5.00	**9**	16.2			

* Reproduced by permission of Fisher Scientific Company.

TABLE A-2. TEMPERATURE CONVERSION, °C TO °F TO °C (*Continued*)

0 to 50			50 to 100			100 to 150			150 to 200		
°C		°F	°C		°F	°C		°F	°C		°F
−17.8	**0**	32.0	10.0	**50**	122.0	37.8	**100**	212.0	65.6	**150**	302.0
−17.2	**1**	33.8	10.6	**51**	123.8	38.3	**101**	213.8	66.1	**151**	303.8
−16.7	**2**	35.6	11.1	**52**	125.6	38.9	**102**	215.6	66.7	**152**	305.6
−16.1	**3**	37.4	11.7	**53**	127.4	39.4	**103**	217.4	67.2	**153**	307.4
−15.6	**4**	39.2	12.2	**54**	129.2	40.0	**104**	219.2	67.8	**154**	309.2
−15.0	**5**	41.0	12.8	**55**	131.0	40.6	**105**	221.0	68.3	**155**	311.0
−14.4	**6**	42.8	13.3	**56**	132.8	41.1	**106**	222.8	68.9	**156**	312.8
−13.9	**7**	44.6	13.9	**57**	134.6	41.7	**107**	224.6	69.4	**157**	314.6
−13.3	**8**	46.4	14.4	**58**	136.4	42.2	**108**	226.4	70.0	**158**	316.4
−12.8	**9**	48.2	15.0	**59**	138.2	42.8	**109**	228.2	70.6	**159**	318.2
−12.2	**10**	50.0	15.6	**60**	140.0	43.3	**110**	230.0	71.1	**160**	320.0
−11.7	**11**	51.8	16.1	**61**	141.8	43.9	**111**	231.8	71.7	**161**	321.8
−11.1	**12**	53.6	16.7	**62**	143.6	44.4	**112**	233.6	72.2	**162**	323.6
−10.6	**13**	55.4	17.2	**63**	145.4	45.0	**113**	235.4	72.8	**163**	325.4
−10.0	**14**	57.2	17.8	**64**	147.2	45.6	**114**	237.2	73.3	**164**	327.2
−9.4	**15**	59.0	18.3	**65**	149.0	46.1	**115**	239.0	73.9	**165**	329.0
−8.9	**16**	60.8	18.9	**66**	150.8	46.7	**116**	240.8	74.4	**166**	330.8
−8.3	**17**	62.6	19.4	**67**	152.6	47.2	**117**	242.6	75.0	**167**	332.6
−7.8	**18**	64.4	20.0	**68**	154.4	47.8	**118**	244.4	75.6	**168**	334.4
−7.2	**19**	66.2	20.6	**69**	156.2	48.3	**119**	246.2	76.1	**169**	336.2
−6.7	**20**	68.0	21.1	**70**	158.0	48.9	**120**	248.0	76.7	**170**	338.0
−6.1	**21**	69.8	21.7	**71**	159.8	49.4	**121**	249.8	77.2	**171**	339.8
−5.6	**22**	71.6	22.2	**72**	161.6	50.0	**122**	251.6	77.8	**172**	341.6
−5.0	**23**	73.4	22.8	**73**	163.4	50.6	**123**	253.4	78.3	**173**	343.4
−4.4	**24**	75.2	23.3	**74**	165.2	51.1	**124**	255.2	78.9	**174**	345.2
−3.9	**25**	77.0	23.9	**75**	167.0	51.7	**125**	257.0	79.4	**175**	347.0
−3.3	**26**	78.8	24.4	**76**	168.8	52.2	**126**	258.8	80.0	**176**	348.8
−2.8	**27**	80.6	25.0	**77**	170.6	52.8	**127**	260.6	80.6	**177**	350.6
−2.2	**28**	82.4	25.6	**78**	172.4	53.3	**128**	262.4	81.1	**178**	352.4
−1.7	**29**	84.2	26.1	**79**	174.2	53.9	**129**	264.2	81.7	**179**	354.2
−1.1	**30**	86.0	26.7	**80**	176.0	54.4	**130**	266.0	82.2	**180**	356.0
−0.6	**31**	87.8	27.2	**81**	177.8	55.0	**131**	267.8	82.8	**181**	357.8
0.0	**32**	89.6	27.8	**82**	179.6	55.6	**132**	269.6	83.3	**182**	359.6
+0.6	**33**	91.4	28.3	**83**	181.4	56.1	**133**	271.4	83.9	**183**	361.4
1.1	**34**	93.2	28.9	**84**	183.2	56.7	**134**	273.2	84.4	**184**	363.2
1.7	**35**	95.0	29.4	**85**	185.0	57.2	**135**	275.0	85.0	**185**	365.0
2.2	**36**	96.8	30.0	**86**	186.8	57.8	**136**	276.8	85.6	**186**	366.8
2.8	**37**	98.6	30.6	**87**	188.6	58.3	**137**	278.6	86.1	**187**	368.6
3.3	**38**	100.4	31.1	**88**	190.4	58.9	**138**	280.4	86.7	**188**	370.4
3.9	**39**	102.2	31.7	**89**	192.2	59.4	**139**	282.2	87.2	**189**	372.2
4.4	**40**	104.0	32.2	**90**	194.0	60.0	**140**	284.0	87.8	**190**	374.0
5.0	**41**	105.8	32.8	**91**	195.8	60.6	**141**	285.8	88.3	**191**	375.8
5.6	**42**	107.6	33.3	**92**	197.6	61.1	**142**	287.6	88.9	**192**	377.6
6.1	**43**	109.4	33.9	**93**	199.4	61.7	**143**	289.4	89.4	**193**	379.4
6.7	**44**	111.2	34.4	**94**	201.2	62.2	**144**	291.2	90.0	**194**	381.2
7.2	**45**	113.0	35.0	**95**	203.0	62.8	**145**	293.0	90.6	**195**	383.0
7.8	**46**	114.8	35.6	**96**	204.8	63.3	**146**	294.8	91.1	**196**	384.8
8.3	**47**	116.6	36.1	**97**	206.6	63.9	**147**	296.6	91.7	**197**	386.6
8.9	**48**	118.4	36.7	**98**	208.4	64.4	**148**	298.4	92.2	**198**	388.4
9.4	**49**	120.2	37.2	**99**	210.2	65.0	**149**	300.2	92.8	**199**	390.2
10.0	**50**	122.0	37.8	**100**	212.0	65.6	**150**	302.0	93.3	**200**	392.0

Interpolation Factors

0.50	**1**	1.8	2.22	**4**	7.2	3.89	**7**	12.6	5.56	**10**	18.0
1.11	**2**	3.6	2.78	**5**	9.0	4.44	**8**	14.4			
1.67	**3**	5.4	3.33	**6**	10.8	5.00	**9**	16.2			

TABLE A–2. TEMPERATURE CONVERSION, °C TO °F TO °C (*Continued*)

200 to 250			250 to 300			300 to 350			350 to 400		
°C		°F	°C		°F	°C		°F	°C		°F
93.3	**200**	392.0	121.1	**250**	482.0	148.9	**300**	572.0	176.7	**350**	662.0
93.9	**201**	393.8	121.7	**251**	483.8	149.4	**301**	573.8	177.2	**351**	663.8
94.4	**202**	395.6	122.2	**252**	485.6	150.0	**302**	575.6	177.8	**352**	665.6
95.0	**203**	397.4	122.8	**253**	487.4	150.6	**303**	577.4	178.3	**353**	667.4
95.6	**204**	399.2	123.3	**254**	489.2	151.1	**304**	579.2	178.9	**354**	669.2
96.1	**205**	401.0	123.9	**255**	491.0	151.7	**305**	581.0	179.4	**355**	671.0
96.7	**206**	402.8	124.4	**256**	492.8	152.2	**306**	582.8	180.0	**356**	672.8
97.2	**207**	404.6	125.0	**257**	494.6	152.8	**307**	584.6	180.6	**357**	674.6
97.8	**208**	406.4	125.6	**258**	496.4	153.3	**308**	586.4	181.1	**358**	676.4
98.3	**209**	408.2	126.1	**259**	498.2	153.9	**309**	588.2	181.7	**359**	678.2
98.9	**210**	410.0	126.7	**260**	500.0	154.4	**310**	590.0	182.2	**360**	680.0
99.4	**211**	411.8	127.2	**261**	501.8	155.0	**311**	591.8	182.8	**361**	681.8
100.0	**212**	413.6	127.8	**262**	503.6	155.6	**312**	593.6	183.3	**362**	683.6
100.6	**213**	415.4	128.3	**263**	505.4	156.1	**313**	595.4	183.9	**363**	685.4
101.1	**214**	417.2	128.9	**264**	507.2	156.7	**314**	597.2	184.4	**364**	687.2
101.7	**215**	419.0	129.4	**265**	509.0	157.2	**315**	599.0	185.0	**365**	689.9
102.2	**216**	420.8	130.0	**266**	510.8	157.8	**316**	600.8	185.6	**366**	690.8
102.8	**217**	422.6	130.6	**267**	512.6	158.3	**317**	602.6	186.1	**367**	692.6
103.3	**218**	424.4	131.1	**268**	514.4	158.9	**318**	604.4	186.7	**368**	694.4
103.9	**219**	426.2	131.7	**269**	516.2	159.4	**319**	606.2	187.2	**369**	696.2
104.4	**220**	428.0	132.2	**270**	518.0	160.0	**320**	608.0	188.2	**370**	698.0
105.0	**221**	429.8	132.8	**271**	519.8	160.6	**321**	609.8	188.3	**371**	699.8
105.6	**222**	431.6	133.3	**272**	521.6	161.1	**322**	611.6	188.9	**372**	701.6
106.1	**223**	433.4	133.9	**273**	523.4	161.7	**323**	613.4	189.4	**373**	703.4
106.7	**224**	435.2	134.4	**274**	525.2	162.2	**324**	615.2	190.0	**374**	705.2
107.2	**225**	437.0	135.0	**275**	527.0	162.8	**325**	617.0	190.6	**375**	707.0
107.8	**226**	438.8	135.6	**276**	528.8	163.3	**326**	618.8	191.1	**376**	708.8
108.3	**227**	440.6	136.1	**277**	530.6	163.9	**327**	620.6	191.7	**377**	710.6
108.9	**228**	442.4	136.7	**278**	532.4	164.4	**328**	622.4	192.2	**378**	712.4
109.4	**229**	444.2	137.2	**279**	534.2	165.0	**329**	624.2	192.8	**379**	714.2
110.0	**230**	446.0	137.8	**280**	536.0	165.6	**330**	626.0	193.3	**380**	716.0
110.6	**231**	447.8	138.3	**281**	537.8	166.1	**331**	627.8	193.9	**381**	717.8
111.1	**232**	449.6	138.9	**282**	539.6	166.7	**332**	629.6	194.4	**382**	719.6
111.7	**233**	451.4	139.4	**283**	541.4	167.2	**333**	631.4	195.0	**383**	721.4
112.2	**234**	453.2	140.0	**284**	543.2	167.8	**334**	633.2	195.6	**384**	723.2
112.8	**235**	455.0	140.6	**285**	545.0	168.3	**335**	635.0	196.1	**385**	725.0
113.3	**236**	456.8	141.1	**286**	546.8	168.9	**336**	636.8	196.7	**386**	726.8
113.9	**237**	458.6	141.7	**287**	548.6	169.4	**337**	638.6	197.2	**387**	728.6
114.4	**238**	460.4	142.2	**288**	550.4	170.0	**338**	640.4	197.8	**388**	730.4
115.0	**239**	462.2	142.8	**289**	552.2	170.6	**339**	642.2	198.3	**389**	732.2
115.6	**240**	464.0	143.3	**290**	554.0	171.1	**340**	644.0	198.9	**390**	734.0
116.1	**241**	465.8	143.9	**291**	555.8	171.7	**341**	645.8	199.4	**391**	735.8
116.7	**242**	467.6	144.4	**292**	557.6	172.2	**342**	647.6	200.0	**392**	737.6
117.2	**243**	469.4	145.0	**293**	559.4	172.8	**343**	649.4	200.6	**393**	739.4
117.8	**244**	471.2	145.6	**294**	561.2	173.3	**344**	651.2	201.1	**394**	741.2
118.3	**245**	473.0	146.1	**295**	563.0	173.9	**345**	653.0	201.7	**395**	743.0
118.9	**246**	474.8	146.7	**296**	564.8	174.4	**346**	654.8	202.2	**396**	744.8
119.4	**247**	476.6	147.2	**297**	566.6	175.0	**347**	656.6	202.8	**397**	746.6
120.0	**248**	478.4	147.8	**298**	568.4	175.6	**348**	658.4	203.3	**398**	748.4
120.6	**249**	480.2	148.3	**299**	570.2	176.1	**349**	660.2	203.9	**399**	750.2
121.1	**250**	482.0	148.9	**300**	572.0	176.7	**350**	662.0	204.4	**400**	752.0

Interpolation Factors

°C		°F	°C		°F	°C		°F	°C		°F
0.56	**1**	1.8	2.22	**4**	7.2	3.89	**7**	12.6	5.56	**10**	18.0
1.11	**2**	3.6	2.78	**5**	9.0	4.44	**8**	14.4			
[illegible]67	**3**	5.4	3.33	**6**		5.00	**9**	16.2			

TABLE A-2. TEMPERATURE CONVERSION, °C TO °F TO °C (*Continued*)

400 to 450			450 to 500			500 to 550			550 to 1000		
°C		°F	°C		°F	°C		°F	°C		°F
204.4	**400**	752.0	232.2	**450**	842.0	260.0	**500**	932.0	288	**550**	1022
205.0	**401**	753.8	232.8	**451**	843.8	260.6	**501**	933.8	293	**560**	1040
205.6	**402**	755.6	233.3	**452**	845.6	261.1	**502**	935.6	299	**570**	1058
206.1	**403**	757.4	233.9	**453**	847.4	261.7	**503**	937.4	304	**580**	1076
206.7	**404**	759.2	234.4	**454**	849.2	262.3	**504**	939.2	310	**590**	1094
207.2	**405**	761.0	235.0	**455**	851.0	262.8	**505**	941.0	316	**600**	1112
207.8	**406**	762.8	235.6	**456**	852.8	263.4	**506**	942.8	321	**610**	1130
208.3	**407**	764.6	236.1	**457**	854.6	263.9	**507**	944.6	327	**620**	1148
208.9	**408**	766.4	236.7	**458**	856.4	264.5	**508**	946.4	332	**630**	1166
209.4	**409**	768.2	237.2	**459**	858.2	265.0	**509**	948.2	338	**640**	1184
210.0	**410**	770.0	237.8	**460**	860.0	265.6	**510**	950.0	343	**650**	1202
210.6	**411**	771.8	238.3	**461**	861.8	266.1	**511**	951.8	349	**660**	1220
211.1	**412**	773.6	238.9	**462**	863.6	266.7	**512**	953.6	354	**670**	1238
211.7	**413**	775.4	239.4	**463**	865.4	267.3	**513**	955.4	360	**680**	1256
212.2	**414**	777.2	240.0	**464**	867.2	267.8	**514**	957.2	866	**690**	1274
212.8	**415**	779.0	240.6	**465**	869.0	268.4	**515**	959.0	371	**700**	1292
213.3	**416**	780.8	241.1	**466**	870.8	268.9	**516**	960.8	377	**710**	1310
213.9	**417**	782.6	241.7	**467**	872.6	269.5	**517**	962.6	382	**720**	1328
214.4	**418**	784.4	242.2	**468**	874.4	270.0	**518**	964.4	388	**730**	1346
215.0	**419**	786.2	242.8	**469**	876.2	270.6	**519**	966.2	393	**740**	1364
215.6	**420**	788.0	243.3	**470**	878.0	271.1	**520**	968.0	399	**750**	1382
216.1	**421**	789.8	243.9	**471**	879.8	271.7	**521**	969.8	404	**760**	1400
216.7	**422**	791.6	244.4	**472**	881.6	272.3	**522**	971.6	410	**770**	1418
217.2	**423**	793.4	245.0	**473**	883.4	272.8	**523**	973.4	416	**780**	1436
217.8	**424**	795.2	245.6	**474**	885.2	273.4	**524**	975.2	421	**790**	1454
218.3	**425**	797.0	246.1	**475**	887.0	273.9	**525**	977.0	427	**800**	1472
218.9	**426**	798.8	246.7	**476**	888.8	274.5	**526**	978.8	432	**810**	1490
219.4	**427**	800.6	247.2	**477**	890.6	275.0	**527**	980.6	438	**820**	1508
220.0	**428**	802.4	247.8	**478**	892.4	275.6	**528**	982.4	443	**830**	1526
220.6	**429**	804.2	248.3	**479**	894.2	276.1	**529**	984.2	449	**840**	1544
221.1	**430**	806.0	248.9	**480**	896.0	276.7	**530**	986.0	454	**850**	1562
221.7	**431**	807.8	249.4	**481**	897.8	277.3	**531**	987.8	460	**860**	1580
222.2	**432**	809.6	250.0	**482**	899.6	277.8	**532**	989.6	466	**870**	1598
222.8	**433**	811.4	250.6	**483**	901.4	278.4	**533**	991.4	471	**880**	1616
223.3	**434**	813.2	251.1	**484**	903.2	278.9	**534**	993.2	477	**890**	1634
223.9	**435**	815.0	251.7	**485**	905.0	279.5	**535**	995.0	482	**900**	1652
224.4	**436**	816.8	252.2	**486**	906.8	280.0	**536**	996.8	488	**910**	1670
225.0	**437**	818.6	252.8	**487**	908.6	280.6	**537**	998.6	493	**920**	1688
225.6	**438**	820.4	253.3	**488**	910.4	281.1	**538**	1000.4	499	**930**	1706
226.1	**439**	822.2	253.9	**489**	912.2	281.7	**539**	1002.2	504	**940**	1724
226.7	**440**	824.0	254.4	**490**	914.0	282.3	**540**	1004.0	510	**950**	1742
227.2	**441**	825.8	255.0	**491**	915.8	282.8	**541**	1005.8	516	**960**	1760
227.8	**442**	827.6	255.6	**492**	917.6	283.4	**542**	1007.6	521	**970**	1778
228.3	**443**	829.4	256.1	**493**	919.4	283.9	**543**	1009.4	527	**980**	1796
228.9	**444**	831.2	256.7	**494**	921.2	284.5	**544**	1011.2	532	**990**	1814
229.4	**445**	833.0	257.2	**495**	923.0	285.0	**545**	1013.0	538	**1000**	1832
230.0	**446**	834.8	257.8	**496**	924.8	285.6	**546**	1014.8			
230.6	**447**	836.6	258.3	**497**	926.6	286.1	**547**	1016.6			
231.1	**448**	838.4	258.9	**498**	928.4	286.7	**548**	1018.4			
231.7	**449**	840.2	259.4	**499**	930.2	287.3	**549**	1020.2			
232.2	**450**	842.0	260.0	**500**	932.0	287.8	**550**	1022.0			

Interpolation Factors

°C		°F	°C		°F	°C		°F	°C		°F
0.56	**1**	1.8	2.22	**4**	7.2	3.89	**7**	12.6	5.56	**10**	18.0
1.11	**2**	3.6	2.78	**5**	9.0	4.44	**8**	14.4			
1.67	**3**	5.4	3.33	**6**	10.8	5.00	**9**	16.2			

Table A-2. Temperature Conversion, °C to °F to °C (*Continued*)

1000 to 1500			1500 to 2000			2000 to 2500			2500 to 3000		
°C		°F	°C		°F	°C		°F	°C		°F
538	**1000**	1832	816	**1500**	2732	1093	**2000**	3632	1371	**2500**	4532
543	**1010**	1850	821	**1510**	2750	1099	**2010**	3650	1377	**2510**	4550
549	**1020**	1868	827	**1520**	2768	1104	**2020**	3668	1382	**2520**	4568
554	**1030**	1886	832	**1530**	2786	1110	**2030**	3686	1388	**2530**	4586
560	**1040**	1904	838	**1540**	2804	1116	**2040**	3704	1393	**2540**	4604
566	**1050**	1922	843	**1550**	2822	1121	**2050**	3722	1399	**2550**	4622
571	**1060**	1940	849	**1560**	2840	1127	**2060**	3740	1404	**2560**	4640
577	**1070**	1958	854	**1570**	2858	1132	**2070**	3758	1410	**2570**	4658
582	**1080**	1976	860	**1580**	2876	1138	**2080**	3776	1416	**2580**	4676
588	**1090**	1994	866	**1590**	2894	1143	**2090**	3794	1421	**2590**	4694
593	**1100**	2012	871	**1600**	2912	1149	**2100**	3812	1427	**2600**	4712
599	**1110**	2030	877	**1610**	2930	1154	**2110**	3830	1432	**2610**	4730
604	**1120**	2048	882	**1620**	2948	1160	**2120**	3848	1438	**2620**	4748
610	**1130**	2066	888	**1630**	2966	1166	**2130**	3866	1443	**2630**	4766
616	**1140**	2084	893	**1640**	2984	1171	**2140**	3884	1449	**2640**	4784
621	**1150**	2102	899	**1650**	3002	1177	**2150**	3902	1454	**2650**	4802
627	**1160**	2120	904	**1660**	3020	1182	**2160**	3920	1460	**2660**	4820
632	**1170**	2138	910	**1670**	3038	1188	**2170**	3938	1466	**2670**	4838
638	**1180**	2156	916	**1680**	3056	1193	**2180**	3956	1471	**2680**	4856
643	**1190**	2174	921	**1690**	3074	1199	**2190**	3974	1477	**2690**	4874
649	**1200**	2192	927	**1700**	3092	1204	**2200**	3992	1482	**2700**	4892
654	**1210**	2210	932	**1710**	3110	1210	**2210**	4010	1488	**2710**	4910
660	**1220**	2228	938	**1720**	3128	1216	**2220**	4028	1493	**2720**	4928
666	**1230**	2246	943	**1730**	3146	1221	**2230**	4046	1499	**2730**	4946
671	**1240**	2264	949	**1740**	3164	1227	**2240**	4064	1504	**2740**	4964
677	**1250**	2282	954	**1750**	3182	1232	**2250**	4082	1510	**2750**	4982
682	**1260**	2300	960	**1760**	3200	1238	**2260**	4100	1516	**2760**	5000
688	**1270**	2318	966	**1770**	3218	1243	**2270**	4118	1521	**2770**	5018
693	**1280**	2336	971	**1780**	3236	1249	**2280**	4136	1527	**2780**	5036
699	**1290**	2354	977	**1790**	3254	1254	**2290**	4154	1532	**2790**	5054
704	**1300**	2372	982	**1800**	3272	1260	**2300**	4172	1538	**2800**	5072
710	**1310**	2390	988	**1810**	3290	1266	**2310**	4190	1543	**2810**	5090
716	**1320**	2408	993	**1820**	3308	1271	**2320**	4208	1549	**2820**	5108
721	**1330**	2426	999	**1830**	3326	1277	**2330**	4226	1554	**2830**	5126
727	**1340**	2444	1004	**1840**	3344	1282	**2340**	4244	1560	**2840**	5144
732	**1350**	2462	1010	**1850**	3362	1288	**2350**	4262	1566	**2850**	5162
738	**1360**	2480	1016	**1860**	3380	1293	**2360**	4280	1571	**2860**	5180
743	**1370**	2498	1021	**1870**	3398	1299	**2370**	4298	1577	**2870**	5198
749	**1380**	2516	1027	**1880**	3416	1304	**2380**	4316	1582	**2880**	5216
754	**1390**	2534	1032	**1890**	3434	1310	**2390**	4334	1588	**2890**	5234
760	**1400**	2552	1038	**1900**	3452	1316	**2400**	4352	1593	**2900**	5252
766	**1410**	2570	1043	**1910**	3470	1321	**2410**	4370	1599	**2910**	5270
771	**1420**	2588	1049	**1920**	3488	1327	**2420**	4388	1604	**2920**	5288
777	**1430**	2606	1054	**1930**	3506	1332	**2430**	4406	1610	**2930**	5306
782	**1440**	2624	1060	**1940**	3524	1338	**2440**	4424	1616	**2940**	5324
788	**1450**	2642	1066	**1950**	3542	1343	**2450**	4442	1621	**2950**	5342
793	**1460**	2660	1071	**1960**	3560	1349	**2460**	4460	1627	**2960**	5360
799	**1470**	2678	1077	**1970**	3578	1354	**2470**	4478	1632	**2970**	5378
804	**1480**	2696	1082	**1980**	3596	1360	**2480**	4496	1638	**2980**	5396
810	**1490**	2714	1088	**1990**	3614	1366	**2490**	4514	1643	**2990**	5414
816	**1500**	2732	1093	**2000**	3632	1371	**2500**	4532	1649	**3000**	5432

Interpolation Factors

0.56	**1**	1.8	2.22	**4**	7.2	3.89	**7**	12.6	5.56	**10**	18.0
1.11	**2**	3.6	2.78	**5**	9.0	4.44	**8**	14.4			
1.67			3.33	**6**	10.8	5.00	**9**	16.2			

TABLE A-3. CHARACTERISTICS OF EXPANSION-THERMOMETER LIQUIDS *

Liquid	Freezing Point, °F	Boiling Point, °F	Coeff. of Volumetric Expansion $V = V_0(1 + BT)$, $V = V_0(1 + \alpha T + \beta T^2 + \gamma T^3)$, °C: $B \times 10^3$ at 20°C	$\alpha \times 10^3$	$\beta \times 10^6$	$\gamma \times 10^8$	Specific Heat at 50–80°F, Btu/lb °F	Weight Density at 50–80°F, lb/ft³	Viscosity at 50–80°F, centipoises
Acetone	−139	133	1.487	1.324	3.81	−0.880	0.528	49.4	0.32
Ethyl alcohol	−175	172	1.12	1.012	2.20	....	0.581	49.4	1.2
Ethyl ether	−177	95	1.656	1.513	2.36	4.00	0.547	45.9	0.22
Mercury	−38	675	0.1819	0.1818	0.0078	....	0.033	849.0	1.5
Pentane	−203	97	1.608	1.465	3.093	1.61	0.527	39.4	0.24
Toluene	−134	230	1.224	....	......	....	0.421	55.0	0.57
Water	32	212	0.207	−0.0643	8.505	−6.79	1.00	62.4	1.0

* Data from *Handbook of Chemistry and Physics* (1947).

TABLE A-4. CHARACTERISTICS OF EXPANSION-THERMOMETER GASES *

Gas	Molecular Weight	Boiling Point, °F	Critical Temperature, °F	Critical Pressure, Atmosphere Absolute	Specific Heat C_v at ~200°F, Btu/lb °F	Specific Heat Ratio, $V = C_p/C_v$, at ~200°F	Viscosity at ~70°F, Poises
Carbon dioxide	44.00	−109	88	73.0	0.156	1.27	146×10^{-6}
Helium	4.002	−452	−449	2.3	0.76	1.62	197
Hydrogen	2.016	−423	−400	12.8	2.46	1.43	89
Nitrogen	28.02	−321	−233	32.5	0.175	1.40	171

* Data from *Handbook of Chemistry and Physics*, 1947, and Eshbach, *Handbook of Engineering Fundamentals*, 1936.

TABLE A-5. CHARACTERISTICS OF VAPOR-THERMOMETER FLUIDS *

Fluid	Melting Point, °F	Boiling Point, °F	Critical Temperature, °F	Maximum Operating Temperature, °F	Temperature, °F, to Produce a Vapor Pressure — Vapor Press., psig: 50	100	150	200	250	Average Coefficient of Cubical Expansion at B.P., per °F
Aniline	−21	363	799	550	482	544	...	...	...	0.59×10^{-3}
Butane (*n*)	−211	31	307	400	114	156	186	211	...	1.07
Ethyl alcohol	−177	173	469	250	250	...	...	...	...	0.76
Ethyl chloride	−218	54	369	350	140	181	211	235	256	0.95
Ethyl ether	−182	94	381	250	184	228	259	...	...	1.02
Hexane (*n*)	−138	156	455	500	250	299	336	367	392	0.78
Methyl chloride	−144	−12	289	400	63	99	126	147	164	1.05
Propane	−310	−44	204	400	29	65	91	112	129	1.24
Sulfur dioxide	−99	14	315	800	85	119	143	162	178	0.99
Toluene	−139	231	610	500	342	398	439	471	498	0.73
Water	32	212	705	...	297	338	366	388	407	0.40

* Data from *Handbook of Chemistry and Physics*, 1947.

TABLE A-6. THERMAL CHARACTERISTICS OF THERMOMETER AND THERMOCOUPLE MATERIALS *

Material	Thermal Conductivity, Btu/°F hr ft	Specific Heat, Btu/lb °F	Weight Density, lb/ft³	Thermal Capacity, Btu/ft³ °F	Thermal Diffusivity, ft²/hr	Thermal Coefficient Linear Expansion, 1/°F
Ambrac	22	0.13	550	72	0.31	9.1×10^{-6}
Aluminum	125	0.23	170	39	2.9	15.4
Brass 70-30	70	0.094	530	50	1.4	11.0
Chromel	13	0.12	470	56	0.23	8.3
Constantan	14	0.13	550	72	0.20	8.3
Copper	225	0.093	560	52	4.3	10.0
Inconel	8.5	0.11	530	58	0.15	....
Iron, cast	27	0.12	440	53	0.51	4.7
Iron, wrought	34	0.13	490	64	0.53	6.3
Lead	20	0.03	670	20	1.0	16.1
Manganin	15	0.12	530	64	0.23	....
Monel	14	0.13	560	73	0.19	8.9
Nichrome	8.5	0.12	510	61	0.14	9.1
Nickel	30	0.11	550	61	0.49	7.8
Platinum	33	0.03	1350	41	0.81	5.0
Steel, carbon	23	0.15	490	74	0.31	7.5
Steel, 18-8	16	0.13	500	65	0.25	11.1
Asbestos	0.044	0.20	150	30	0.0015	....
Glass	0.44	0.20	170	34	0.013	3.3
Porcelain	1.0	0.26	150	39	0.026	2.3
Rubber	.90	0.48	100	48	0.019	38
Quartz	4.9	0.17	160	27	0.18	0.32
Silicon carbide	10	0.18	200	36	0.28	....
Sillimanite	1.4	0.18	190	34	0.041	....

* Data from *Handbook of Chemistry and Physics*, 1947; Eshbach, *Handbook of Engineering Fundamentals*, 1936; *Smithsonian Physical Tables*, 1933; Marks, *Mechanical Engineers Handbook*, 1941.

The data is given for a commercial form of the material and at an average operating temperature, insofar as possible.

TABLE A-7. PHYSICAL CHARACTERISTICS OF THERMOMETER AND THERMOCOUPLE MATERIALS

Material	Composition	Approx. Limit., °F	May Be Used for
Aluminum	97 Al, 1 Mg	500	Air and water
Ambrac	75 Cu, 20 Ni, 5 Zn	1000	Sulfur-free gases and liquids
Brass	80 Cu, 20 Zn	800	Air
Bronze	90 Cu, 10 Sn	900	Steam and water and air
Chromax	55 Ni, 29 Fe, 15 Cr	2000	
Chromel	47 Fe, 32 Ni, 20 Cr	2000	Chemicals and sulfur-free gases
Inconel	78 Ni, 13 Cr, 8 Fe	2000	Sulfur-free furnace gases
Iron, cal.	99 Fe, 1 C	1200	Furnace gases
cast	95 Fe, 4 C	1600	Chemicals and reducing atm.
14 Cr	85 Fe, 14 Cr	1500	Chemicals and reducing atm.
28 Cr	71 Fe, 28 Cr	1800	Furnace gases
wrought	100 Fe	1300	Oxygen-free gases
Lead	99 Pb	150	Water and some chemicals
Monel	69 Ni, 28 Cu	900	Steam and some acids
Nichrome	62 Ni, 24 Fe, 13 Cr	2000	High mechanical strength
Nickel	100 Ni	2000	Sulfur-free gases
Platinum	100 Pt	2800	Molten glass
Steel, cal.	99 Fe, 1 C	1200	Oxygen-free gases
18-8	63 Fe, 18 Ni, 8 Cr	1500	Corrosive gases and oils and foods and steam
Firebrick	53 SiO_2, 43 Al_2O_3	2600	Secondary well in acid fluxes
Mullite	72 Al_2O_3, 28 SiO_2	2700	Secondary well in acid fluxes
Porcelain	71 SiO_2 + oxides	2400	High-temp. furnaces and acid fluxes
Quartz	SiO_2	2000	Oxidizing atm. and acid fluxes
Rubber	....	150	Air and some oils
Silica	95 SiO_2 + oxides	2800	Oxidizing atm.
Silicon carbide	SiC	3000	Furnace gases
Sillimanite	63 Al_2O_3, 37 SiO_2	2800	Furnace gases and flames
Vycor	A glass	1600	Furnace gases. Passes radiation

TABLE A–8. AVERAGE THERMOELECTRIC COEFFICIENTS *

Referred to Lead

Metal	kT_c, micro-volts/°C	k, micro-volts/deg.^{2} C
Bismuth	+74.42	−0.0320
Cadmium	−12.00	−0.3238
Cobalt	+17.32	+0.0780
Constantan	+34.76	+0.0794
Copper	−2.777	−0.0097
Magnesium	+0.095	−0.0001
Tin	−0.230	−0.0086
Zinc	−3.047	+0.0099

* Data from P. W. Bridgeman, *The Thermodynamics of Electrical Phenomena in Metals.*

Table A–9. Thermal Emf of Various Metals versus Platinum *
Referred to Platinum

Metal	−100	0	+100	200	300	400	500	600	700	800	900	1000	1100
	Millivolts with Measuring Junction at °C												
Brass (yellow)	......	0	+0.60	1.49	2.58	3.85	5.30	6.96	.....	.....	.....	.....	.....
Copper	−0.37	0	+0.76	1.83	3.15	4.68	6.41	8.34	10.49	12.84	15.41	18.20	.....
Iron	−1.94	0	+1.98	3.69	5.03	6.08	7.00	8.02	9.34	11.09	13.10	14.64	.....
Manganin	......	0	+0.61	1.55	2.77	4.25	5.95	7.84	.....	.....	.....	.....	.....
Nickel	+1.22	0	−1.48	3.10	4.59	5.45	6.16	7.04	8.10	9.35	10.69	12.13	13.62
Phos. bronze	......	0	+0.55	1.34	2.34	3.50	4.81	6.30	.....	.....	.....	.....	.....
Silver	−0.39	0	+0.74	1.77	3.05	4.57	6.36	8.41	10.75	13.36	16.20	.....	.....
Steel (18-8)	......	0	+0.44	1.04	1.76	2.60	3.56	4.67	5.93	7.37	8.99	.....	.....
Tungsten	−0.15	0	+1.12	2.62	4.48	6.70	9.30	12.26	15.60	19.30	23.36	27.80	32.60
80 Ni, 20 Cr	......	0	+1.14	2.62	4.34	6.25	8.31	10.53	12.91	15.44	18.11	.....	.....
60 Ni, 24 Fe, 16 Cr	......	0	+0.85	2.01	3.41	5.00	6.76	8.68	10.78	13.06	15.50	.....	.....

* Data from *Temperature, Its Measurement and Control in Science and Industry*, 1941. A positive sign denotes that the given emf results in a current from the metal to platinum at the reference junction. The reference junction is at 0°C.

Table A-10. Thermal Emf for Iron–Constantan Thermocouple *

Reference Junction 32°F

°F	0	100	200	300	400	500	600	700	800	900	1000
						Millivolts					
0	−0.922	1.96	4.92	7.95	11.02	14.09	17.16	20.23	23.31	26.41	29.54
5	−0.778	2.11	5.07	8.10	11.18	14.24	17.31	20.38	23.46	26.57	29.70
10	−0.634	2.26	5.22	8.26	11.32	14.40	17.47	20.54	23.62	26.72	29.86
15	−0.490	2.40	5.37	8.41	11.48	14.55	17.62	20.69	23.77	26.88	30.02
20	−0.346	2.55	5.53	8.56	11.63	14.70	17.77	20.85	23.93	27.04	30.18
25	−0.202	2.70	5.68	8.72	11.79	14.86	17.93	21.00	24.08	27.19	30.34
30	−0.058	2.85	5.83	8.87	11.94	15.01	18.08	21.15	24.24	27.35	30.50
35	0.088	3.00	5.98	9.02	12.09	15.16	18.23	21.31	24.39	27.50	30.66
40	0.228	3.14	6.13	9.18	12.25	15.32	18.39	21.46	24.55	27.66	30.82
45	0.378	3.29	6.28	9.33	12.40	15.47	18.54	21.62	24.70	27.82	30.98
50	0.518	3.44	6.44	9.48	12.55	15.62	18.69	21.77	24.86	27.97	31.14
55	0.658	3.59	6.58	9.64	12.71	15.78	18.85	21.92	25.01	28.13	31.30
60	0.808	3.74	6.74	9.79	12.86	15.93	19.00	22.08	25.17	28.29	31.46
65	0.948	3.88	6.89	9.94	13.02	16.08	19.15	22.23	25.32	28.44	31.62
70	1.10	4.03	7.04	10.10	13.17	16.24	19.31	22.39	25.48	28.60	31.78
75	1.24	4.18	7.19	10.25	13.32	16.39	19.46	22.54	25.63	28.76	31.94
80	1.38	4.33	7.34	10.40	13.48	16.55	19.62	22.69	25.79	28.91	32.10
85	1.53	4.48	7.50	10.55	13.63	16.70	19.77	22.85	25.94	29.07	32.26
90	1.67	4.62	7.65	10.71	13.78	16.85	19.92	23.00	26.10	29.23	32.42
95	1.82	4.77	7.80	10.86	13.94	17.01	20.08	23.16	26.25	29.38	32.58
100	1.96	4.92	7.95	11.02	14.09	17.16	20.23	23.31	26.41	29.54	32.74
Mv per °F	0.0288	0.0296	0.0303	0.0307	0.0307	0.0307	0.0307	0.0308	0.0310	0.0313	0.0320

°F	1100	1200	1300	1400	1500	1600	1700	1800	1900	2000	2100
						Millivolts					
0	32.74	36.04	39.46	42.96	46.48	50.00	53.52	57.04	60.56	64.08	67.60
5	32.90	36.21	39.64	43.14	46.66	50.18	53.70	57.22	60.74	64.26	67.78
10	33.07	36.38	39.81	43.31	46.83	50.35	53.87	57.39	60.91	64.43	67.95
15	33.23	36.55	39.98	43.49	47.01	50.53	54.05	57.57	61.09	64.61	68.13
20	33.40	36.72	40.16	43.66	47.18	50.70	54.22	57.74	61.26	64.78	68.30
25	33.56	36.89	40.34	43.84	47.36	50.88	54.40	57.92	61.44	64.96	68.48
30	33.73	37.07	40.51	44.02	47.54	51.06	54.58	58.10	61.62	65.14	68.66
35	33.89	37.24	40.68	44.19	47.71	51.23	54.75	58.27	61.79	65.31	68.83
40	34.06	37.41	40.86	44.37	47.89	51.41	54.93	58.45	61.97	65.49	69.01
45	34.22	37.58	41.03	44.54	48.06	51.58	55.10	58.62	62.14	65.66	69.18
50	34.39	37.75	41.21	44.72	48.24	51.76	55.28	58.80	62.32	65.84	69.36
55	34.55	37.92	41.38	44.90	48.42	51.94	55.46	58.98	62.50	66.02	69.54
60	34.72	38.09	41.56	45.07	48.59	52.11	55.63	59.15	62.67	66.19	69.71
65	34.88	38.26	41.73	45.25	48.77	52.29	55.81	59.33	62.85	66.37	69.89
70	35.05	38.43	41.91	45.42	48.94	52.46	55.98	59.50	63.02	66.54	70.06
75	35.21	38.60	42.08	45.60	49.12	52.64	56.16	59.68	63.20	66.72	70.24
80	35.38	38.78	42.26	45.78	49.30	52.82	56.34	59.86	63.38	66.90	70.42
85	35.54	38.95	42.43	45.95	49.47	52.99	56.51	60.03	63.55	67.07	70.59
90	35.71	39.12	42.61	46.13	49.65	53.17	56.69	60.21	63.73	67.25	70.77
95	35.87	39.29	42.78	46.30	49.82	53.34	56.86	60.38	63.90	67.42	70.94
100	36.04	39.46	42.96	46.48	50.00	53.52	57.04	60.56	64.08	67.60	71.12
Mv per °F	0.0330	0.0342	0.0350	0.0352	0.0352	0.0352	0.0352	0.0352	0.0352	0.0352	0.0352

* Form of table supplied by The Brown Instrument Company. Data do not conform to National Bureau of Standards specifications but are those in use at the present time by nearly all the large instrument companies.

TABLE A–11. THERMAL EMF FOR COPPER–CONSTANTAN THERMOCOUPLE *
Reference Junction at 32°F

°F	−200	−100	−0	+0	+100	+200	+300	+400	+500
					Millivolts				
0	−4.11	−2.56	−0.671	−0.671	1.52	3.97	6.64	9.52	12.57
5	−4.18	−2.65	−0.773	−0.568	1.63	4.09	6.78	9.67	12.73
10	−4.25	−2.73	−0.874	−0.465	1.75	4.22	6.92	9.82	12.88
15	−4.31	−2.82	−0.975	−0.360	1.87	4.35	7.06	9.97	13.04
20	−4.38	−2.90	−1.07	−0.255	1.99	4.48	7.21	10.12	13.20
25	−4.43	−2.98	−1.17	−0.149	2.11	4.61	7.35	10.27	13.36
30	−4.50	−3.06	−1.27	−0.043	2.23	4.75	7.49	10.42	13.52
35	−4.57	−3.14	−1.37	0.064	2.35	4.88	7.63	10.57	13.67
40	−4.63	−3.22	−1.47	0.172	2.47	5.01	7.77	10.72	13.83
45	−4.69	−3.30	−1.56	0.281	2.59	5.15	7.92	10.87	13.99
50	−4.75	−3.38	−1.66	0.389	2.71	5.28	8.06	11.03	14.15
55	−4.80	−3.46	−1.75	0.499	2.83	5.41	8.20	11.18	14.31
60	−4.86	−3.53	−1.84	0.609	2.96	5.55	8.35	11.33	14.47
65	−4.92	−3.61	−1.94	0.720	3.08	5.68	8.49	11.49	14.63
70	−4.97	−3.68	−2.03	0.832	3.21	5.82	8.64	11.64	14.79
75	−5.03	−3.76	−2.12	0.944	3.33	5.96	8.79	11.79	14.96
80	−5.08	−3.83	−2.21	1.06	3.46	6.09	8.93	11.95	15.12
85	−5.13	−3.90	−2.30	1.17	3.58	6.23	9.08	12.10	15.28
90	−5.18	−3.97	−2.39	1.29	3.71	6.37	9.23	12.26	15.44
95	−5.23	−4.04	−2.47	1.40	3.84	6.51	9.37	12.41	15.61
100	−5.28	−4.11	−2.56	1.52	3.97	6.64	9.52	12.57	15.77
Mv per °F	0.0117	0.0155	0.0189	0.0219	0.0245	0.0268	0.0287	0.0305	0.0320

* Form of table supplied by The Brown Instrument Company. Data conform to National Bureau of Standards specifications.

TABLE A–12. THERMAL EMF FOR CHROMEL–ALUMEL THERMOCOUPLE *

Reference Junction at 32°F

°F	0	100	200	300	400	500	600	700	800	900	1000	1100	1200
							Millivolts						
0	−0.68	1.52	3.82	6.09	8.31	10.56	12.85	15.18	17.52	19.88	22.25	24.62	26.98
5	−0.58	1.63	3.93	6.20	8.42	10.67	12.96	15.29	17.63	20.00	22.37	24.74	27.10
10	−0.47	1.74	4.05	6.31	8.53	10.79	13.08	15.41	17.75	20.12	22.49	24.85	27.21
15	−0.37	1.85	4.16	6.42	8.64	10.90	13.19	15.52	17.87	20.24	22.60	24.97	27.33
20	−0.26	1.97	4.28	6.53	8.76	11.02	13.31	15.64	17.99	20.36	22.72	25.00	27.45
25	−0.15	2.08	4.39	6.64	8.87	11.13	13.43	15.76	18.10	20.47	22.84	25.21	27.57
30	−0.04	2.20	4.51	6.75	8.98	11.25	13.55	15.88	18.22	20.59	22.96	25.33	27.68
35	0.07	2.31	4.62	6.86	9.09	11.36	13.67	16.00	18.34	20.71	23.08	25.45	27.80
40	0.18	2.43	4.74	6.98	9.20	11.47	13.78	16.11	18.46	20.83	23.20	25.57	27.92
45	0.29	2.54	4.85	7.09	9.31	11.58	13.89	16.23	18.58	20.95	23.32	25.69	28.04
50	0.40	2.66	4.97	7.20	9.43	11.70	14.01	16.35	18.70	21.07	23.43	25.80	28.15
55	0.51	2.77	5.08	7.31	9.55	11.81	14.12	16.47	18.81	21.18	23.55	25.92	28.27
60	0.62	2.89	5.19	7.42	9.66	11.93	14.24	16.58	18.93	21.30	23.67	26.04	28.39
65	0.73	3.00	5.30	7.53	9.77	12.04	14.36	16.70	19.05	21.42	23.79	26.16	28.51
70	0.84	3.12	5.42	7.64	9.88	12.16	14.48	16.82	19.17	21.54	23.91	26.27	28.62
75	0.95	3.24	5.53	7.75	9.99	12.27	14.60	16.93	19.29	21.66	24.02	26.39	28.74
80	1.06	3.36	5.64	7.87	10.11	12.39	14.71	17.05	19.41	21.78	24.14	26.51	28.86
85	1.17	3.48	5.75	7.98	10.22	12.50	14.83	17.17	19.52	21.89	24.26	26.63	28.98
90	1.29	3.59	5.87	8.09	10.33	12.62	14.94	17.29	19.64	22.01	24.38	26.74	29.09
95	1.40	3.70	5.98	8.20	10.44	12.73	15.06	17.40	19.76	22.13	24.50	26.86	29.21
100	1.52	3.82	6.09	8.31	10.56	12.85	15.18	17.52	19.88	22.25	24.62	26.98	29.33
Mv per °F	0.0220	0.0230	0.0227	0.0222	0.0225	0.0229	0.0233	0.0234	0.0236	0.0237	0.0237	0.0236	0.0235

°F	1300	1400	1500	1600	1700	1800	1900	2000	2100	2200	2300	2400
						Millivolts						
0	29.33	31.65	33.94	36.20	38.43	40.62	42.77	44.89	46.97	49.01	51.00	52.95
5	29.45	31.77	34.06	36.31	38.54	40.73	42.88	45.00	47.08	49.11	51.10	53.05
10	29.56	31.88	34.17	36.42	38.65	40.83	42.98	45.10	47.18	49.21	51.20	53.14
15	29.68	32.00	34.29	36.54	38.76	40.94	43.09	45.20	47.28	49.31	51.30	53.24
20	29.79	32.11	34.40	36.65	38.87	41.05	43.20	45.31	47.38	49.41	51.39	53.33
25	29.91	32.23	34.51	36.76	38.98	41.16	43.31	45.41	47.49	49.51	51.49	53.43
30	30.02	32.34	34.62	36.87	39.09	41.27	43.41	45.52	47.59	49.61	51.59	53.52
35	30.14	32.46	34.74	36.99	39.20	41.38	43.52	45.62	47.69	49.71	51.69	53.62
40	30.26	32.57	34.85	37.10	39.31	41.48	43.62	45.73	47.79	49.81	51.78	53.71
45	30.38	32.69	34.97	37.21	39.42	41.59	43.73	45.83	47.89	49.91	51.88	53.81
50	30.49	32.80	35.08	37.32	39.53	41.70	43.83	45.93	47.99	50.01	51.98	53.90
55	30.61	32.92	35.19	37.43	39.64	41.81	43.94	46.04	48.10	50.11	52.08	54.00
60	30.72	33.03	35.30	37.54	39.75	41.91	44.04	46.14	48.20	50.21	52.17	54.09
65	30.84	33.15	35.42	37.65	39.86	42.02	44.15	46.25	48.30	50.31	52.27	54.19
70	30.96	33.26	35.53	37.76	39.96	42.13	44.26	46.35	48.40	50.41	52.37	54.28
75	31.08	33.38	35.64	37.88	40.07	42.24	44.37	46.46	48.51	50.51	52.47	54.38
80	31.19	33.49	35.75	37.99	40.18	42.34	44.47	46.56	48.61	50.61	52.56	54.47
85	31.31	33.60	35.87	38.10	40.29	42.45	44.58	46.66	48.71	50.71	52.66	54.57
90	31.42	33.71	35.98	38.21	40.40	42.56	44.68	46.76	48.81	50.80	52.75	54.66
95	31.54	33.83	36.09	38.32	40.51	42.67	44.79	46.87	48.91	50.90	52.85	54.76
100	31.65	33.94	36.20	38.43	40.62	42.77	44.89	46.97	49.01	51.00	52.95	54.85
Mv per °F	0.0232	0.0229	0.0226	0.0223	0.0219	0.0215	0.0212	0.0208	0.0204	0.0199	0.0195	0.0190

* Form of table supplied by The Brown Instrument Company. Data conform to National Bureau of Standards specifications.

TABLE A-13. THERMAL EMF FOR PLATINUM–PLATINUM + 10% RHODIUM THERMOCOUPLE *

Reference Junction at 32°F

°F	0	100	200	300	400	500	600	700	800	900	1000	1100	1200	1300	1400	1500
								Millivolts								
0	−0.0920	0.221	0.595	1.016	1.473	1.956	2.457	2.975	3.505	4.044	4.594	5.155	5.725	6.307	6.898	7.500
5	−0.0778	0.239	0.615	1.038	1.497	1.980	2.482	3.001	3.532	4.071	4.621	5.183	5.754	6.336	6.928	7.530
10	−0.0636	0.257	0.635	1.060	1.521	2.005	2.508	3.028	3.559	4.098	4.649	5.211	5.783	6.366	6.958	7.561
15	−0.0494	0.274	0.655	1.082	1.544	2.030	2.534	3.054	3.585	4.125	4.677	5.239	5.812	6.395	6.987	7.591
20	−0.0351	0.292	0.676	1.105	1.568	2.055	2.560	3.081	3.612	4.153	4.705	5.268	5.841	6.424	7.017	7.621
25	−0.0207	0.310	0.696	1.127	1.592	2.080	2.585	3.107	3.639	4.180	4.733	5.296	5.869	6.453	7.047	7.651
30	−0.0060	0.328	0.717	1.150	1.616	2.105	2.611	3.133	3.667	4.208	4.761	5.324	5.898	6.483	7.077	7.682
35	0.0090	0.346	0.737	1.173	1.640	2.130	2.637	3.159	3.694	4.235	4.789	5.352	5.927	6.512	7.107	7.713
40	0.0243	0.365	0.758	1.196	1.664	2.155	2.663	3.186	3.720	4.263	4.817	5.381	5.956	6.542	7.137	7.744
45	0.0398	0.383	0.779	1.218	1.688	2.180	2.689	3.212	3.747	4.290	4.845	5.409	5.985	6.571	7.167	7.774
50	0.0555	0.401	0.800	1.241	1.712	2.205	2.715	3.239	3.774	4.318	4.873	5.438	6.015	6.601	7.198	7.805
55	0.0714	0.420	0.821	1.264	1.736	2.230	2.741	3.266	3.801	4.345	4.901	5.466	6.044	6.630	7.228	7.835
60	0.0875	0.439	0.843	1.287	1.760	2.255	2.767	3.293	3.828	4.373	4.929	5.495	6.073	6.660	7.258	7.866
65	0.104	0.458	0.864	1.310	1.784	2.280	2.793	3.319	3.855	4.400	4.957	5.524	6.102	6.690	7.288	7.897
70	0.120	0.477	0.886	1.333	1.808	2.305	2.819	3.346	3.882	4.428	4.985	5.553	6.132	6.720	7.318	7.928
75	0.137	0.496	0.907	1.356	1.832	2.330	2.845	3.372	3.909	4.455	5.013	5.581	6.161	6.749	7.348	7.958
80	0.153	0.516	0.929	1.380	1.857	2.356	2.871	3.399	3.936	4.483	5.042	5.610	6.190	6.779	7.379	7.989
85	0.170	0.535	0.950	1.403	1.881	2.381	2.897	3.425	3.963	4.511	5.070	5.638	6.219	6.808	7.409	8.020
90	0.187	0.555	0.972	1.426	1.906	2.406	2.923	3.452	3.990	4.539	5.098	5.667	6.249	6.838	7.439	8.051
95	0.204	0.575	0.994	1.449	1.931	2.431	2.949	3.478	4.017	4.566	5.126	5.696	6.278	6.868	7.469	8.081
100	0.221	0.595	1.016	1.473	1.956	2.457	2.975	3.505	4.044	4.594	5.155	5.725	6.307	6.898	7.500	8.112
Mv per °F	0.00325	0.00374	0.00421	0.00457	0.00483	0.00501	0.00518	0.00530	0.00539	0.00550	0.00561	0.00570	0.00582	0.00591	0.00602	0.00612

°F	1600	1700	1800	1900	2000	2100	2200	2300	2400	2500	2600	2700	2800	2900	3000	3100
								Millivolts								
0	8.112	8.734	9.365	10.007	10.657	11.316	11.977	12.642	13.305	13.968	14.629	15.288	15.943	16.596	17.247	17.892
5	8.143	8.765	9.397	10.039	10.690	11.349	12.010	12.675	13.339	14.001	14.662	15.321	15.976	16.629	17.279	17.920
10	8.174	8.796	9.429	10.071	10.723	11.382	12.043	12.708	13.372	14.034	14.695	15.353	16.009	16.661	17.311	17.957
15	8.205	8.827	9.461	10.103	10.756	11.415	12.076	12.741	13.405	14.067	14.728	15.386	16.042	16.694	17.344	17.989
20	8.236	8.859	9.493	10.136	10.789	11.448	12.110	12.775	13.438	14.100	14.761	15.418	16.074	16.726	17.376	18.021
25	8.267	8.890	9.525	10.168	10.822	11.481	12.143	12.808	13.472	14.133	14.794	15.451	16.107	16.759	17.408	......
30	8.298	8.922	9.557	10.201	10.855	11.514	12.177	12.841	13.505	14.166	14.826	15.484	16.139	16.791	17.440	......
35	8.329	8.953	9.589	10.233	10.887	11.547	12.210	12.874	13.538	14.200	14.859	15.517	16.172	16.824	17.473	......
40	8.360	8.895	9.621	10.266	10.920	11.580	12.243	12.907	13.571	14.233	14.892	15.550	16.205	16.856	17.505	......
45	8.391	9.016	9.653	10.298	10.953	11.613	12.276	12.941	13.604	14.266	14.925	15.583	16.238	16.889	17.538	......
50	8.422	9.048	9.685	10.331	10.986	11.646	12.310	12.974	13.637	14.299	14.958	15.615	16.270	16.922	17.570	......
55	8.453	9.079	9.717	10.363	11.019	11.679	12.343	13.007	13.670	14.332	14.991	15.648	16.303	16.955	17.602	......
60	8.484	9.111	9.749	10.396	11.052	11.712	12.376	13.040	13.703	14.365	15.024	15.680	16.335	16.987	17.634	......
65	8.515	9.143	9.781	10.428	11.085	11.745	12.409	13.074	13.737	14.398	15.057	15.713	16.368	17.020	17.667	......
70	8.546	9.175	9.813	10.461	11.118	11.778	12.442	13.107	13.770	14.431	15.090	15.746	16.401	17.052	17.699	......
75	8.577	9.206	9.845	10.493	11.151	11.811	12.475	13.140	13.803	14.464	15.123	15.779	16.434	17.085	17.731	......
80	8.609	9.238	9.877	10.526	11.184	11.844	12.509	13.173	13.836	14.497	15.156	15.812	16.466	17.117	17.763	......
85	8.640	9.270	9.909	10.558	11.217	11.877	12.542	13.206	13.869	14.530	15.189	15.845	16.499	17.150	17.796	......
90	8.671	9.302	9.942	10.591	11.250	11.911	12.575	13.239	13.902	14.563	15.222	15.878	16.531	17.182	17.828	......
95	8.702	9.333	9.974	10.624	11.283	11.944	12.608	13.272	13.935	14.596	15.255	15.911	16.564	17.215	17.860	......
100	8.734	9.365	10.007	10.657	11.316	11.977	12.642	13.305	13.968	14.629	15.288	15.943	16.596	17.247	17.892	......
Mv per °F	0.00622	0.00631	0.00642	0.00650	0.00659	0.00661	0.00665	0.00663	0.00663	0.00661	0.00659	0.00655	0.00653	0.00651	0.00645	......

* Form of table supplied by The Brown Instrument Company. Data conform to National Bureau of Standards specifications.

TABLE A-14. THERMAL EMF FOR PLATINUM–PLATINUM + 13% RHODIUM THERMOCOUPLE *
Reference Junction at 32°F

°F	0	100	200	300	400	500	600	700	800	900	1000	1100	1200	1300	1400	1500
								Millivolts								
0	−0.0890	0.220	0.596	1.030	1.504	2.012	2.546	3.102	3.675	4.263	4.867	5.486	6.122	6.773	7.438	8.118
5	−0.0756	0.237	0.616	1.052	1.529	2.038	2.573	3.130	3.704	4.293	4.897	5.517	6.154	6.806	7.472	8.152
10	−0.0621	0.255	0.637	1.075	1.553	2.064	2.601	3.159	3.733	4.323	4.928	5.548	6.187	6.839	7.505	8.186
15	−0.0484	0.273	0.658	1.098	1.578	2.090	2.628	3.187	3.762	4.352	4.959	5.579	6.219	6.872	7.539	8.220
20	−0.0346	0.291	0.679	1.121	1.603	2.116	2.656	3.215	3.792	4.382	4.990	5.611	6.251	6.905	7.573	8.255
25	−0.0205	0.309	0.700	1.144	1.628	2.142	2.683	3.243	3.821	4.412	5.021	5.643	6.283	6.938	7.607	8.289
30	−0.0060	0.327	0.721	1.167	1.653	2.169	2.711	3.272	3.851	4.442	5.052	5.675	6.316	6.972	7.641	8.324
35	0.0090	0.345	0.742	1.190	1.678	2.196	2.739	3.300	3.880	4.472	5.083	5.706	6.348	7.005	7.674	8.359
40	0.0244	0.364	0.764	1.214	1.703	2.223	2.767	3.329	3.910	4.502	5.114	5.738	6.381	7.038	7.708	8.394
45	0.0399	0.382	0.785	1.237	1.728	2.249	2.794	3.357	3.939	4.532	5.144	5.770	6.413	7.071	7.742	8.428
50	0.0555	0.400	0.807	1.261	1.753	2.276	2.822	3.386	3.969	4.562	5.175	5.802	6.446	7.104	7.776	8.463
55	0.0712	0.419	0.829	1.285	1.778	2.303	2.850	3.415	3.998	4.592	5.206	5.834	6.478	7.137	7.810	8.498
60	0.0871	0.439	0.851	1.309	1.804	2.330	2.878	3.444	4.028	4.623	5.237	5.866	6.511	7.170	7.845	8.533
65	0.103	0.458	0.873	1.333	1.830	2.356	2.906	3.472	4.057	4.653	5.268	5.898	6.544	7.203	7.879	8.567
70	0.119	0.477	0.895	1.357	1.856	2.383	2.934	3.501	4.087	4.684	5.299	5.930	6.577	7.237	7.913	8.602
75	0.136	0.496	0.917	1.381	1.881	2.410	2.962	3.530	4.116	4.714	5.330	5.962	6.609	7.270	7.947	8.637
80	0.152	0.516	0.939	1.405	1.907	2.438	2.990	3.559	4.145	4.745	5.361	5.994	6.642	7.303	7.981	8.672
85	0.169	0.536	0.961	1.429	1.933	2.465	3.018	3.588	4.174	4.775	5.392	6.026	6.674	7.337	8.015	8.706
90	0.186	0.556	0.984	1.454	1.959	2.492	3.046	3.617	4.204	4.806	5.423	6.058	6.707	7.371	8.049	8.741
95	0.203	0.576	1.007	1.479	1.985	2.519	3.074	3.646	4.233	4.836	5.454	6.090	6.740	7.404	8.083	8.776
100	0.220	0.596	1.030	1.504	2.012	2.546	3.102	3.675	4.263	4.867	5.486	6.122	6.773	7.438	8.118	8.811
Mv per °F	0.00323	0.00376	0.00434	0.00474	0.00508	0.00534	0.00556	0.00573	0.00588	0.00604	0.00619	0.00636	0.00651	0.00665	0.00680	0.00693

°F	1600	1700	1800	1900	2000	2100	2200	2300	2400	2500	2600	2700	2800	2900	3000
								Millivolts							
0	8.811	9.518	10.237	10.970	11.720	12.478	13.242	14.010	14.777	15.543	16.309	17.073	17.833	18.588	19.342
5	8.845	9.553	10.273	11.007	11.758	12.516	13.280	14.048	14.815	15.581	16.347	17.111	17.871	18.626	19.380
10	8.880	9.589	10.310	11.045	11.796	12.554	13.318	14.086	14.853	15.619	16.385	17.149	17.909	18.664	19.417
15	8.915	9.624	10.346	11.082	11.833	12.592	13.356	14.124	14.892	15.657	16.424	17.187	17.947	18.702	19.455
20	8.951	9.660	10.383	11.119	11.871	12.631	13.395	14.163	14.930	15.695	16.462	17.225	17.985	18.739	19.492
25	8.986	9.695	10.419	11.156	11.909	12.669	13.433	14.201	14.968	15.734	16.501	17.264	18.023	18.777	19.530
30	9.021	9.731	10.456	11.194	11.947	12.707	13.472	14.240	15.006	15.772	16.539	17.302	18.060	18.815	19.567
35	9.056	9.767	10.492	11.231	11.985	12.745	13.510	14.278	15.045	15.811	16.577	17.340	18.098	18.853	19.605
40	9.092	9.803	10.529	11.269	12.023	12.783	13.549	14.316	15.083	15.849	16.615	17.377	18.135	18.891	19.642
45	9.127	9.839	10.565	11.306	12.061	12.821	13.587	14.355	15.122	15.887	16.654	17.415	18.173	18.929	19.680
50	9.162	9.875	10.602	11.344	12.099	12.860	13.625	14.393	15.160	15.925	16.692	17.453	18.211	18.966	19.717
55	9.197	9.911	10.638	11.382	12.136	12.898	13.663	14.432	15.199	15.964	16.731	17.492	18.249	19.004	19.755
60	9.233	9.947	10.675	11.420	12.174	12.936	13.702	14.470	15.237	16.002	16.769	17.530	18.286	19.041	19.793
65	9.268	9.983	10.712	11.457	12.212	12.974	13.740	14.509	15.275	16.041	16.807	17.568	18.324	19.079	19.830
70	9.304	10.019	10.749	11.495	12.250	13.012	13.779	14.547	15.313	16.079	16.845	17.605	18.362	19.117	19.867
75	9.339	10.055	10.785	11.532	12.288	13.050	13.817	14.586	15.351	16.117	16.883	17.643	18.400	19.155	19.905
80	9.375	10.091	10.822	11.570	12.326	13.089	13.856	14.624	15.389	16.155	16.921	17.682	18.437	19.192	19.942
85	9.410	10.127	10.859	11.607	12.364	13.127	13.894	14.663	15.428	16.194	16.959	17.720	18.475	19.230	19.980
90	9.446	10.164	10.896	11.645	12.402	13.166	13.933	14.701	15.466	16.232	16.997	17.758	18.513	19.267	20.017
95	9.482	10.200	10.933	11.682	12.440	13.204	13.971	14.739	15.505	16.271	17.035	17.796	18.551	19.305	20.055
100	9.518	10.237	10.970	11.720	12.478	13.242	14.010	14.777	15.543	16.309	17.073	17.833	18.588	19.342	20.093
Mv per °F	0.00707	0.00719	0.00733	0.00750	0.00758	0.00764	0.00768	0.00767	0.00766	0.00766	0.00764	0.00760	0.00755	0.00754	0.00751

* Form of table supplied by The Brown Instrument Company. Data conform to National Bureau of Standards specifications.

Table A–15. Resistance (versus Temperature) of Various Metals *

Metal	Resistivity, microhm-cm	−200	−100	0	+100	200	300	400	500	600	700	800	900
		Relative Resistance (R_t/R_0) at °C											
Alumel	28.1	.....	.....	1.000	1.239	1.428	1.537	1.637	1.726	1.814	1.899	1.982	2.066
Copper	1.56	0.117	0.557	1.000	1.431	1.862	2.299	2.747	3.210	3.695	4.208	4.752	5.334
Iron	8.57	.....	.....	1.000	1.650	2.464	3.485	4.716	6.162	7.839	9.790	12.009	12.790
Manganin	48.2	.....	.....	1.000	1.002	0.996	0.991	0.983	.....	.....	.....	.....	.....
Nickel	6.38	.....	.....	1.000	1.663	2.501	3.611	4.847	5.398	5.882	6.327	6.751	7.156
Platinum	9.83	0.177	0.599	1.000	1.392	1.773	2.142	2.499	2.844	3.178	3.500	3.810	4.109
Pt 90, Rh 10	18.4	.....	.....	1.000	1.166	1.330	1.490	1.646	1.798	1.947	2.093	2.234	2.370
Silver	1.50	0.176	0.596	1.000	1.408	1.827	2.256	2.698	3.150	3.616	4.094	4.586	5.091

* Data from *Temperature, Its Measurement and Control in Science and Industry*, 1941.
Data for temperatures below zero not given for same sample as data for temperatures above zero.

Table A–16. Total and Spectral Emissivity of Open Surfaces *

Material	Type Surface	Emissivity Total, e_t	Emissivity Spectral, e_n, $n = 0.65$	Material	Type Surface	Emissivity Total, e_t	Emissivity Spectral, e_n, $n = 0.65$
Alumel	Clean	....	0.37	Iron, wrought	Dull, oxidized	0.94	.....
Aluminum	Clean	0.03	....	Lead	Clean	0.05	.....
	Oxidized	0.19	0.30		Oxidized	0.63	.....
Brass	Clean	0.04	....	Monel	Clean	.....	0.37
	Oxidized	0.61	....		Oxidized	0.43	.....
Carbon	....	0.08	0.90	Nichrome	Clean	.....	0.35
Chromel	Oxidized	0.82	0.87		Oxidized	0.87	0.90
Chromium	Clean	0.08	0.34	Nickel	Clean	0.12	0.36
	Oxidized	....	0.70		Oxidized	0.85	0.90
Constantan	Clean	....	0.35	Platinum	Clean	0.15	0.30
	Oxidized	....	0.84	Porcelain	....	.....	0.40
Copper	Clean	0.02	0.10	Silica brick	....	0.85	.....
	Clean, molten	0.15	0.15	Silver	Clean	0.04	0.07
	Oxidized	0.60	0.70	Steel	Clean	0.08	0.35
	Calorized	0.26	....		Clean, molten	0.28	0.37
	Caloriz., oxid.	0.19	....		Oxidized	0.79	0.80
Firebrick	....	0.75	....		Caloriz., oxid.	0.57	.....
Glass	....	0.94	....		Rough plate	0.97	.....
Iron	Clean	0.05	0.35	Steel (18-8)	Oxidized	0.97	0.85
	Clean, molten	....	0.37	Tungsten	300°K	0.032	0.472
	Rusted	0.65	....		1000°K	0.114	0.458
	Oxidized	0.84	0.70		1500	0.192	0.448
Iron, cast	Clean	0.21	0.37		2000	0.260	0.438
	Clean, molten	0.29	0.40		2500	0.303	0.428
	Oxidized	0.78	0.70		3000	0.334	0.418
	Strongly oxid.	0.95	....		3500	0.351	0.408

* Data from *Temperature, Its Measurement and Control in Science and Industry*, 1941.
Emissivity of oxidized surfaces may vary by 30% and depends on roughness of surface. Values are given at operating temperatures, wherever possible.

TABLE A–17. CORRECTIONS FOR OPTICAL–PYROMETER READINGS FOR LOW-TRANSMISSION AND EMISSIVITY FACTORS

At 0.665 microns

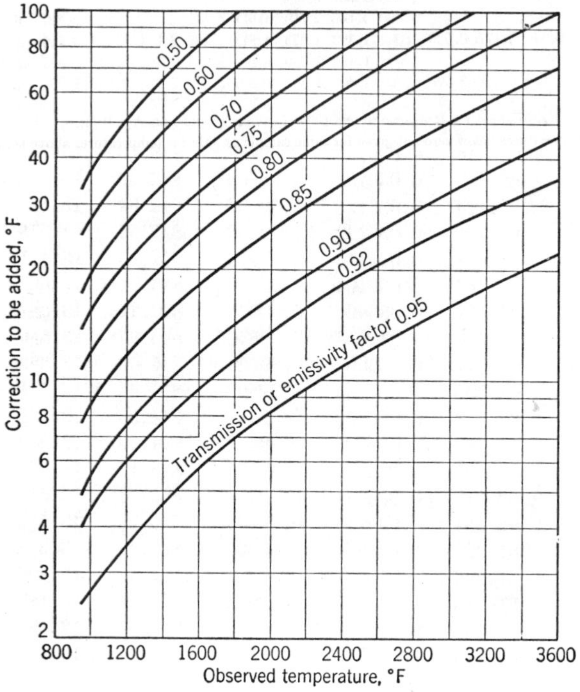

Table A-18. Weight Density of Mercury and Water

Temp., °F	Water Lb/ft³	Water Lb/in.³	Water Relative Density	Mercury Lb/ft³	Mercury Specific Gravity *	Mercury Relative Density
20	......	....	....	849.72	13.612	1.00406
25	......	....	....	849.29	13.605	1.00354
30	......	....	....	848.85	13.598	1.00302
(32)	62.416	0.036121	0.99987	848.72	13.596	1.00288
35	62.422	0.036124	0.99996	848.47	13.592	1.00258
40	62.424	0.036125	1.00000	848.04	13.585	1.00206
45	62.420	0.036122	0.99992	847.60	13.578	1.00155
50	62.408	0.036116	0.99973	847.16	13.571	1.00103
55	62.388	0.036104	0.99942	846.73	13.564	1.00052
60	62.364	0.036090	0.99903	846.29	13.557	1.00000
65	62.335	0.036073	0.99856	845.85	13.550	0.99948
(68)	62.314	0.036061	0.99823	845.60	13.546	0.99919
70	62.300	0.036053	0.99800	845.42	13.543	0.99897
75	62.258	0.036029	0.99734	844.98	13.536	0.99845
80	62.214	0.036003	0.99662	844.60	13.530	0.99801
85	62.165	0.035975	0.99585	844.17	13.523	0.99749
90	62.112	0.035944	0.99499	843.73	13.516	0.99698
95	62.054	0.035911	0.99406	843.35	13.510	0.99653
100	61.991	0.035875	0.99306	842.92	13.503	0.99602
105	61.926	0.035837	0.99201	842.48	13.496	0.99550
110	61.858	0.035797	0.99092	842.04	13.489	0.99498
115	61.787	0.035756	0.98978	841.61	13.482	0.99447
120	61.710	0.035712	0.98856	841.23	13.476	0.99402

* Referred to water at 39.2°F.

The density of water is 62.4245 pcf at 4°C (39.2°F). The density of mercury is 848.692 lb per ft³ at 0°C (32°F).

TABLE A–19. CONVERSION OF PRESSURE UNITS *

To Obtain → Multiply by ↘	Atmospheres	Cm Mercury at 32°F	In. Mercury at 32°F	In. Water at 40°F	Ft Water at 40°F	Kg per sq meter	Psf	Psi
Atmospheres	1.000	76.00	29.92	406.8	33.90	1.033×10^{4}	2117	14.70
Cm mercury at 32°F	1.316×10^{-2}	1.000	3.937×10^{-1}	5.354	4.460×10^{-1}	136.0	27.85	1.934×10^{-1}
In. mercury at 32°F	3.342×10^{-2}	2.540	1.000	13.60	1.133	345.3	70.73	4.912×10^{-1}
In. water at 40°F	2.458×10^{-3}	1.868×10^{-1}	7.355×10^{-2}	1.000	8.333×10^{-2}	25.40	5.204	3.613×10^{-2}
Ft water at 40°F	2.950×10^{-2}	2.232	8.826×10^{-1}	12.00	1.000	304.8	62.43	4.335×10^{-1}
Kg per sq meter	9.678×10^{-5}	7.356×10^{-3}	2.896×10^{-3}	3.937×10^{-2}	3.281×10^{-3}	1.000	2.048×10^{-1}	1.422×10^{-3}
Psf	4.725×10^{-4}	3.591×10^{-2}	1.414×10^{-2}	1.922×10^{-1}	1.602×10^{-2}	4.882	1.000	6.944×10^{-3}
Psi	6.804×10^{-2}	5.171	2.036	27.68	2.307	703.1	144.0	1.000

* Reproduced by permission from O. W. Eshbach, *Handbook of Engineering Fundamentals*, John Wiley & Sons, 1936.

TABLE A-20. CHARACTERISTICS OF MANOMETER FLUIDS

Liquid	Freezing Point, °F	Boiling Point, °F	Specific Gravity at 68°F	Viscosity, centipoises	Coefficient Volumetric Expansion × 10^3 per °F	Vapor Pressure at 68°F, mm Hg	Surface Tension, dynes/cm	Capillary Rise Relative to Mercury
Benzene	42	176	0.879	0.645	1.237	....	28.9	0.93
Carbon tetrachloride	−9	177	1.595	0.969	1.236	91	26.8	0.47
Ethyl alcohol	−175	172	0.791	1.20	1.12	43.9	22.3	0.78
Kerosene	.....	~250	0.78–0.82	2.0	0.95	....	23–32	0.96
Mercury	−38	675	13.55	1.55	0.1819	0.0012	487	1.00
Toluene	−134	230	0.882	0.590	1.224	....	28.4	0.89
Water	32	212	1.000	1.00	0.207	17.54	72.7	2.0

Tables A-21, A-22, A-23

These tables show orifice-flow coefficients K_O including the velocity-of-approach factor $1/\sqrt{1 - \beta^4}$. The Reynolds number is indicated in the heading at each column. The diameter ratio β is shown in the left hand column. The values of flow coefficient are accurate to within the following tolerances:

Flange taps, ± 0.5 per cent for $0.15 < \beta < 0.70$
Vena-contracta taps, ± 0.4 per cent for $0.15 < \beta < 0.75$
Pipe taps, ± 0.75 per cent for $0.20 < \beta < 0.67$

Double these tolerances apply outside the given limits of β, and double these tolerances apply for figures in italics.

These data reprinted from *Fluid Meters, Their Theory and Application*, Part I, ASME, 1937.

Table A-21. Orifice-Flow Coefficients for Flange Taps

For 2-in. Pipe

β	10,000	15,000	25,000	35,000	50,000	75,000	100,000	150,000	250,000	500,000	10,000,000
0.100	0.6134	0.6101	0.6075	0.6064	0.6056	0.6049	0.6046	0.6043	0.6040	0.6039	0.6036
0.150	0.6112	0.6072	0.6040	0.6027	0.6017	0.6009	0.6005	0.6001	0.5999	0.5995	0.5993
0.200	0.6104	0.6060	0.6026	0.6011	0.6000	0.5991	0.5987	0.5983	0.5979	0.5976	0.5974
0.250	0.6115	0.6070	0.6033	0.6017	0.6005	0.5997	0.5992	0.5987	0.5983	0.5981	0.5978
0.300	0.6150	0.6100	0.6062	0.6055	0.6032	0.6023	0.6018	0.6013	0.6009	0.6006	0.6003
0.350	0.6204	0.6149	0.6106	0.6088	0.6074	0.6063	0.6058	0.6053	0.6049	0.6045	0.6042
0.400	0.6278	0.6216	0.6144	0.6129	0.6117	0.6111	0.6104	0.6099	0.6096	0.6094	0.6090
0.450	0.6381	0.6307	0.6248	0.6222	0.6203	0.6188	0.6180	0.6173	0.6167	0.6163	0.6159
0.500	0.6521	0.6430	0.6357	0.6326	0.6302	0.6284	0.6275	0.6266	0.6259	0.6253	0.6247
0.550	0.6708	0.6592	0.6501	0.6461	0.6433	0.6410	0.6398	0.6386	0.6377	0.6371	0.6363
0.600	0.6945	0.6801	0.6687	0.6637	0.6601	0.6572	0.6558	0.6544	0.6532	0.6524	0.6515
0.625	0.7090	0.6930	0.6801	0.6746	0.6705	0.6672	0.6656	0.6641	0.6628	0.6618	0.6608
0.650	0.7251	0.7071	0.6927	0.6866	0.6819	0.6783	0.6765	0.6748	0.6732	0.6722	0.6712
0.675	0.7431	0.7230	0.7070	0.7000	0.6949	0.6908	0.6888	0.6868	0.6852	0.6840	0.6829
0.700	0.7630	0.7406	0.7227	0.7156	0.7092	0.7047	0.7025	0.7003	0.6985	0.6972	0.6959

For 3-in. Pipe

β	10,000	15,000	25,000	35,000	50,000	75,000	100,000	250,000	500,000	10,000,000
0.100	0.6120	0.6076	0.6041	0.6026	0.6014	0.6005	0.6001	0.5993	0.5990	0.5988
0.150	0.6108	0.6056	0.6014	0.5997	0.5985	0.5973	0.5967	0.5959	0.5956	0.5952
0.200	0.6119	0.6063	0.6018	0.5999	0.5984	0.5974	0.5968	0.5958	0.5955	0.5951
0.250	0.6146	0.6089	0.6043	0.6022	0.6008	0.5996	0.5990	0.5980	0.5977	0.5973
0.300	0.6183	0.6120	0.6073	0.6053	0.6037	0.6025	0.6019	0.6009	0.6005	0.6001
0.350	0.6236	0.6170	0.6116	0.6094	0.6078	0.6064	0.6058	0.6046	0.6042	0.6038
0.400	0.6314	0.6239	0.6178	0.6152	0.6132	0.6117	0.6109	0.6096	0.6092	0.6087
0.450	0.6425	0.6333	0.6260	0.6228	0.6206	0.6187	0.6178	0.6162	0.6156	0.6151
0.500	0.6580	0.6465	0.6374	0.6335	0.6305	0.6283	0.6271	0.6251	0.6244	0.6237
0.550	0.6788	0.6641	0.6525	0.6475	0.6438	0.6409	0.6394	0.6368	0.6358	0.6350
0.600	0.7056	0.6871	0.6722	0.6659	0.6610	0.6573	0.6555	0.6521	0.6510	0.6499
0.625	0.7218	0.7009	0.6840	0.6769	0.6714	0.6673	0.6651	0.6613	0.6601	0.6589
0.650	0.7400	0.7164	0.6974	0.6893	0.6833	0.6785	0.6761	0.6718	0.6705	0.6692
0.675	0.7602	0.7336	0.7123	0.7031	0.6963	0.6910	0.6883	0.6835	0.6820	0.6804
0.700	0.7828	0.7530	0.7290	0.7188	0.7113	0.7052	0.7022	0.6968	0.6951	0.6934

For 4-in. Pipe

β	10,000	15,000	25,000	50,000	75,000	100,000	250,000	500,000	10,000,000
0.100	0.6118	0.6064	0.6020	0.5988	0.5976	0.5971	0.5962	0.5959	0.5955
0.150	0.6128	0.6065	0.6014	0.5976	0.5963	0.5956	0.5945	0.5941	0.5937
0.200	0.6151	0.6083	0.6030	0.5990	0.5977	0.5970	0.5958	0.5954	0.5950
0.250	0.6178	0.6111	0.6056	0.6015	0.6001	0.5992	0.5982	0.5978	0.5974
0.300	0.6212	0.6142	0.6086	0.6044	0.6030	0.6023	0.6010	0.6006	0.6002
0.350	0.6263	0.6188	0.6127	0.6082	0.6067	0.6060	0.6046	0.6042	0.6037
0.400	0.6343	0.6257	0.6188	0.6135	0.6118	0.6110	0.6094	0.6089	0.6084
0.450	0.6463	0.6357	0.6273	0.6210	0.6189	0.6178	0.6160	0.6153	0.6147
0.500	0.6631	0.6497	0.6391	0.6310	0.6285	0.6271	0.6247	0.6239	0.6231
0.550	0.6834	0.6690	0.6551	0.6447	0.6413	0.6395	0.6364	0.6354	0.6344
0.600	0.7158	0.6936	0.6757	0.6623	0.6578	0.6556	0.6516	0.6503	0.6490
0.625	0.7340	0.7087	0.6883	0.6731	0.6680	0.6655	0.6609	0.6594	0.6580
0.650	0.7540	0.7253	0.7023	0.6850	0.6793	0.6764	0.6712	0.6696	0.6679
0.675	0.7749	0.7425	0.7165	0.6971	0.6906	0.6874	0.6815	0.6795	0.6777
0.700	0.7985	0.7621	0.7328	0.7109	0.7037	0.7001	0.6935	0.6913	0.6892
0.725	0.8298	0.7887	0.7558	0.7311	0.7228	0.7197	0.7113	0.7088	0.7065
0.750	0.8645	0.8182	0.7807	0.7531	0.7438	0.7391	0.7307	0.7279	0.7253

TABLE A-21. ORIFICE-FLOW COEFFICIENTS FOR FLANGE TAPS (*Continued*)

For 6-in. Pipe

β	25,000	35,000	50,000	75,000	100,000	150,000	250,000	500,000	1,000,000	10,000,000
0.100	0.6018	0.5992	0.5973	0.5957	0.5950	0.5942	*0.5936*	*0.5932*	*0.5929*	*0.5927*
0.150	0.6039	0.6010	0.5987	0.5970	0.5961	0.5952	*0.5946*	*0.5940*	*0.5937*	*0.5935*
0.200	0.6061	0.6030	0.6008	0.5990	0.5981	0.5972	0.5965	*0.5959*	*0.5957*	*0.5954*
0.250	0.6082	0.6051	0.6029	0.6011	0.6002	0.5994	0.5987	*0.5981*	*0.5978*	*0.5976*
0.300	0.6107	0.6077	0.6055	0.6037	0.6028	0.6020	0.6013	0.6007	*0.6004*	*0.6002*
0.350	0.6149	0.6117	0.6094	0.6075	0.6066	0.6056	0.6049	0.6043	*0.6041*	*0.6038*
0.400	0.6209	0.6173	0.6145	0.6124	0.6113	0.6103	0.6095	0.6089	0.6085	*0.6082*
0.450	0.6302	0.6256	0.6222	0.6197	0.6183	0.6170	0.6160	0.6152	0.6148	*0.6144*
0.500	*0.6430*	0.6371	0.6328	0.6294	0.6276	0.6260	0.6246	0.6236	0.6231	*0.6226*
0.550	*0.6606*	0.6529	0.6470	0.6425	0.6403	0.6380	0.6362	0.6349	0.6342	*0.6336*
0.600	*0.6836*	*0.6735*	0.6658	0.6599	0.6569	0.6540	0.6515	0.6498	0.6488	*0.6481*
0.625	*0.6976*	*0.6859*	0.6772	0.6704	0.6669	0.6635	0.6607	0.6587	0.6578	*0.6568*
0.650	*0.7132*	*0.6998*	0.6898	0.6803	0.6782	0.6743	0.6711	0.6688	0.6677	*0.6666*
0.675	*0.7308*	*0.7156*	*0.7041*	0.6953	0.6909	0.6865	0.6829	0.6803	0.6790	*0.6777*
0.700	*0.7500*	*0.7328*	*0.7199*	0.7099	0.7048	0.6998	0.6958	0.6938	0.6913	*0.6899*
0.725	*0.7721*	*0.7527*	*0.7381*	*0.7267*	0.7211	0.7153	0.7108	0.7074	0.7058	*0.7041*
0.750	*0.8017*	*0.7788*	*0.7616*	*0.7482*	0.7417	0.7348	0.7294	0.7255	0.7235	*0.7217*

For 8-in. Pipe

β	25,000	35,000	50,000	75,000	100,000	150,000	250,000	500,000	1,000,000	10,000,000
0.100	0.6035	0.6002	0.5977	0.5958	0.5949	0.5939	*0.5932*	*0.5926*	*0.5923*	*0.5920*
0.150	0.6067	0.6030	0.6002	0.5980	0.5969	0.5958	*0.5949*	*0.5943*	*0.5940*	*0.5936*
0.200	0.6088	0.6050	0.6022	0.6000	0.5989	0.5978	0.5969	*0.5963*	*0.5960*	*0.5956*
0.250	0.6103	0.6067	0.6040	0.6018	0.6008	0.5997	0.5989	*0.5983*	*0.5979*	*0.5976*
0.300	0.6127	0.6091	0.6065	0.6045	0.6034	0.6024	0.6016	0.6009	*0.6006*	*0.6003*
0.350	0.6167	0.6129	0.6101	0.6079	0.6068	0.6058	0.6049	0.6043	*0.6039*	*0.6036*
0.400	0.6229	0.6186	0.6155	0.6130	0.6118	0.6105	0.6095	0.6088	0.6085	*0.6081*
0.450	0.6325	0.6272	0.6232	0.6202	0.6186	0.6171	0.6159	0.6150	0.6146	*0.6140*
0.500	*0.6466*	0.6396	0.6343	0.6303	0.6284	0.6263	0.6246	0.6235	0.6229	*0.6224*
0.550	*0.6661*	0.6568	0.6497	0.6441	0.6414	0.6389	0.6364	0.6349	0.6340	*0.6333*
0.600	*0.6912*	*0.6786*	0.6693	0.6619	0.6583	0.6546	0.6517	0.6496	0.6484	*0.6474*
0.625	*0.7067*	*0.6922*	0.6813	0.6730	0.6687	0.6645	0.6611	0.6586	0.6572	*0.6561*
0.650	*0.7239*	*0.7073*	0.6947	0.6851	0.6803	0.6754	0.6715	0.6687	0.6672	*0.6658*
0.675	*0.7432*	*0.7243*	*0.7100*	0.6990	0.6934	0.6879	0.6834	0.6801	0.6784	*0.6768*
0.700	*0.7646*	*0.7430*	*0.7269*	0.7142	0.7079	0.7016	0.6966	0.6928	0.6910	*0.6891*
0.725	*0.7889*	*0.7643*	*0.7459*	*0.7315*	0.7243	0.7173	0.7115	0.7072	0.7052	*0.7032*
0.750	*0.8214*	*0.7923*	*0.7706*	*0.7536*	0.7452	0.7367	0.7299	0.7249	0.7223	*0.7200*

For 10-in. Pipe

β	25,000	35,000	50,000	75,000	100,000	150,000	250,000	500,000	1,000,000	10,000,000
0.100	0.6058	0.6018	0.5998	0.5965	0.5953	0.5941	*0.5932*	*0.5925*	*0.5922*	*0.5918*
0.150	0.6095	0.6050	0.6016	0.5990	0.5975	0.5964	*0.5954*	*0.5946*	*0.5942*	*0.5938*
0.200	0.6112	0.6068	0.6035	0.6009	0.5996	0.5983	0.5972	*0.5965*	*0.5961*	*0.5957*
0.250	0.6125	0.6083	0.6051	0.6027	0.6018	0.6002	0.5991	*0.5985*	*0.5982*	*0.5978*
0.300	0.6145	0.6104	0.6074	0.6051	0.6039	0.6027	0.6018	0.6011	*0.6008*	*0.6004*
0.350	0.6182	0.6139	0.6109	0.6085	0.6073	0.6061	0.6051	0.6044	*0.6041*	*0.6037*
0.400	0.6246	0.6198	0.6163	0.6135	0.6122	0.6108	0.6097	0.6089	0.6085	*0.6081*
0.450	0.6348	0.6288	0.6244	0.6209	0.6192	0.6174	0.6160	0.6150	0.6145	*0.6140*
0.500	*0.6500*	0.6421	0.6361	0.6315	0.6291	0.6268	0.6249	0.6235	0.6229	*0.6223*
0.550	*0.6713*	0.6603	0.6521	0.6457	0.6425	0.6393	0.6368	0.6348	0.6338	*0.6330*
0.600	*0.6988*	*0.6840*	0.6728	0.6643	0.6600	0.6556	0.6521	0.6496	0.6483	*0.6471*
0.625	*0.7135*	*0.6970*	0.6846	0.6749	0.6706	0.6653	0.6613	0.6585	0.6571	*0.6561*
0.650	*0.7346*	*0.7148*	0.6999	0.6883	0.6826	0.6769	0.6722	0.6687	0.6671	*0.6655*
0.675	*0.7557*	*0.7330*	*0.7159*	0.7027	0.6961	0.6894	0.6842	0.6802	0.6781	*0.6764*
0.700	*0.7792*	*0.7533*	*0.7337*	0.7186	0.7110	0.7035	0.6974	0.6929	0.6906	*0.6885*
0.725	*0.8057*	*0.7761*	*0.7540*	*0.7367*	0.7281	0.7194	0.7126	0.7074	0.7048	*0.7025*
0.750	*0.8414*	*0.8064*	*0.7800*	*0.7597*	0.7494	0.7391	0.7310	0.7247	0.7218	*0.7190*

Table A-22. Orifice-Flow Coefficients for Vena-Contracta Taps

For 2-in. Pipe

β	10,000	15,000	25,000	35,000	50,000	75,000	100,000	150,000	250,000	500,000
0.100	0.6195	0.6148	0.6106	0.6083	0.6061	0.6041	0.6031	......	......	......
0.150	0.6133	0.6097	0.6059	0.6039	0.6022	0.6004	0.5995	......	......	......
0.200	0.6098	0.6067	0.6035	0.6020	0.6004	0.5989	0.5981	0.5971	......	......
0.250	0.6090	0.6062	0.6035	0.6020	0.6007	0.5995	0.5988	0.5978	......	......
0.300	0.6109	0.6082	0.6056	0.6041	0.6029	0.6018	0.6016	0.6001	0.5993	......
0.350	0.6150	0.6123	0.6096	0.6081	0.6068	0.6055	0.6048	0.6039	0.6030	......
0.400	0.6214	0.6185	0.6154	0.6138	0.6125	0.6111	0.6103	0.6093	0.6084	0.6074
0.450	......	0.6261	0.6229	0.6212	0.6197	0.6182	0.6173	0.6165	0.6155	0.6144
0.500	......	0.6361	0.6327	0.6308	0.6293	0.6277	0.6268	0.6257	0.6247	0.6236
0.550	......	......	0.6454	0.6433	0.6418	0.6401	0.6390	0.6379	0.6368	0.6357
0.600	......	......	......	0.6601	0.6582	0.6565	0.6553	0.6541	0.6530	0.6516
0.625	......	......	......	......	0.6684	0.6666	0.6654	0.6639	0.6628	0.6613
0.650	......	......	......	......	0.6802	0.6782	0.6768	0.6753	0.6740	0.6724
0.675	......	......	......	......	0.6938	0.6916	0.6900	0.6884	0.6870	0.6852
0.700	......	......	......	......	0.7095	0.7070	0.7054	0.7034	0.7018	0.7000
0.725	......	......	......	......	0.7280	0.7250	0.7233	0.7212	0.7191	0.7169
0.750	......	......	......	......	0.7496	0.7460	0.7440	0.7417	0.7392	0.7368
0.775	......	......	......	......	......	0.7709	0.7687	0.7661	0.7634	0.7607
0.800	......	......	......	......	......	0.8012	0.7988	0.7960	0.7930	0.7900
0.825	......	......	......	......	......	0.8395	0.8370	0.8340	0.8306	0.8273

For 3-in. Pipe

β	15,000	25,000	35,000	50,000	75,000	100,000	150,000	250,000	500,000	1,000,000
0.100	0.6081	0.6045	0.6026	0.6009	0.5993	0.5984	0.5972	......	......	......
0.150	0.6048	0.6018	0.6002	0.5987	0.5974	0.5967	0.5957	......	......	......
0.200	0.6037	0.6010	0.5995	0.5983	0.5971	0.5963	0.5955	0.5947	......	......
0.250	0.6050	0.6026	0.6011	0.5998	0.5988	0.5981	0.5973	0.5964	......	......
0.300	0.6081	0.6054	0.6039	0.6027	0.6015	0.6008	0.6000	0.5991	0.5982	......
0.350	0.6123	0.6096	0.6081	0.6067	0.6055	0.6047	0.6039	0.6029	0.6021	......
0.400	0.6185	0.6154	0.6139	0.6125	0.6111	0.6103	0.6092	0.6084	0.6074	0.6067
0.450	......	0.6230	0.6213	0.6197	0.6183	0.6174	0.6164	0.6154	0.6144	0.6137
0.500	......	0.6327	0.6309	0.6292	0.6277	0.6268	0.6257	0.6246	0.6236	0.6228
0.550	......	......	0.6434	0.6418	0.6402	0.6391	0.6379	0.6368	0.6356	0.6348
0.600	......	......	0.6601	0.6581	0.6564	0.6553	0.6542	0.6530	0.6517	0.6507
0.625	......	......	......	0.6685	0.6666	0.6655	0.6642	0.6628	0.6613	0.6604
0.650	......	......	......	0.6806	0.6783	0.6770	0.6756	0.6739	0.6724	0.6713
0.675	......	......	......	0.6946	0.6920	0.6904	0.6886	0.6868	0.6849	0.6837
0.700	......	......	......	0.7107	0.7077	0.7060	0.7039	0.7018	0.6995	0.6980
0.725	......	......	......	0.7294	0.7260	0.7241	0.7215	0.7192	0.7166	0.7149
0.750	......	......	......	0.7513	0.7475	0.7453	0.7424	0.7396	0.7368	0.7349
0.775	......	......	......	0.7773	0.7728	0.7704	0.7673	0.7641	0.7612	0.7589
0.800	......	......	......	......	0.8038	0.8009	0.7975	0.7940	0.7905	0.7882
0.825	......	......	......	......	0.8426	0.8395	0.8355	0.8317	0.8277	0.8251

For 4-in. Pipe

β	15,000	25,000	35,000	50,000	75,000	100,000	150,000	250,000	500,000	1,000,000
0.100	0.6063	0.6029	0.6011	0.5994	0.5980	0.5971	0.5959	......	......	......
0.150	0.6039	0.6010	0.5994	0.5980	0.5967	0.5960	0.5950	......	......	......
0.200	0.6034	0.6007	0.5993	0.5981	0.5969	0.5961	0.5954	......	......	......
0.250	0.6049	0.6025	0.6010	0.5998	0.5987	0.5980	0.5972	0.5964	......	......
0.300	0.6081	0.6054	0.6039	0.6027	0.6015	0.6008	0.6000	0.5991	0.5983	......
0.350	0.6123	0.6096	0.6081	0.6067	0.6055	0.6047	0.6039	0.6029	0.6022	......
0.400	0.6185	0.6154	0.6139	0.6125	0.6111	0.6103	0.6092	0.6084	0.6074	0.6067
0.450	......	0.6229	0.6213	0.6197	0.6183	0.6174	0.6164	0.6154	0.6144	0.6137
0.500	......	......	0.6309	0.6292	0.6277	0.6268	0.6257	0.6246	0.6236	0.6228
0.550	......	......	0.6434	0.6418	0.6402	0.6391	0.6379	0.6368	0.6356	0.6348
0.600	......	......	0.6601	0.6581	0.6564	0.6553	0.6542	0.6530	0.6517	0.6507
0.625	......	......	......	......	0.6667	0.6655	0.6642	0.6628	0.6613	0.6604
0.650	......	......	......	......	0.6785	0.6771	0.6756	0.6739	0.6724	0.6713
0.675	......	......	......	......	......	0.6906	0.6887	0.6869	0.6849	0.6837
0.700	......	......	......	......	......	0.7063	0.7041	0.7019	0.6995	0.6980
0.725	......	......	......	......	......	0.7245	0.7219	0.7193	0.7167	0.7148
0.750	......	......	......	......	......	0.7460	0.7429	0.7399	0.7370	0.7348
0.775	......	......	......	......	......	0.7714	0.7680	0.7645	0.7613	0.7587
0.800	......	......	......	......	......	0.8022	0.7983	0.7945	0.7906	0.7881
0.825	......	......	......	......	......	0.8408	0.8364	0.8321	0.8277	0.8249

TABLE A-22. ORIFICE-FLOW COEFFICIENTS FOR VENA-CONTRACTA TAPS (*Continued*)

For 6-in. Pipe

β	15,000	25,000	35,000	50,000	75,000	100,000	150,000	250,000	500,000	1,000,000
0.100	0.6024	0.5995	0.5978	0.5963	0.5951	0.5942	......	......	......	......
0.150	0.6019	0.5992	0.5978	0.5965	0.5953	0.5946	......	......	......	......
0.200	0.6028	0.6001	0.5988	0.5977	0.5965	0.5958	0.5951	......	......	......
0.250	0.6048	0.6022	0.6009	0.5997	0.5986	0.5979	0.5971	0.5963	......	......
0.300	0.6080	0.6053	0.6039	0.6026	0.6014	0.6007	0.6000	0.5990	0.5984	0.5978
0.350	0.6123	0.6096	0.6082	0.6068	0.6055	0.6047	0.6040	0.6030	0.6023	0.6015
0.400	0.6185	0.6155	0.6139	0.6125	0.6111	0.6104	0.6093	0.6083	0.6074	0.6067
0.450	......	0.6229	0.6213	0.6197	0.6183	0.6175	0.6165	0.6154	0.6143	0.6136
0.500	......	0.6327	0.6309	0.6293	0.6277	0.6269	0.6258	0.6247	0.6235	0.6228
0.550	......	......	0.6434	0.6418	0.6402	0.6391	0.6379	0.6368	0.6356	0.6347
0.600	......	......	0.6601	0.6581	0.6565	0.6554	0.6542	0.6529	0.6517	0.6506
0.625	......	......	......	......	0.6668	0.6656	0.6642	0.6629	0.6614	0.6604
0.650	......	......	......	......	0.6788	0.6774	0.6757	0.6740	0.6725	0.6713
0.675	......	......	......	......	......	0.6910	0.6890	0.6870	0.6850	0.6837
0.700	......	......	......	......	......	0.7069	0.7044	0.7021	0.6996	0.6980
0.725	......	......	......	......	......	0.7254	0.7226	0.7196	0.7168	0.7147
0.750	......	......	......	......	......	0.7475	0.7441	0.7406	0.7373	0.7346
0.775	......	......	......	......	......	0.7734	0.7694	0.7654	0.7614	0.7584
0.800	......	......	......	......	......	0.8048	0.8001	0.7956	0.7909	0.7878
0.825	......	......	......	......	......	0.8435	0.8383	0.8330	0.8276	0.8244

For 8-in. Pipe

β	15,000	25,000	35,000	50,000	75,000	100,000	150,000	250,000	500,000	1,000,000	1,500,000
0.100	0.6023	0.5996	0.5980	0.5964	0.5952	0.5944	......	......	......	......	......
0.150	0.6017	0.5993	0.5980	0.5966	0.5954	0.5947	......	......	......	......	......
0.200	0.6029	0.6002	0.5989	0.5978	0.5966	0.5959	0.5952	......	......	......	......
0.250	0.6051	0.6023	0.6009	0.5998	0.5986	0.5980	0.5971	0.5964	......	......	......
0.300	0.6083	0.6054	0.6039	0.6027	0.6015	0.6008	0.6001	0.5991	0.5984	......	......
0.350	0.6123	0.6096	0.6082	0.6068	0.6055	0.6047	0.6040	0.6030	0.6022	......	......
0.400	0.6189	0.6155	0.6139	0.6125	0.6111	0.6104	0.6093	0.6083	0.6074	0.6068	......
0.450	......	......	0.6213	0.6197	0.6183	0.6175	0.6165	0.6154	0.6143	0.6136	......
0.500	......	......	0.6309	0.6293	0.6277	0.6269	0.6258	0.6247	0.6235	0.6228	......
0.550	......	......	0.6434	0.6418	0.6402	0.6391	0.6379	0.6368	0.6356	0.6347	......
0.600	......	......	......	0.6581	0.6565	0.6554	0.6542	0.6529	0.6517	0.6506	......
0.625	......	......	......	......	0.6669	0.6656	0.6642	0.6630	0.6615	0.6604	......
0.650	......	......	......	......	0.6790	0.6777	0.6759	0.6743	0.6726	0.6714	0.6708
0.675	......	......	......	......	......	0.6915	0.6894	0.6873	0.6853	0.6838	0.6833
0.700	......	......	......	......	......	0.7076	0.7050	0.7025	0.6999	0.6981	0.6974
0.725	......	......	......	......	......	0.7265	0.7233	0.7202	0.7170	0.7148	0.7139
0.750	......	......	......	......	......	0.7487	0.7447	0.7411	0.7373	0.7346	0.7335
0.775	......	......	......	......	......	0.7748	0.7701	0.7659	0.7614	0.7582	0.7570
0.800	......	......	......	......	......	0.8063	0.8010	0.7961	0.7909	0.7875	0.7859
0.825	......	......	......	......	......	0.8454	0.8393	0.8336	0.8277	0.8239	0.8219

For 10-in. Pipe

β	25,000	35,000	50,000	75,000	100,000	150,000	250,000	500,000	1,000,000	1,500,000
0.100	0.5998	0.5982	0.5966	0.5954	0.5946	0.5936	......	......	......	......
0.150	0.5995	0.5982	0.5968	0.5956	0.5949	0.5940	0.5934	......	......	......
0.200	0.6004	0.5991	0.5979	0.5967	0.5961	0.5953	0.5946	......	......	......
0.250	0.6024	0.6010	0.6000	0.5987	0.5981	0.5972	0.5965	......	......	......
0.300	0.6055	0.6040	0.6029	0.6016	0.6010	0.6002	0.5992	0.5985	......	......
0.350	0.6097	0.6082	0.6068	0.6055	0.6048	0.6039	0.6030	0.6022	......	......
0.400	0.6155	0.6139	0.6124	0.6111	0.6103	0.6092	0.6082	0.6074	0.6069	......
0.450	......	0.6212	0.6196	0.6183	0.6174	0.6164	0.6154	0.6143	0.6136	......
0.500	......	0.6308	0.6292	0.6277	0.6268	0.6257	0.6246	0.6235	0.6228	......
0.550	......	0.6434	0.6417	0.6401	0.6390	0.6379	0.6368	0.6356	0.6347	......
0.600	......	......	0.6581	0.6565	0.6554	0.6542	0.6530	0.6517	0.6507	......
0.625	......	......	......	0.6670	0.6657	0.6643	0.6631	0.6616	0.6605	......
0.650	......	......	......	0.6792	0.6780	0.6761	0.6746	0.6728	0.6716	0.6710
0.675	......	......	......	0.6936	0.6921	0.6899	0.6877	0.6856	0.6840	0.6835
0.700	......	......	......	0.7108	0.7086	0.7056	0.7030	0.7002	0.6983	0.6975
0.725	......	......	......	......	0.7276	0.7240	0.7208	0.7173	0.7149	0.7139
0.750	......	......	......	......	0.7499	0.7454	0.7417	0.7374	0.7345	0.7333
0.775	......	......	......	......	0.7762	0.7709	0.7664	0.7614	0.7580	0.7567
0.800	......	......	......	......	0.8079	0.8020	0.7966	0.7910	0.7871	0.7853
0.825	......	......	......	......	0.8473	0.8404	0.8342	0.8279	0.8234	0.8210

Table A–23. Orifice–Flow Coefficients for Pipe Taps

For 2-in. Pipe

β	10,000	15,000	20,000	25,000	35,000	50,000	75,000	100,000	150,000	250,000	500,000	10,000,000
0.100	0.6172	0.6134	0.6116	0.6104	0.6091	0.6081	0.6073	0.6070	*0.6066*	*0.6063*	*0.6060*	*0.6057*
0.150	0.6212	0.6164	0.6140	0.6126	0.6109	0.6097	0.6087	0.6083	*0.6078*	*0.6074*	*0.6071*	*0.6068*
0.200	0.6291	0.6236	0.6209	0.6192	0.6173	0.6159	0.6148	0.6143	0.6137	*0.6133*	*0.6129*	*0.6126*
0.250	0.6412	0.6350	0.6319	0.6301	0.6280	0.6264	0.6251	0.6245	0.6239	*0.6233*	*0.6230*	*0.6226*
0.300	0.6565	0.6496	0.6461	0.6440	0.6417	0.6399	0.6385	0.6379	0.6372	0.6366	0.6362	*0.6358*
0.350	0.6758	0.6680	0.6640	0.6617	0.6590	0.6570	0.6553	0.6545	0.6538	0.6532	0.6527	*0.6522*
0.400	*0.7003*	0.6910	0.6864	0.6836	0.6810	0.6781	0.6762	0.6752	0.6744	0.6736	0.6730	*0.6725*
0.425	*0.7154*	0.7052	0.7001	0.6970	0.6936	0.6909	0.6889	0.6879	0.6869	0.6861	0.6855	*0.6848*
0.450	*0.7320*	0.7209	0.7152	0.7119	0.7080	0.7051	0.7029	0.7018	0.7006	0.6997	0.6991	*0.6984*
0.475	*0.7510*	*0.7386*	0.7323	0.7286	0.7243	0.7211	0.7187	0.7173	0.7161	0.7152	0.7144	*0.7136*
0.500	*0.7724*	*0.7585*	0.7515	0.7473	0.7425	0.7390	0.7362	0.7347	0.7334	0.7323	0.7314	*0.7306*
0.525	*0.7968*	*0.7812*	0.7753	0.7686	0.7632	0.7592	0.7561	0.7545	0.7530	0.7517	0.7508	*0.7498*
0.550	*0.8245*	*0.8068*	*0.7979*	0.7927	0.7866	0.7821	0.7786	0.7768	0.7750	0.7737	0.7725	*0.7716*
0.575	*0.8555*	*0.8356*	*0.8256*	0.8197	0.8129	0.8078	0.8038	0.8018	0.7999	0.7983	0.7970	*0.7960*
0.600	*0.8909*	*0.8682*	*0.8571*	*0.8503*	0.8426	0.8368	0.8323	0.8301	0.8278	0.8260	0.8246	*0.8234*
0.625	*0.9310*	*0.9054*	*0.8926*	*0.8850*	0.8762	0.8697	0.8646	0.8621	0.8595	0.8574	0.8559	*0.8545*
0.650	*0.9765*	*0.9475*	*0.9330*	*0.9243*	0.9144	0.9069	0.9012	0.8983	0.8953	0.8930	0.8914	*0.8897*
0.675	*1.0318*	*0.9958*	*0.9793*	*0.9694*	*0.9581*	0.9496	0.9431	0.9397	0.9365	0.9339	0.9317	*0.9300*
0.700	*1.0881*	*1.0506*	*1.0319*	*1.0205*	*1.0077*	0.9981	0.9906	0.9868	0.9830	0.9801	0.9779	*0.9757*

For 3-in. Pipe

β	10,000	15,000	20,000	25,000	35,000	50,000	75,000	100,000	150,000	250,000	500,000	10,000,000
0.100	0.6171	0.6124	0.6100	0.6085	0.6069	0.6057	0.6047	0.6042	*0.6038*	*0.6034*	*0.6031*	*0.6028*
0.150	0.6223	0.6165	0.6136	0.6118	0.6098	0.6083	0.6071	0.6065	*0.6060*	*0.6055*	*0.6052*	*0.6048*
0.200	0.6307	0.6242	0.6210	0.6191	0.6169	0.6151	0.6139	0.6133	0.6126	*0.6121*	*0.6117*	*0.6113*
0.250	0.6424	0.6354	0.6320	0.6299	0.6276	0.6257	0.6243	0.6236	0.6229	*0.6223*	*0.6219*	*0.6215*
0.300	0.6576	0.6502	0.6463	0.6440	0.6415	0.6395	0.6380	0.6373	0.6365	0.6359	*0.6354*	*0.6350*
0.350	0.6771	0.6686	0.6644	0.6619	0.6590	0.6568	0.6551	0.6542	0.6534	0.6527	*0.6522*	*0.6517*
0.400	*0.7022*	0.6923	0.6873	0.6842	0.6809	0.6784	0.6763	0.6753	0.6744	0.6736	0.6730	*0.6724*
0.425	*0.7175*	0.7065	0.7010	0.6977	0.6939	0.6912	0.6889	0.6878	0.6868	0.6859	0.6853	*0.6845*
0.450	*0.7349*	0.7226	0.7165	0.7129	0.7087	0.7055	0.7031	0.7019	0.7006	0.6996	0.6989	*0.6982*
0.475	*0.7547*	0.7410	0.7341	0.7300	0.7253	0.7217	0.7190	0.7176	0.7163	0.7152	0.7143	*0.7135*
0.500	*0.7770*	*0.7615*	0.7537	0.7491	0.7438	0.7398	0.7367	0.7351	0.7336	0.7324	0.7314	*0.7305*
0.525	*0.8026*	*0.7850*	0.7762	0.7709	0.7648	0.7604	0.7568	0.7550	0.7533	0.7518	0.7508	*0.7498*
0.550	*0.8319*	*0.8117*	0.8016	0.7955	0.7886	0.7833	0.7793	0.7774	0.7753	0.7737	0.7725	*0.7714*
0.575	*0.8646*	*0.8415*	*0.8301*	0.8232	0.8152	0.8093	0.8048	0.8024	0.8001	0.7983	0.7969	*0.7957*
0.600	*0.9020*	*0.8756*	*0.8623*	0.8545	0.8454	0.8387	0.8335	0.8308	0.8283	0.8260	0.8244	*0.8229*
0.625	*0.9446*	*0.9143*	*0.8991*	0.8900	0.8797	0.8719	0.8658	0.8627	0.8597	0.8573	0.8555	*0.8537*
0.650	*0.9929*	*0.9581*	*0.9407*	*0.9303*	0.9184	0.9095	0.9024	8.8990	0.8954	0.8927	0.8906	*0.8886*
0.675	*1.0480*	*1.0081*	*0.9881*	*0.9761*	0.9624	0.9521	0.9458	0.9402	0.9361	0.9330	0.9305	*0.9283*
0.700	*1.1112*	*1.0653*	*1.0423*	*1.0285*	*1.0127*	1.0009	0.9918	0.9872	0.9826	0.9789	0.9762	*0.9735*

For 4-in. Pipe

β	10,000	15,000	20,000	25,000	35,000	50,000	75,000	100,000	150,000	250,000	500,000	10,000,000
0.100	0.6181	0.6124	0.6096	0.6077	0.6059	0.6044	0.6033	0.6027	0.6021	*0.6018*	*0.6014*	*0.6010*
0.150	0.6243	0.6174	0.6141	0.6120	0.6097	0.6079	0.6065	0.6059	0.6052	*0.6046*	*0.6042*	*0.6038*
0.200	0.6326	0.6253	0.6216	0.6194	0.6169	0.6150	0.6135	0.6128	0.6121	0.6115	*0.6110*	*0.6106*
0.250	0.6442	0.6365	0.6327	0.6304	0.6278	0.6257	0.6241	0.6234	0.6226	0.6220	*0.6215*	*0.6211*
0.300	0.6591	0.6510	0.6468	0.6444	0.6420	0.6396	0.6379	0.6372	0.6363	0.6356	0.6352	*0.6347*
0.350	0.6785	0.6694	0.6649	0.6622	0.6592	0.6569	0.6550	0.6541	0.6532	0.6525	0.6519	*0.6514*
0.400	0.7035	0.6935	0.6882	0.6850	0.6813	0.6786	0.6765	0.6754	0.6744	0.6736	0.6729	*0.6723*
0.425	*0.7197*	0.7080	0.7021	0.6986	0.6945	0.6915	0.6891	0.6880	0.6868	0.6859	0.6852	*0.6844*
0.450	*0.7375*	0.7244	0.7178	0.7138	0.7093	0.7060	0.7033	0.7021	0.7007	0.6997	0.6989	*0.6981*
0.475	*0.7551*	0.7433	0.7358	0.7313	0.7261	0.7223	0.7194	0.7179	0.7164	0.7152	0.7143	*0.7134*
0.500	*0.7815*	0.7646	0.7560	0.7508	0.7451	0.7407	0.7373	0.7357	0.7339	0.7326	0.7315	*0.7305*
0.525	*0.8084*	*0.7888*	0.7791	0.7733	0.7665	0.7614	0.7575	0.7556	0.7536	0.7520	0.7510	*0.7498*
0.550	*0.8392*	*0.8164*	0.8051	0.7983	0.7905	0.7847	0.7802	0.7780	0.7757	0.7739	0.7725	*0.7713*
0.575	*0.8734*	*0.8475*	0.8344	0.8266	0.8177	0.8109	0.8057	0.8032	0.8006	0.7984	0.7969	*0.7954*
0.600	*0.9129*	*0.8829*	0.8678	0.8588	0.8484	0.8407	0.8347	0.8316	0.8286	0.8262	0.8244	*0.8227*
0.625	*0.9581*	*0.9231*	*0.9056*	0.8952	0.8833	0.8741	0.8672	0.8638	0.8602	0.8574	0.8555	*0.8534*
0.650	*1.0090*	*0.9687*	*0.9485*	0.9364	0.9225	0.9121	0.9040	0.9000	0.8960	0.8927	0.8904	*0.8880*
0.675	*1.0675*	*1.0208*	*0.9975*	0.9834	0.9675	0.9553	0.9460	0.9414	0.9367	0.9330	0.9302	*0.9275*
0.700	*1.1344*	*1.0804*	*1.0534*	*1.0372*	1.0186	1.0047	0.9939	0.9887	0.9832	0.9789	0.9756	*0.9725*
0.725	*1.2149*	*1.1522*	*1.1208*	*1.1020*	1.0806	1.0645	1.0520	1.0457	1.0395	1.0344	1.0307	*1.0272*
0.750	*1.3094*	*1.2344*	*1.1968*	*1.1743*	1.1485	1.1293	1.1142	1.1067	1.0993	1.0931	1.0888	*1.0844*

TABLE A-23. ORIFICE-FLOW COEFFICIENTS FOR PIPE TAPS (*Continued*)

For 6-in. Pipe

β	25,000	35,000	50,000	75,000	100,000	150,000	250,000	500,000	1,000,000	10,000,000
0.100	0.6084	0.6058	0.6038	0.6023	0.6015	0.6007	0.6001	0.5997	0.5994	0.5992
0.150	0.6134	0.6104	0.6080	0.6063	0.6054	0.6044	0.6038	0.6032	0.6029	0.6027
0.200	0.6210	0.6178	0.6154	0.6136	0.6126	0.6117	0.6110	0.6104	0.6101	0.6099
0.250	0.6318	0.6286	0.6261	0.6243	0.6234	0.6225	0.6217	0.6212	0.6208	0.6206
0.300	0.6452	0.6420	0.6396	0.6377	0.6368	0.6359	0.6351	0.6346	0.6343	0.6340
0.350	0.6631	0.6597	0.6571	0.6550	0.6539	0.6529	0.6521	0.6515	0.6512	0.6509
0.400	0.6864	0.6823	0.6792	0.6769	0.6756	0.6744	0.6735	0.6728	0.6724	0.6721
0.425	0.7004	0.6957	0.6924	0.6897	0.6883	0.6870	0.6859	0.6852	0.6847	0.6843
0.450	0.7162	0.7110	0.7071	0.7040	0.7025	0.7010	0.6998	0.6989	0.6984	0.6978
0.475	0.7344	0.7283	0.7238	0.7203	0.7187	0.7169	0.7155	0.7144	0.7139	0.7134
0.500	0.7547	0.7479	0.7426	0.7387	0.7366	0.7345	0.7329	0.7316	0.7311	0.7306
0.525	0.7779	0.7698	0.7637	0.7590	0.7567	0.7543	0.7524	0.7511	0.7504	0.7497
0.550	0.8043	0.7949	0.7877	0.7822	0.7793	0.7766	0.7744	0.7727	0.7719	0.7712
0.575	0.8341	0.8231	0.8147	0.8083	0.8051	0.8018	0.7993	0.7972	0.7963	0.7954
0.600	0.8678	0.8548	0.8450	0.8375	0.8337	0.8299	0.8268	0.8245	0.8234	0.8224
0.625	0.9064	0.8911	0.8797	0.8707	0.8662	0.8618	0.8582	0.8556	0.8543	0.8530
0.650	0.9498	0.9320	0.9185	0.9081	0.9029	0.8976	0.8935	0.8904	0.8888	0.8874
0.675	0.9996	0.9787	0.9630	0.9508	0.9447	0.9386	0.9338	0.9302	0.9283	0.9267
0.700	1.0565	1.0321	1.0139	0.9997	0.9924	0.9854	0.9798	0.9755	0.9734	0.9714
0.725	1.1217	1.0933	1.0721	1.0556	1.0474	1.0390	1.0324	1.0275	1.0250	1.0228
0.750	1.2031	1.1686	1.1426	1.1225	1.1125	1.1023	1.0944	1.0889	1.0855	1.0826

For 8-in. Pipe

β	25,000	35,000	50,000	75,000	100,000	150,000	250,000	500,000	1,000,000	10,000,000
0.100	0.6097	0.6065	0.6040	0.6020	0.6011	0.6001	0.5993	0.5988	0.5985	0.5982
0.150	0.6153	0.6115	0.6087	0.6065	0.6054	0.6043	0.6034	0.6028	0.6025	0.6021
0.200	0.6229	0.6191	0.6162	0.6140	0.6130	0.6118	0.6109	0.6103	0.6100	0.6096
0.250	0.6336	0.6299	0.6271	0.6248	0.6237	0.6227	0.6218	0.6212	0.6208	0.6205
0.300	0.6469	0.6433	0.6405	0.6384	0.6373	0.6362	0.6354	0.6347	0.6344	0.6341
0.350	0.6648	0.6609	0.6580	0.6556	0.6545	0.6534	0.6525	0.6517	0.6514	0.6511
0.400	0.6881	0.6835	0.6801	0.6774	0.6760	0.6748	0.6737	0.6729	0.6725	0.6721
0.425	0.7023	0.6973	0.6934	0.6903	0.6888	0.6874	0.6862	0.6853	0.6848	0.6843
0.450	0.7186	0.7127	0.7082	0.7048	0.7032	0.7014	0.7001	0.6990	0.6985	0.6980
0.475	0.7374	0.7306	0.7254	0.7214	0.7194	0.7174	0.7159	0.7146	0.7140	0.7134
0.500	0.7856	0.7505	0.7445	0.7399	0.7375	0.7351	0.7333	0.7318	0.7312	0.7306
0.525	0.7829	0.7735	0.7663	0.7608	0.7581	0.7553	0.7531	0.7514	0.7506	0.7498
0.550	0.8106	0.7971	0.7908	0.7843	0.7809	0.7777	0.7749	0.7731	0.7721	0.7712
0.575	0.8417	0.8285	0.8184	0.8104	0.8069	0.8030	0.7999	0.7976	0.7964	0.7953
0.600	0.8772	0.8615	0.8497	0.8405	0.8360	0.8314	0.8276	0.8249	0.8235	0.8223
0.625	0.9178	0.8992	0.8852	0.8744	0.8689	0.8635	0.8592	0.8559	0.8543	0.8527
0.650	0.9637	0.9405	0.9253	0.9126	0.9063	0.8999	0.8948	0.8909	0.8890	0.8873
0.675	1.0165	0.9907	0.9714	0.9564	0.9488	0.9413	0.9352	0.9307	0.9285	0.9265
0.700	1.0765	1.0462	1.0236	1.0060	0.9971	0.9884	0.9813	0.9762	0.9710	0.9708
0.725	1.1460	1.1106	1.0840	1.0634	1.0530	1.0428	1.0344	1.0283	1.0252	1.0224
0.750	1.2299	1.1868	1.1543	1.1294	1.1165	1.1041	1.0939	1.0863	1.0826	1.0791

For 10-in. Pipe

β	25,000	35,000	50,000	75,000	100,000	150,000	250,000	500,000	1,000,000	10,000,000
0.100	0.6114	0.6074	0.6044	0.6021	0.6009	0.5997	0.5988	0.5981	0.5978	0.5974
0.150	0.6175	0.6129	0.6096	0.6069	0.6056	0.6043	0.6033	0.6025	0.6021	0.6017
0.200	0.6249	0.6205	0.6172	0.6145	0.6132	0.6119	0.6109	0.6101	0.6097	0.6093
0.250	0.6351	0.6308	0.6277	0.6251	0.6238	0.6226	0.6216	0.6208	0.6205	0.6201
0.300	0.6484	0.6442	0.6411	0.6387	0.6375	0.6363	0.6354	0.6346	0.6343	0.6339
0.350	0.6661	0.6617	0.6585	0.6559	0.6546	0.6534	0.6524	0.6517	0.6513	0.6509
0.400	0.6895	0.6846	0.6807	0.6778	0.6763	0.6748	0.6736	0.6728	0.6724	0.6719
0.425	0.7044	0.6987	0.6943	0.6911	0.6893	0.6877	0.6864	0.6854	0.6848	0.6843
0.450	0.7210	0.7143	0.7094	0.7055	0.7037	0.7018	0.7002	0.6991	0.6984	0.6980
0.475	0.7403	0.7326	0.7269	0.7223	0.7201	0.7178	0.7160	0.7146	0.7140	0.7134
0.500	0.7624	0.7532	0.7464	0.7410	0.7383	0.7358	0.7336	0.7320	0.7312	0.7305
0.525	0.7878	0.7769	0.7687	0.7624	0.7592	0.7560	0.7535	0.7515	0.7506	0.7497
0.550	0.8169	0.8038	0.7940	0.7864	0.7826	0.7787	0.7757	0.7734	0.7722	0.7712
0.575	0.8495	0.8339	0.8224	0.8132	0.8087	0.8042	0.8005	0.7979	0.7965	0.7952
0.600	0.8868	0.8683	0.8544	0.8437	0.8382	0.8328	0.8286	0.8253	0.8236	0.8221
0.625	0.9297	0.9076	0.8911	0.8782	0.8718	0.8655	0.8602	0.8564	0.8545	0.8527
0.650	0.9780	0.9519	0.9324	0.9173	0.9097	0.9019	0.8960	0.8914	0.8894	0.8870
0.675	1.0334	1.0027	0.9797	0.9617	0.9528	0.9438	0.9366	0.9312	0.9285	0.9260
0.700	1.0972	1.0608	1.0337	1.0126	1.0021	0.9915	0.9832	0.9767	0.9736	0.9707
0.725	1.1704	1.1279	1.0960	1.0711	1.0587	1.0463	1.0364	1.0289	1.0252	1.0224
0.750	1.2636	1.2113	1.1721	1.1415	1.1263	1.1110	1.0988	1.0898	1.0852	1.0791

TABLE A–24. WEIR-DISCHARGE COEFFICIENTS

	Rectangular: Without Side Contraction	Rectangular: With Side Contraction	V Notch	Trapezoidal (Cipoletti)
Discharge coefficient	$C_{RW} = 0.644 \pm 5\%$	Use width $= b - 0.2h$	$C_{VW} = 0.605 \pm 6\%$	$C_{TW} = 0.63 \pm 3\%$
Channel-width limitations	b	$B - b > 4h$ $B > 3b$	....	....
Channel-depth limitations	$H > 2h$		....	....
Weir-width limitations	....	$0.5 \text{ ft} < b < 4 \text{ ft}$	....	$b > 5h$ $b > 0.5 \text{ ft}$
Head limitations	$0.02 \text{ ft} < h < 2.6 \text{ ft}$	$0.08 \text{ ft} < h < 2.5 \text{ ft}$	....	....
Angle limitation	....	....	$35° < \theta < 120°$	$\theta = 2 \tan^{-1} \frac{1}{4}$

INDEX